Making Space HAPPEN

Making Space HAPPEN

Private Space Ventures and the Visionaries Behind Them

Paula Berinstein

Plexus Publishing, Inc.
Medford, New Jersey

First printing, 2002

Making Space Happen: Private Space Ventures and the Visionaries Behind Them

Published by:
Plexus Publishing, Inc.
143 Old Marlton Pike
Medford, New Jersey 08055
U.S.A.

Library of Congress Cataloging-in-Publication Data

Making space happen : private space ventures and the visionaries behind them / Paula Berinstein.
p. cm.
Includes bibliographical references and index.
ISBN 0-9666748-3-9 (pbk.)
1. Space industrialization. 2. Outer space--Civilian use--Economic aspects. 3. Space tourism. 4. United States. National Aeronautics and Space Administration. I. Berinstein, Paula.
HD9711.75.A2 M35 2002
338.4'7910919--dc21

2002001564

Printed and bound in The United States of America

Publisher: Thomas H. Hogan, Sr.
Editor-in-Chief: John B. Bryans
Managing Editor: Deborah R. Poulson
Production Manager: M. Heide Dengler
Sales Manager: Pat Palatucci
Copy Editor: Robert Saigh
Book Designer: Kara Mia Jalkowski
Cover Designer: Lisa Boccadutre

Dedication

For Alan

Table of Contents

Sidebars

Foreword

I am five years old. I am walking hand in hand with my grandmother in the country outside Milan. Our bare feet pad across the soft, damp grass. Suddenly my grandmother halts and points above us to the heavens.

"Look!" she says. "That star—the brightest one—she is looking at us."

I laugh, but she continues, seriously. "Si, little one, all the stars watch us. Look carefully."

I do. And in that instant I feel the star return my gaze. I feel as if it is a stellar heart that beats with mine. For a moment, all the loneliness of my childhood evaporates. I feel peaceful; I feel a oneness with the Universe.

I often think about that first extraterrestrial gaze, how it made me quiver with awe. How it made me feel both like the center of the Universe and an invisible microdot lost in incomprehensible space. I felt both magnificently empowered by this magical array of stellar jewelry and terribly humbled by the infinite vastness of it all. At that moment, I knew not a thing about quasars and black holes and brown dwarves. I did not even know that radio telescopes existed, let alone that I would spend years peering through one, that one day I would gather the waves they perceive and transform them into mysterious sounds, the song of the Universe. All I knew was that the sky had suddenly opened up to me and I would never be the same.

That night my grandmother not only introduced me to the wonders of the cosmos, she helped me discover that more than one can ever dream becomes possible when we allow imagination to excite exploration. I am not alone in this realization and the passion that springs forth from it. Throughout the ages, space and all it contains have inspired not only awe, but an insatiable hunger to know it intimately, to become part of it and make it a part of us. However, it is only in the last half-century that we have begun to satisfy this hunger, and only in the last decade that the ability to do so has begun to approach reality for the average person.

Oh yes, the time for you and me to go into space is nearly here. Soon we will be able to look at the Earth from the blackness that surrounds it, to revel and cavort in weightlessness, to gaze upon the stars at last unobstructed by haze. There will be thousands to thank for this gift—restless darers who have dreamed, theorized, calculated, built, tested, trained—who have broken pencil tips and defied Earth's gravity well in a single go. Through vision, will, and ceaseless hard work, these giants helped humanity enter space for the first time.

But there are others, giants in the making, who are working even as I write this to make a home for us in space once and for all. I know these people—they are creative, thoughtful, dedicated, even sublime—and I want you to know them too. Rick Tumlinson and James George of the Space Frontier Foundation, out-of-the-box thinkers who helped a private company and the Russian space agency come together to turn space station Mir into a commercial enterprise. Never before had the world witnessed such a near miracle—a joint effort between American private citizens and a Russian public-private partnership via a company based in the Netherlands to open space to the human race! My friend Dennis Tito became the first direct beneficiary, even though his trip to Mir never materialized. When funds ran short to keep the station aloft, Dennis made history by visiting the International Space Station instead. As the first self-financed citizen space explorer, he spent a whirlwind week aboard the little structure, changing forever the way we think about civilians taking trips to space.

You've got to meet Denise Norris. She must be made of pure energy! Watch and listen to her and you'll be spellbound. Denise is planning to take us back to the Moon as virtual explorers, but with a twist. Applied Space Resources, which she heads, doesn't make rockets or instruments. They leave that to others. Her company is a marketing company! Inspired by an article called "Buck Rogers, CEO," Denise decided to build her business not around science or hardware, but a defined market with a palpable need. As you read this book you will discover that that is not the way space enterprise has traditionally been conducted, but you will certainly wonder why not once you have finished. Everything Denise is planning is based around market need—the need for those of us on Earth to interact personally with goings-on in space. Denise Norris will engage us and personalize our experiences up there through electronically assisted interaction like video feeds and the Internet. She will land a rover on the Moon. You'll watch as special instruments pick up extraterrestrial dirt just for you (you will be able to purchase it raw or encased in a sparkling crystal); you'll click your own pictures of the lunar landscape; and you'll have your own HAL—e-mail the onboard computer and get customized data back.

And then there is Buzz Aldrin, the second man to walk on Luna, our beautiful and mysterious Moon. Buzz is that rare breed, visionary cum doer. In perpetual motion, not only does he work to design and build a space transportation infrastructure, but at the same time urges us to transcend our earthly lives and strive for so much more. Going into space will stretch us, vitalize our brains and our hearts, for the final frontier will not be space alone, but human relationships. As we move outward, we will meet not just the universe, but ourselves. Perhaps in the process we will begin to understand why we continually want to combat each other, why we are obsessed with our own satisfaction at the expense of fellow humans, and learn to live in harmony with each other.

These people will change our lives by making space happen. They will take us forward into a world we have never known, a world in which many of the assumptions by which we live our daily lives will be turned on their heads. When humanity moves out into space for good, Earth will change. We will become richer, for the resources of space are boundless. We will gain new perspective from viewing our planet from the outside. We will make new scientific discoveries. Our engineering will advance, as will our medicine and our ability to gather and use clean energy. Our economy will grow in new and exciting ways, for there will be jobs and careers aplenty, new goods and services, new infrastructure.

You will meet these people in this extraordinary book, *Making Space Happen*. In order to write it, author Paula Berinstein spent nearly three years talking to people in both the private and public space arenas, understanding what makes them tick. At the same time, she threw herself body and soul into defining and dissecting the issues involved in making space happen—why we should or shouldn't go in the first place, where the money will come from, how safe it will be for travelers and settlers, what a space trip will be like, what the effect on international relations might be, and more. Berinstein is the ideal person to write such a book. Coming at the topic from the outside the industry, she approaches it without preconception, without bogging down in the technicalities that are so hotly debated on the inside. As a woman, she brings a different perspective from that found in most books on space. She approaches the topic not from a mechanistic point of view, talking of hardware and equations and orbits, but from a personal one—what the opening of space means for you and me. You will find her style engaging and readable, her personal observations fascinating. This is a compelling work: be prepared to be seduced … and surprised.

However there is more to this book, and more to making space happen. Berinstein knows it, and I know it. By moving outward, we evolve. It is not just our external world that will be transformed, but our inner one as well. We must let each new discovery swim in our imaginations and open our hearts to new ways of thinking and feeling about life.

I'm looking forward to the day when we all hear how the music of the spheres resonates with the music of our hearts.

Dr. Fiorella Terenzi

Preface

In the fall of 1998, I made a startling discovery. I went to a conference given by a group I'd never heard of—the Space Frontier Foundation—and found that people outside NASA, ordinary people, were trying to bring about a permanent human presence in space! I mean serious, nonkooks who had established companies, cooked up business and technical plans, obtained funding, lobbied Congress, and established names for themselves in the mainstream press. Why hadn't I, a person with a passionate interest in space, heard of them? They were talking about private missions to asteroids, settlements on the Moon and Mars, lowering the cost of space transportation by a factor of 100, and space tourism for the general public. This wasn't science fiction. No transporters, replicators, tractor beams, or time travel. Much of the technology being talked about was already proven. It was a challenge that could be met if only for one thing: money. Where would it come from, and what would justify its being spent? Would governments fund or subsidize these kinds of projects, as the U.S. government did back in Apollo days and the Russian government has done since *Sputnik* in 1957, or would private funding and commercial markets be the drivers of this new trend?

I had long ago given up on the dream of going to space personally. I couldn't conceive of a realistic scenario under which that would happen. Only highly trained and carefully vetted astronauts and cosmonauts got to go. Even if ordinary civilians got to go again after teacher Christa McAuliffe's death aboard the space shuttle *Challenger*, they wouldn't be me. The odds were so far against my being chosen—or your being chosen, dear reader—that I may as well continue projecting myself into *Star Trek* episodes and at least get some vicarious gratification.

But here were people holding out new hope. They filled not only the conference sessions: they filled my heart. They were doers, people who couldn't sit by while the world took its own directions. They were going to make something happen: space!

I had come to the conference looking for book ideas. I left with a ream of them, and with such excitement and enthusiasm that I worked like a fiend to put together a book proposal as quickly as possible. The story of private efforts to put people into space had to be told, and I wanted to tell it. Who better? I was enthusiastic about space, but despite having worked for Rocketdyne, the company that designs and makes the space shuttle's main engine, I was an outsider. I didn't know from space tourism or mining asteroids, and I certainly didn't know anything about the factors that might affect the ability to pull off such feats. I wasn't a techie and had no interest in being one. I knew that most other people weren't techies either, but I was certain that they'd be interested in hearing about new developments that would change their lives forever. I would bring this under-the-radar movement to the screen.

There was a lot to learn. I had to understand what was happening! I was lost, I was confused, I didn't know who was involved in what, and I wasn't even sure I knew which topics I should cover or whom to profile. I read incessantly. I talked to everyone I could browbeat into taking a few minutes to reveal their secrets. I asked friends, colleagues, and anyone I'd met whether they personally would want to go into space. (Everyone did!) I went over and over the writings of the people I had set my sights on interviewing, underlining, taking notes, repeating ideas to myself until I was sure I understood them well enough to ask questions. But it was all rote at the beginning. I was a sponge that holds its water, never discharging it. And then something began to happen.

I began to develop a point of view. I started to form opinions. Finally I could see that what this person was telling me made infinite sense, and what that person said was missing the boat. But even more than that, I began to see what was necessary and to evaluate proposals critically. I could separate the fluff from the substance. I also began to realize that the people in the field were not working from a common set of facts. Before recognizing this, I blamed myself for not understanding. Now I know that space is no different from medicine or science or education or politics. One person quotes one set of facts and statistics, while another person presents something totally contradictory. One's experience teaches different things from another's. Of course—that's the way the world works. How could I expect differently? Not even NASA sings with one voice, so why should others?

Of course this sorting out and clarifying happens to anyone who begins to get a grasp of her subject, but even knowing that, I was exhilarated that the clouds were beginning to part. And I was emotionally involved, which I swore I wouldn't be. I'm telling you all this, dear reader, because I want you to know that as much as I've tried to be objective and fair, some bias may have crept into my telling of the story. If so, I apologize.

Let me tell you about the structure and content of the book. Most of what you will read is centered around specific people and the issues they face, issues like funding, environmental concerns, safety, market size, international relations, ethics, and the effects of space on the human body. Each chapter treats a different aspect of the "private space movement," such as tourism or Mars or political issues, and the people involved in that aspect. Nevertheless, you will find some overlap, since certain ideas and people are integral to multiple areas.

At the end of the book you will find a number of appendices that expand on topics touched on in the book. You'll also find a helpful bibliography and glossary—features that no nonfiction book should lack.

A word about the difference between "private" and "commercial." This book is about *private* space efforts. Many of them are commercial as well, but not all. So you'll find discussions of going to Mars, which no one can figure out how to make profitable, and lunar bases, which some people *think* they can make profitable, but lots don't.

You will find a diversity of opinions and approaches among people who are all more or less trying to tackle the same problem. Don't be surprised—there is no unified view coming from private space activists. However, I hope you will get such a view from me. Since I have no vested interest in any particular project or approach, I think that's possible.

Much of what you see and hear about space is technical. "The gizmo of the whothingee spins at a rate of four somethings per nanosecond, which gives you the capability to slurge through the bumhoop 5 percent faster than if you use the yurtlenut." The appeal of *Star Trek*-style technobabble notwithstanding, you won't find it in this book. There are a few technical write-ups in the appendices, where the interested can read about them; the uninterested can keep the flow of the book going without having to get bogged down.

You may discover that your favorite person or issue has been left out. Believe me, my initial proposal included everything and everyone except the kitchen sink and the Grinch. Had I kept to it, I would have written a twelve-volume set rather than the book you are holding. It's not that the omitted people and topics aren't interesting and important. Space solar power, for one thing, is a big deal among some space advocates. (I'm referring to satellites that beam solar energy to Earth in the form of microwaves; the energy is converted to electricity on Earth.) It's also a huge topic, quite technical, and unlikely ever to happen (it's just too massively expensive to consider in these political times, though if Earth faces some future crisis, governments may be willing to bite the bullet—but I digress. I suppose that little statement will inspire a few protesting e-mails). So it's out of the book.

Almost everyone I spotlight in the book has enemies. All have detractors, and all have flaws. It is not my purpose to present gossip or to "trash" anyone. There are many complex relationships among these players. Some actually hate and disrespect each other, but you'll hear very little from me on that subject. Find me a field in which the same is not true of the participants, and I'll show you a nonhuman species.

You won't find much about NASA and the other national space agencies in this book either. Nor will you find much about the big aerospace companies. Heresy? Not really. That's not what this book is about. This book is about the underdog, the nonestablishment, the people who are toiling behind the scenes. If NASA or the Russian space agency ends up getting the general public into space, great, but frankly, I don't think it's going to happen, and anyway, there are other books you can read about NASA. (Actually, it isn't quite true that the Russians won't get us there. Of all the strange developments, the Russians—*not* NASA—are the ones who are taking paying private individuals to *Space Station Alpha*, and wouldn't it be ironic if we all ended up getting to go because of the Russians? How's that for an "evil empire?")

Of course, there are the people I *couldn't* talk to, though not for lack of trying. I was absolutely dying to talk to Richard Branson, who has started a company called Virgin Galactic Airways. I think Branson is the bee's knees. He's just so creative and clever and driven and unusual, and if anyone can make space tourism happen, it's he. I tried. I received a gracious letter from him—a letter I have up in my office I'm so proud of it—turning me down. Lack of time or some such thing. Oh well. He wasn't the only one.

So here, dear reader, is my Bransonless book for you to consider. Its purpose is to inform and astonish. You'll let me know if it succeeds.

Acknowledgments

This is the part of every book I enjoy the most: thanking all the people who helped me. For this is not my work alone. Without the gracious participation of all the interviewees and all the people who furnished facts, explanations, analysis, and opinion, there would be no book. And so, I'd like to thank ...

All the people I interviewed for the book. There are too many to name, but you'll find them mentioned throughout these pages. Many of them I profiled; some of them I did not, but they contributed significantly to the book just the same.

My wonderful editor-in-chief at Plexus, John Bryans, who has been with me through thick and thin. John has believed in me from the beginning. His support and encouragement have been unwavering. He is just one great guy!

Jack Flannery, former executive director of the U.S. Space Foundation. Jack has made invaluable suggestions along the way and directed me to helpful people.

James George and Jason Klassi, who double-checked my facts for me and provided perspective and encouragement.

My sister, Jan Berinstein, who copy edited the manuscript and kept me going when I was in the throes of end-of-book syndrome.

And ...

Greg Arsenault, who offered extremely helpful suggestions about content and organization, and who provided an outsider's look at the subject.

My friends Tim Robertson, Mike Chibnik, John and Wendy Calderon, Dave Holland, Linda and Gerry Seck, and all the fine people at the Ventura County Astronomical Society.

Seth Potter, who tossed around ideas with me and helped me sort out who was doing what.

John Nelson, who gave me the guts to ask really nosy questions.

Martin Sebborn and Carol Wilmot, good friends who always provide insight as well as fun.

Tim Glauert, who liked *Mission to Mars* even though I didn't.

Gary and Barb Cooper, who love everything spacey.

Patricia Barnes-Svarney, fellow writer and space fan.

Alan Jones, who advised on the physics of the Martian winds.

Bev Olson, fellow space advocate and provider of valuable information about Mars issues.

Charlie Carr, who didn't live to see the dream realized.

John Spencer, space tourism visionary.

Barbara Brown, who offered insights into the issues of women and space as well as space medicine.

Jeff Krukin, who hooked me into this private space thing in the first place.

Jim Cornelius, longtime friend who's provided support and space news.

Mark Marcus, Howard and Karen Resnick, Barbara Javor, Marina Javor, and Kathy and Scott Watters, who didn't have much to do with this particular book but have been my friends for years and years.

Andrew Lea, who's also trying to make space happen.

My neighbors, who always ask about "the book."

My mother.

My father.

Alex Chaney, my games consultant.

Ellie Chaney, my artistic consultant.

But more than any of them, my wonderful husband Alan, who provided so much support, insight, love, and inspiration that he really should share the credit for this book.

Cast of Characters

Buzz Aldrin, former Apollo astronaut and space entrepreneur: You need a public-private partnership.

David Ashford, aeronautical engineer and X PRIZE contestant: Money is the critical issue.

Greg Bennett, rocket engineer: We go because it's fun.

Jim Benson, entrepreneur: "Baby steps." His office is in a former shooting gallery.

Patrick Collins, economist: The market is there. Space tourism will make for a more prosperous Earth.

Peter Diamandis, organizer of a major prize for space transportation: "Make it a race." The fastest way to become a millionaire in space is to start as a billionaire.

Alonzo Fyfe, ethicist: Give value for value.

Kirby Ikin, space insurer: Insurance reassures investors. You can also do tricks with it.

Ron Jones, space entrepreneur: Mass travel is the only way to get the cost down.

John Lewis, chemist and planetary geologist: "Fill in the question marks." He believes the solar system can and should support trillions of people.

Wendell Mendell, NASA planetary scientist: "Mr. Goldin, are you mad at *me*?"

Charles Miller, entrepreneur and former head of lobbying organization: "Is one week of your life worth $3 million?" Don't wear Vulcan ears on Capitol Hill.

Denise Norris, entrepreneur: "We're a marketing company, not an aerospace company."

Tom Olson, Space investment fund creator: "I just thought different."

Kim Stanley Robinson, science fiction author: "The privatization of space makes me shudder."

Tom Rogers, head of space tourism organization (and the granddaddy of space tourism): "What I want to do in space is none of your business."

Rene Schaad, scientist and space buff: Europeans don't want to spend money on space.

Dennis Tito, the first civilian to pay his own way to space: It was worth the $20 million.

Rick Tumlinson and James George, heads of space advocacy organization: Space is a place, not a program.

Alan Wasser, space property rights advocate: Maybe some day, a Mt. Wasser on the Moon.

Harvey Wichman, space psychology professor and ergonomics specialist: Weightlessness is much more trouble than you'd think.

Robert Zubrin, Head of the Mars Society: "See you on Mars."

Making Space HAPPEN

INTRODUCTION

Why Aren't We There Already?

Despite the ambitious pictures painted in the last hundred years of science fiction, humans living and working in space is not a given. We thought that would happen back in the Apollo Moon landing days. Going to the Moon as we did would comprise our grand beginning, followed by a base there, then on to Mars, the other planets in our solar system, and the galaxy itself. Whether we'd be walking around in silver lamé suits, well, maybe, but that wasn't the point. Space would advance us technologically. It would move humans outward beyond the familiar. Our civilization would change, molded by the new wisdom we'd find through exploration and perhaps contact with extraterrestrial life. How else could we unlock the mysteries of the universe?

To some extent, we have grown and learned. The fabulous Hubble space telescope has shown us galaxies twelve billion years old. We've just begun to explore Jupiter's moon Europa, which we think might harbor life. We've identified planets around other suns. And a few of us have endured long exposure to space, principally on the Russian space station *Mir*, which has taught us about how the human body reacts to zero gravity over time.

But we haven't been back to the Moon since 1974. Our Mars exploration has proceeded excruciatingly slowly, and a human mission to the Red Planet is still a pipe dream. At the beginning of the 21st century, only 400 humans have ever been to space. No, we are not a spacefaring civilization.

It never occurred to me until recently that anyone would question the inevitability of our going into space. The possibility of exporting our problems, yes, that anti-space argument had occurred to me, but I knew we'd go eventually. I could not conceive of us being earthbound forever.

I was sheltered. Many people question the wisdom of moving off the planet, and in ways that never would have occurred to me.

So it seems that inevitability is not enough of a reason, especially given the massive expenses involved.

This book tackles the big questions: Should we go, and if so why? If not, why not? If we do, how much will it cost and where will the money come from? What are the obstacles we face in getting there, and those we'll encounter once we do go? How will life change, how will society change, if we go? *Can* we go, given the limitations of the human body?

WHAT SPACE IS

Cold—usually. Temperatures depend on whether or not a spot is hit by direct sunlight. Near the earth and the Moon, sunlit objects can heat up to about 250° F, 121° C. Where no sun shines, temperatures can be -250° F, -156° C.

Radiation-filled. Natural emissions in space take the form of sporadic bursts caused by solar flares and background radiation from cosmic rays. The earth is protected from this radiation by its surrounding magnetic field, but space travelers and visitors to the Moon and Mars are not. Solar flares can be instantly deadly. Exposure to cosmic rays builds up radiation levels in the body, ultimately leading to a much greater risk of cancer than normal.

Airless. There's no air in space, and none on many bodies in space. That means you have to take your air with you.

Riddled with interstellar gas and photons (packets of light). The density of the gas is so low—less than a trillionth of that at the Earth's surface—that you don't have to worry about its obscuring your view of the cosmos.

Without gravity. You have to have significant mass in order to have gravity. The Earth has such mass, as do Mars and the Moon, though considerably less than Earth.

Space stations and spacecraft don't, unless we make it ourselves by spinning them. So far, no one has actually done this in space.

The Apollo program cost Americans $25 billion ($100 billion in today's dollars). For our money, we got twenty-one humans to the Moon over a period of six years; twelve actually set foot on the surface. It was a technologically and spiritually stunning set of accomplishments, but it was really about the Cold War and two countries fighting for ascendancy. After the winner was declared, the American media and the U.S. government lost interest. Only when *Apollo 13* went wildly wrong did the program make news again. And only when NASA's huge workforce was threatened with layoffs was the U.S. Congress motivated to fund the agency.

What will it cost us to go back into space to stay? I'm not referring to the space shuttle, which flies a measly three or four times a year. I don't mean *Space Station Alpha*, the NASA-led international project so plagued by political, technical, and financial problems that I can't bear to look. I'm not speaking of planetary probes and exploratory spacecraft, as exciting and important as they are, for they don't carry humans, though to some extent they're preparing us for human settlement. Nor do I mean communications satellites, which in no way represent humans getting into space. I mean a permanent human presence. People living and working in space. This book looks at some of the possibilities.

What will we do when we get there? Parade around in tight-fitting uniforms and meet interesting aliens who all look and dress alike? Probably not. What is there to do on that old dead Moon and arid red Mars anyway? And does anyone seriously want to live in orbit where you can never go outside? Since we can't zip around at warp speed and explore other star systems, does it really matter whether we get to space or not? Good questions. This book offers up a number of answers.

Who will get to go? Will it be still more astronauts with specialized training, mostly men between the ages of thirty-five and forty-five? Surely you and I won't get to go. The odds are worse than those of the Irish Sweepstakes. (The chances of getting to be an astronaut are about the same as those of playing one in the movies—maybe slightly worse when you consider how many films feature astronauts.) Or are they? Ever heard of space tourism? You can't build a space tourism industry on middle-aged men alone. If such a delightful prospect ever materializes, you and I will be the target market. This book explores how and when a robust space tourism industry might take shape.

What will it take to get us there? Money, yes, we know that. Where will that come from? Technology, same question: Where do we get it? Permission from the Federal Aviation Administration? We'll undoubtedly need that. Markets, incentives, organization, public interest, and medical advancement all will be required. They never talk about most of that in *Star Trek*. I guess that's all taken for granted in the 24th century, but our "enterprising" friends had to think about it at some point or it never would have happened. By setting up all that infrastructure, we 21st century humans are paving the way for our descendants. This book explores the obstacles, challenges, and steps that need to be taken to get from here to there.

WHY IT'S SO HARD TO GET US INTO SPACE

Financial constraints. The main barrier to human entry into space has been financial rather than technical. Time lag between investment and payoff holds back development of commercial customers for *Space Station Alpha* and other endeavors. If the government subsidizes a company's competition, potential investors may be scared away. Space is expensive because of the low volume of flights. (There are currently fewer than 100 commercial launches per year.) Even some of the established companies are stagnating and having trouble maintaining their capital bases.

Technical factors. Lack of infrastructure holds back development of commercial markets. Rocket technology hasn't advanced in any major way in thirty years. We have no reusable rockets or vehicles other than NASA's space shuttle.

Industry structure. The space industry in the U.S. is dominated by NASA and the military. In other countries, like France, governments subsidize launch vehicle manufacturers like Arianespace.

THE COMMERCIAL SPACE INDUSTRY

The commercial space industry has not been faring well of late. Looking beyond the major market—satellites—one sees an almost-industry struggling to be born. Rocket entrepreneurs are trying like crazy to build cost-effective vehicles to take tourists aloft, vehicles to deliver packages at space flight speeds, or vehicles to supply *Space Station Alpha* or maintain satellites. We have no such vehicles. We have four very expensive NASA space shuttles—far too costly to operate commercially, and not robust enough to fly often. All the others, the ones that lob satellites into space, are expendable, thrown away after use. The reason? It's cheaper to build a new expendable vehicle ($50 to $200 million, about the cost of an airliner) than to develop, build, and maintain a reusable one (not that the shuttle is any example, but the orbiter *Endeavour* cost $1.7 billion).

In the late 1990s, the aerospace giant Lockheed Martin began work on a reusable prototype—the X-33—that would be a precursor to a larger vehicle, VentureStar. The X-33, which had always been intended as an unmanned vehicle, was supposed to cost $7 billion. However, debacle after debacle and cost overrun after cost overrun caused some observers to shake their heads in disgust and conclude that spending all that money on an uneconomic vehicle that doesn't carry people is no way to get humans into space. In 2001, the U.S. Congress cancelled the program.

In the mid-1990s a lot of would-be rocket entrepreneurs opened up shop in the hopes of beating the odds. At that time, the satellite industry looked as though it would be capable of sucking lots and lots of vehicles into use—some estimates came in at more than 1,500 satellites to be launched by 2010. Teledesic's (now merged with formerly bankrupt ICO Global Communications) and Iridium's futures looked bright and ample. As we now know, it did not turn out that way. Iridium went into bankruptcy before the U.S. military decided to snap the company's satellites up for its own use. The other companies fared worse.

But before things got worse, they got quite a bit better. The best known of the companies—Kelly Space and Technology, Kistler Aerospace, and Rotary Rocket—were preparing to secure large sums of money through private bond and equity offerings; Rotary, which is now defunct, even considered an IPO. But in mid-1998, the satellite market began to crack, and the companies' fortunes flagged. Rotary ended up laying off most of its workforce; in 2000 it was ignominiously seized for back property taxes. Kelly and Kistler are still hanging in there, but their prospects have diminished.

Growth of the Commercial Space Industry

The commercial space industry has outstripped government space spending tremendously since the mid-1980s.

In 1986, space commerce, including satellite, launch equipment, and ground equipment sales, plus satellite services, was a $3.2 billion-dollar market.

In 1996, worldwide satellite industry revenue, including manufacturing, services, and launches, reached $44.8 billion. In 1998, it was $65.9 billion. In 2000, it was $81.1 billion.

In 1990, the U.S. commercial space launch business came to $570 million. In 1999 it was $992 million. In 2000, the worldwide launch services industry generated $8.2 billion in revenue.

Government space budgets totaled about $19 billion in 1996, about $18.6 billion in 1998.

Source: A speech to the Space Transportation Assn. conference on September 22, 1998 by Raymond S. Colladay, President Lockheed Martin Astronautics quoting facts and figures which Mel Brashears, the President of Lockheed Martin Space and Strategic Missiles Sector, recently cited.

Source: Futron Corporation, Satellite Industry Guide, October 1999.

Source: Satellite Industry Association press release, April 2001.

In addition to the rocket entrepreneurs, there are companies that want to send probes and cargo into space to entertain people. Some want to land on the Moon and let you drive rovers around remotely, or pick up moon dirt as you watch, then offer it for sale as part of trinkets back on Earth. Or they want to put telescopes on the Moon and let you snap your own picture of deep space objects. It's vicarious entertainment, and it costs a lot less than putting humans themselves into space.

Then there are the entrepreneurs who want to explore space robotically without the help of government, send probes to Mars and asteroids and, again, let you watch as they poke around. Some want to inspire government, or partner with it, to send humans to Mars.

These people hail from only a handful of countries around the world. The U.S. and Russia are the only two with a history of manned space programs, though other countries like France, Canada, and the U.K. have sent astronauts up. A good portion of the space tourism work is being done in Japan. Australia is host to some commercial space activity, and there are firms in the U.K. working on technology for taking humans up. The European Space Agency plans to send out exploratory probes, but the emphasis in Europe is largely upon robotic exploration, not commercial use (other than satellites) or humans in space. A little bit of work is being done in Argentina. The Chinese have announced plans to put a person into space by 2005 and to institute a robust space exploration and commerce program. Though India is a leader in remote sensing and has its own launch vehicle, it isn't putting humans into space or exploring other bodies in the solar system.

Those who would expand human activity in space face multiple obstacles. Investors are uninterested in financing endeavors that will take such a long time to pay back, and in putting up the hundreds of millions of dollars required to get even one product to market. They know that the potential for profitability isn't exactly golden. Moreover, they're worried that the markets aren't there, and that the dangers of space will eat up whatever profits might be generated. They're also concerned that private companies can't compete with governments, or government-backed companies like the major aerospace firms, which are favored with many perks and preferential treatment. Most of the startups that have tried have failed or are struggling to maintain financial advantage. Almost all of them have to depend on funding from wealthy individuals, but there aren't many of those willing to act as the angel, and the patience of those who do is finite. For Americans, there's competition from low-cost Russian expendable launch vehicles that would serve the same markets as reusable vehicles. (This wouldn't have happened before the end of the Cold War, but now Russian hardware is available in the U.S.) And there's still the "giggle factor"—that refusal to take space advocates seriously because of the "Trekkie" image.

But also, people aren't used to thinking of space as profitable. They aren't accustomed to thinking of *using* space. The very idea raises hackles on the backs of environmentalists and others who believe we should solve our problems on Earth and leave space pristine. The arguments for and against the commercial use of space are long and complex, and they haven't been aired in public to any appreciable extent. This book offers an overview of them.

Kim Stanley Robinson, author of the compelling novel, *Red Mars,* believes that likening space to a frontier, as some advocates of space development do, is false advertising. He reminds us that without air, you can't strike out for parts unknown on your own. Using "frontier" to describe those parts implies that you can.

It's not that Robinson is overly picky. It's just that he'd like to see humanity go into space, but not for the wrong reasons. His list of those wrong reasons will knock your socks off. Robinson addresses the issue of Mars, but his reasons apply to other destinations as well.

Ensure human survival

Many people think we should move outward so as not to "keep all our eggs in one basket." The idea is to spread humanity around the solar system, possibly even the galaxy, so that if there's a catastrophic asteroid collision, we blow ourselves up, or pollute ourselves to death, the species will survive. Sounds pretty good on the surface. But, Robinson reminds us, be realistic: who do you think is going to get to leave Earth and who will have to stay behind? Do you really think it will be you setting out for the haven? Do you want to sacrifice yourself for the good of humanity? Escape is a distraction, he believes, and an excuse for not doing the work necessary to survive on Earth. Robinson finds this reason odious because it implies that humanity is more important than other creatures, a value with which he doesn't agree.

Destiny

Some people argue that humanity as a species is destined to fill the galaxy. Robinson calls this reason "Earth as cradle," meaning that humanity was born on this planet, but will necessarily move into other areas as it matures and "fulfills its destiny." This argument falters because, as Robinson sees it, the idea of destiny is a code word for "I think we should do it" and is an attempt to avoid having to make a rational case for moving into space.

Beauty

Another reason to go is that doing so would be a beautiful thing. There actually are people who argue in favor of this position. But

Robinson is practical: This reason won't help raise a dime. As evocative as it is, it also reminds us of times in human history that brought as much pain as glory, such as the era of building beautiful cathedrals.

Economic growth

As for going into space to make money, the problem with that idea is that the required investments are too high and the returns too low, at least in the short term. You won't get rich quick by going into space, so it's difficult to raise private money for space projects.

Save the American aerospace industry

This one may not be the dumbest reason to go into space, says Robinson, but it's quite un-PC to state the rationale so directly. The $400 billion-a-year industry has been floundering ever since the end of the Cold War. Some people think it might be better to use industry resources to settle Mars than to prop up the technically impossible and legally forbidden Star Wars defense system or to spend $22 billion on a military airplane. But we can't say this out loud—it isn't politically palatable—so instead we talk about going into space as something else: a patriotic act, creating a high frontier, or as the salvation of a stagnating civilization.

Dwindling resources

Many space advocates argue that we're outgrowing the Earth, that resources are dwindling and the population is becoming unsupportably large. Not a valid reason for going, says Robinson, even if true. Learn to fix your problems here before moving out. Especially because doing so won't actually ease the population pressures on Earth. For that, hundreds of millions of people, even billions, would have to leave. That's not going to happen any time soon. There just won't be the facilities to support them—not in orbit, not on the Moon, not on Mars.

But there are some reasons to go that Robinson finds valid.

Search for knowledge

"This is what you call Mars as Antarctica or the 'because it is there' reason," says Robinson. "We go to Mars to study it because we study everything. It's science for science, it's humanity at its best." But, he admits, this reason isn't going to help raise a lot of money.

Search for life

"I think the search for life is one of the main reasons for going," says Robinson. "It's one of the best reasons because we don't know the status of life in the universe right now." Perhaps life starts up anywhere you get the right chemicals and the right heat. If we were to find signs of life on Mars, fossil or living, that would be one of the most important discoveries in human history. "The ecological crisis is a life crisis," explains Robinson, "and if we were to find life on Mars it might abash us, make us more survival oriented, make us more respectful of life in general."

Mars will help us survive the ecological crisis

"We are going to have to get involved in climate control and global planetary management," says Robinson, "but we don't know how to do that well enough. Going to Mars is one of the things that will help us understand it better. Comparative planetology is one of the most powerful tools for understanding our planet. Examine the story of Venus. You can say that the ozone layer was discovered in part because people were studying the chemistry of the atmosphere of Venus. Thousands upon thousands of cases of melanoma and premature death have been avoided because people were studying the chemistry of the atmosphere of Venus. This reason would give us more of a constituency—not just the aerospace community, but all environmentalists."

Many people embrace and argue for the reasons Robinson has laid out. However, when you look coldly and honestly at people's real motivations, two camps emerge: the "It's going to be profitable" camp, and the "I just really want to do this" camp. That's what it all boils down to. But I have another reason, something I don't hear from other people. I'm not going to tell you what it is now. You'll just have to wait till the concluding chapter to find out.

The space-as-frontier argument evokes more passion among space activists than just about any other. Advocates from all over the political spectrum embrace it, arguing that without frontiers, i.e., challenges, societies stagnate. Detractors go apoplectic over its arrogance and the implication that because Americans like it, everyone else ought to as well.

The idea of the frontier was popularized by Frederick Jackson Turner, a 19th-century American historian who argued that its existence created the American character and democracy. In 1890, the U.S. Census Bureau declared that the frontier was closed, alarming some who believed the lack of a frontier meant diminished opportunity. In this context, Turner began to think about the frontier's importance; three years later he presented "The Significance of the Frontier in American History" to his colleagues. Turner's argument, though not without controversy, resonated so much that it has become ingrained in American history texts.

The argument goes like this. The frontier was wild. It required people to live by their wits, to adapt, to invent constantly. This fluidity, this "return to primitive conditions on a continually advancing frontier line" rather than along a fixed one caused American social development to begin over and over on many fronts. It left European ways of doing things farther and farther behind, forging a new self-sufficient American culture. The wilderness caused complex society to break down into something simpler, which Turner called "anti-social"; it bred individualism and produced "antipathy to control." Individualism promoted democracy, as one could see in the liberal voting provisions that accompanied frontier states' entrance into the Union. Turner contended that the open land bred competence, while in the East, relative ease made people lazy and complacent.

What the frontier brought to its people, said Turner, was unique intellectual traits: practicality, inventiveness, inquisitiveness, nervous energy, coarseness, and acuteness. It made people strong, exuberant, quick to find expedients. It provided opportunity, "escape from the bondage of the past ... freshness, and confidence, and scorn of older society, impatience of its restraints and its ideas, and indifference to its lessons."

Turner loved the "dominant individualism, working for good and for evil." Most of us don't embrace the negative along with the positive as he did. Those of us who do, such as the space activists who argue that we should go for the frontiers, don't understand those of us who don't. And vice versa.

The frontier concept polarizes people in other ways as well. It's pretty much an American concept, and its advocacy implies that the American character and experience is superior to that of other cultures. Even Europeans and other non-Americans who want us to go into space find the idea ridiculous. A Swiss scientist named Rene Schaad (see Chapter 19) describes the situation this way: "I think it's obvious that Europeans don't buy the frontier argument because there hasn't been a frontier in Europe for 3000 years. The frontier doesn't have a very good reputation in Europe. Many Europeans sympathize with Native Americans."

That's another rub. Much about the frontier in the U.S. was ignoble. Like the fact that there were already people living there, and that those people's way of life was destroyed by the arrival of others. Even if there's no life on Mars or the Moon, this negative implication of "frontier" remains. Schaad maintains that Europeans picture slaughtered Indians and buffaloes when they hear that kind of talk. "Why would you want to do that again?" he asks. "We've grown so much wiser since then. What people these days want to achieve is some kind of balance, sustainability, ecological viability."

Just how analogous *is* the idea of the earthly frontier to that of settling space? Author Kim Stanley Robinson thinks not at all. For him, Mars in particular is much more similar to Antarctica than to any known frontier. But nobody talks about Antarctica as being a frontier, Robinson points out. "It isn't a place where people can pull up their stakes and say, 'I'm sick of the crowds here in California. I'm going to take my family and I'm going to move to Antarctica and homestead my 160 acres of ice plateau.' It would be a ludicrous concept, and it's even more ludicrous to talk about space that way."

Space diverges from earthly frontiers in other ways. Once you're on your way, you can't just stop and say, "I think I'll settle here." There's nothing between destinations, and you can't just bring your ship to a halt and start it up again. There's a point of no return, and beyond that there's absolutely nothing between you and your destination.

Why do people set out for frontiers in the first place? In the U.S., some of them went for the gold. They went to protect others, as when the army built forts. They went to explore on the government's dime. They went for the soil and the grass, as farmers and ranchers expanded into areas more conducive for crop growing and livestock grazing. They went to take advantage of the growing communities that preceded them. They went for the love of adventure and for religious, political, and cultural freedom. Some went toward something and some went away from something.

One of the most passionate advocates of the need for a frontier is Robert Zubrin, a rocket scientist who founded the Mars Society (see Chapter 8). According to Zubrin, nothing is more important for humanity than the creation of a new frontier. Without it, our entire global civilization and its humanistic culture will die. Period. We're going to decline because we're losing the willingness to take risks and to fend or think for ourselves. We're becoming increasingly irrational and superficial, Zubrin argues. Political institutions are impotent bureaucracies constipating our lives. We're losing diversity. With the frontier we can start fresh. If we move out, technological development will accelerate because we'll need heavy technology in order to survive and progress. This development will benefit both Earth and Mars. Back on Earth we'll be motivated to acquire practical educations and to make real contributions. When we get to Mars, intelligence will be prized, as will the labor of each and every settler. If anything, argues Zubrin, Mars will be more humanist and democratic than Earth. I can't wait to find out whether he's right.

In all the years since the first human in space, Yuri Gagarin, circled the earth in 1961, fewer than 500 humans have gone into space. Twelve have been on the Moon, none to Mars. When it's completed in 2005, *Space Station Alpha* will accommodate seven people at any one time (we hope; recent budget woes have led to talk of only three). The Russian space station *Mir,* which has been deorbited, was able to sustain three people.

Stop people on the street and ask them if they'd like to go into space. Chances are they'll tell you they'd drop everything to go. Then they'll probably say something like, "Whatever happened to space anyway? When Neil Armstrong walked on the Moon, I thought we'd keep going back and back and pretty soon we could all go."

It didn't happen that way. In fact, the U.S. destroyed its *Saturn V* rockets and all its moon-going capability after the last Apollo and Skylab missions in the 1970s, which means that in order for us to go back now, we have to start all over again.

What went wrong?

You'd better sit down for this one.

It's a well-kept open secret, and it shocks people when they find out. The U.S. government space program has always been about foreign policy and jobs. At first it was the Cold War. *Sputnik* scared the U.S. half to death, so NASA was created to beat the Russians at their own game. When President Kennedy gave his famous "We're going to the Moon before the decade is out" speech in 1962, he couched his motives in lofty terms: "We choose not to go to the Moon because it is easy, but because it is hard." The truth is that the U.S. went to the Moon to tell the Russians and the rest of the world not to expand human presence in space, and certainly not to settle there.

After Armstrong and Aldrin bounced on the Moon, there were six more U.S. Moon missions, and then nothing. The political and economic climate had changed. Watergate, which exploded a year or two after the last Apollo mission, fomented huge cynicism about government among the American people and turned them inward. At the same time, the U.S. was hit by the OPEC oil embargo and ensuing energy crisis, which squeezed the American economy and sent a wake-up call around the world. Simultaneously, the Vietnam war ended, and suddenly people stopped worrying about our outward face and started focusing on whether or not they had a job. Americans grieved over their war, engaging in intense and painful postmortems and trying to understand how things could have gone so horribly wrong.

NASA no longer had much to do. The massive organization, with all its expertise and resources, had lost its purpose. So, as bureaucracies do, it scrambled around trying to find something to justify its existence. The space shuttle was born. It was supposed to do all sorts of things: house science experiments, launch and repair satellites, provide information on how humans react to space, conduct astronomical observations, and carry components for a space station. When President Richard Nixon announced the shuttle in 1972, he said its purpose was to routinize transportation to and from Earth orbit. It was supposed to "take the astronomical costs out of astronautics." The cause was noble, but thirty years later, space travel has yet to become routine.

> The shuttle cannot go to the Moon. It can't carry enough propellant to get there, and it can't land there because its airplane-like design requires air through which to glide.

But we did get *Skylab*, the first U.S. space station (the Russians had already launched their first one, *Salyut 1*, in 1971), which was supposed to prove that humans could live and work in space for extended periods. Unfortunately, *Skylab* didn't last long. Its active life spanned six months in 1973, after which it was abandoned for lack of funds. It crashed to Earth six years later.

In 1986, the space shuttle *Challenger* exploded, killing all aboard. This disaster caused a reaction so powerful, so devastating, that all government civil space activity in the U.S. stopped. One consequence was that the shuttle was no longer permitted to launch commercial satellites, and production of expendable launch vehicles, which had fallen dramatically, began a climb back to its earlier proportions. At the same time, the European Space Agency's Ariane Launch System came on the market. Its purveyors intended to soak up 50 percent of global commercial launch business, which they did. Prices were initially low enough to undercut NASA in many cases. After *Challenger,* Arianespace, the manufacturer of Ariane rockets, raised its prices, but it soon found new competitors. The Soviets began to offer commercial launch services at a discount, as did China.

Right before the fall of the Soviet Union in 1991, the U.S. Congress passed the 1990 Launch Services Purchase Act prohibiting NASA from launching commercial satellites. Up until that time NASA had been the only organization in the U.S. allowed to launch into space. This ban effectively created the commercial launch industry and began to chip away at NASA's monopoly. Growth in the communications industry zoomed, with large constellations of satellites providing services like voice, fax, and broadband communications. All those satellites needed reliable launchers to get them up there; this demand fueled the production of expendable launch vehicles. Commercial spaceports began to operate at Vandenberg Air Force Base in California and Cape Canaveral in Florida. The industry has grown so much that now the number of commercial launches is greater than that of government launches. Even with that growth, however, today's launch market is still led by the French Arianespace. Other players include China, Japan, and Russia. There are also some international joint ventures: a U.S.-Russian alliance, and a U.S.-Russian-Ukrainian-Norwegian endeavor.

Space Station Alpha, an orbiting science and materials facility to be used by government and industry, was a long time getting off the ground. It was first proposed by President Ronald Reagan in 1984; NASA's planning efforts began in 1985. The planning period stretched out longer and longer, with the date of completion sliding from an original 1992 to 2005. The first module was finally launched in 1998. Cost estimates have risen over the years from $8 billion in 1984 to $30 billion in 2001. Some analysts put the cost much higher, as much as $100 billion—all for a projected life of only ten years.

The demise of the Soviet Union changed everything. As had happened to their counterparts at NASA in the 1970s, Russian scientists and engineers who had focused on defense and aerospace were suddenly at loose ends. The last thing the U.S. wanted was for them to resume their work on "hostile" projects, so the country supported their participation in *Space Station Alpha*. It would soak them up, give them something peaceful to do, and keep them employed. It would further help the Russian economy by funneling American money into the country for work performed. Once again, space served the goddess of foreign policy.

In order to get humans into space, you need vehicles different from those that carry probes, satellites, and other lifeless cargo. Other than the shuttle, which carries six people, we don't have these vehicles. We have the basic technology—it's not a matter of that. However, in order to get them, a number of other things have to happen:

1. Most critical: There has to be an economic justification for them. Right now there isn't.
2. There needs to be an infrastructure to support them.
3. They have to be insurable.
4. They have to be safe.
5. They have to be not only reusable, but also capable of quick turnaround.
6. Regulations and laws have to be in place that support their use.

Today there is a variety of launch vehicles capable of carrying satellites and remote sensing equipment. These vehicles are all expendable, which means that all or part of them are thrown away after one use! Why so wasteful?

Because expendable launch vehicles are a whole lot cheaper than those you can reuse. And we already have the technology. Besides, in today's market, we really don't need reusable launch vehicles—there's not enough to carry. So we build 'em, use 'em once, and throw 'em away.

Important political considerations affect the space industry today as well. One of the most important in the United States is the bugaboo issue of technology transfer.

Following the accusation of Chinese technology espionage in 1999, U.S. foreign policy specialists decided that commercial satellites represented critical technologies that should not fall into the "wrong" hands. As a result, permission to launch was transferred from the liberal and open Commerce Department to the conservative and, according to some, paranoid State Department. That meant that every commercial launch from the United States had to be approved by the State Department or it couldn't take place. Talk about a bottleneck! The transfer caused delays, cancellations, and economic loss.

Another political problem is NASA itself. The agency has a reputation for hostility to commercial space enterprise, though that seems to be changing. Even though various laws, like the Commercial Space Act of 1999, mandate that NASA purchase from and encourage commercial space enterprises, there's still resistance there.

The financial issue is by far the most critical one affecting whether or not humans get into space. Without its solution, none of the other factors matters. The problem is this: Space is tremendously expensive and the payback is slow, small, or nonexistent. There are two choices for overcoming this problem: Either governments finance the projects because they're not profitable, or someone finds a way to make space activities profitable.

It's unlikely that government is going to back large space projects these days. Constituencies throughout the world are calling for lowered government spending. Even if that weren't the case, the results of the last forty-five or so years in space don't encourage them to contribute their tax monies toward such pursuits. What have we gained from the world's space programs anyway? In order to convince governments to spend at high levels again, there would have to be pressing national or global security issues at stake—a major imminent threat, or perhaps verifiable communication from sentient beings on another planet. If space were seen to be fantastically profitable, government wouldn't become involved directly—it would encourage the private sector to do so, perhaps with incentives like tax breaks or grants.

As the recent failure of the $5 billion Iridium mobile satellite service shows, even when there's a potential existing market and demand is estimated to be high, there's no guarantee of success. When there's no market and only a possible pent-up demand, there's little inspiration for investors to plunk down their money.

Even if investors were keen on space, which they aren't, costs would have to drop exponentially for space to become economically attractive. It costs $10,000 a pound to lift something up via the shuttle. Surprisingly, the same is true for expendable launch vehicles, at least those putting satellites into geosynchronous orbit, 24,000 miles above the earth. (At one time the Russians were charging extremely low prices of $600 per pound, but the U.S. State Department protested. They told the Russians that those prices would ruin the U.S. launch industry and if they didn't quote prices similar to those of American launch purveyors, Americans wouldn't be allowed to buy their services. The Russians raised their prices to $3,000 a pound.)

If we could use the analogies of other transportation industries as our guide, we might be able to draw some useful conclusions. The railroads and the aviation industry grew from nothing to major movers of people and goods. How did that happen, and is space likely to follow those models?

THE RAILROADS

Railroads spurred the development of towns and cities and gave people places to go. Even if you traveled on horseback through the wilderness, in effect the same was true. A place might or might not be developed, there might or might not be other people or food there, but there was air and land to stand on, and you could curl up in your tent or wagon or under an overhang and sleep whenever it was convenient. Not so in space. No place in the solar system is hospitable to humans without significant development, except

Earth. As Kim Stanley Robinson says, "America had an atmosphere when people came, you could get there under your own power more or less. You could live on land and off the land, not just sucking your fuel out of the air but in fact going outdoors—farming, pasturing. None of that's possible on Mars. Getting there and what happens once you get there is radically unlike what it was like in America when Europeans arrived."

But leaving that rather important point aside, are there other similarities between the railroads and space? We know that the railroads spurred the development of industrial capitalism by moving goods from factory to customer. Railroads enabled a growing population to spread out and still maintain physical contact with the East, to receive goods, and to travel back and forth on business and pleasure. Local and state governments became heavily involved, sometimes in funding and loans, sometimes not, as the industry matured. They did so because railroads were a matter of economic development. Governments provided incentives, granted monopolies, and otherwise supported private railroad development. Most early money came from private investors, from farmers living along the route, and from merchants in the cities.

Even though passenger travel quickly became affordable, if not immediately comfortable, the real profits in railroads lay in freight transportation. In fact, rates were substantially lower than those for wagon or turnpike, two popular methods of moving cargo. From 1830 to 1840, U.S. railroads grew from nothing to ascendancy.

Even though citizen opposition to railroads existed, governments and businesspeople enthusiastically supported the industry for the most part. The benefits were obvious, and the economics were always strong. Not so with space. There's a lot of difference between getting out of the gravity well to head for nothingness and laying down track to move incrementally from hub to populated rim and back again. Though uncomfortable and difficult, settlement could occur without the railroads; there were several other ways to get to new places. Not so with space. You either get in a space vehicle or you stay put. And once you get there, you can't live on your own—at least not in a practical way.

There's simply too much infrastructure you can't provide for yourself. So, again, it comes down to financing and public support. And the financing required at this stage is a lot more massive than it was for the rails, expensive as they were: $310 million for 9,000 miles of U.S. track by 1850. In today's dollars, that would be $6.4 billion, but you'd need a lot more than that to really get space going. Each shuttle costs nearly $2 billion just to build, and another $1 billion each time it flies, though obviously we can't rely on the shuttle for heavy-duty transportation. *Space Station Alpha* has greatly exceeded its initial projected cost of $8 billion and is expected by some to cost $100 billion over its expected ten-year lifetime. Numbers like $36 billion to develop and build a fleet of space vehicles are being proposed. You can see the level of investment required. It's a lot more than the money needed to develop aviation and/or railroads, and there are more complicated issues to resolve, like safety and traffic control and health and international treaties.

THE AVIATION INDUSTRY

There's no denying it. In the U.S., France, Germany, and Britain, federal governments were responsible for the birth and nurturing of the commercial aviation industry. Without their subsidies, the economics just wouldn't have worked. Without their policies, there wouldn't have been substantial markets.

Between the Wright Brothers' first flights in 1903 and the onset of World War I, aviation was an embryo stuck in the womb. The Great War changed that. Governments threw money at the fledgling industry, and the infant burst forth, quickly becoming a hyperactive child. The world was awash in airplanes and crazy with technical progress.

After the war, you could pick over the surplus and buy a plane for a song. Lots of people did. Flying your own plane in those days must have been a heady experience. You didn't need a license, and your plane didn't need to be certified. You could roam the open skies, scraping together a living as a thrill-ride operator, aerial bus driver, crop duster, or bootlegger. You charged by the transaction, and flew when there was demand.

It may not have been lucrative, but it was something you could do alone or with a couple of people. Fuel was readily available, the technology was adequate, you didn't have to worry about regulations and treaties, you didn't have to fight the gravity well, and you would not be sued blind if you had an accident. Perhaps even more important than those issues was that there was someplace to go and a reason to go there. You didn't need market studies. You just hustled up a little business and made a little money.

You can't do that in space today. There isn't enough of a market to make your flights profitable (largely because there's no place to go, and to develop destinations requires a lot more than just putting up a few buildings). In addition, the world is criss-crossed with regulations that make it impossible just to get in and go. Furthermore, vehicles are too expensive and incapable of carrying enough passengers, and you need to carry too much fuel with you.

In the U.S., it didn't make financial sense to operate a regularly scheduled airline. People took trains, and they sent their cargo by rail. Anyone who tried to run such an enterprise soon found out that the new way wasn't better enough to attract large numbers of customers. In Europe, however, the market for regular flights actually existed. For one thing, the war had decimated Europe's railroads to the point where they were unusable for many purposes. For another, the English Channel that separated London and Paris was unbridgeable by railroads; a regular air route would be just the thing to connect them. A new, improved transportation system could be built, and people would use it.

They did, but the economics didn't work. Planes simply couldn't carry enough to justify the operating costs. So, the governments of Germany, France, and Britain provided financing and subsidies, and a thriving commercial aviation industry focused largely on passengers began to connect western Europe and northern Africa. Ultimately, the U.S. did something similar, but not with passenger flights. The U.S. Post Office supported, regulated, and subsidized airmail. Eventually, Congress passed legislation that caused the government to turn over its airmail routes to private companies, and the industry was off. The government still had a big hand in through licensing, inspecting, building and lighting runways, and the like, but private entrepreneurs and inventors had a lot of freedom to do things their way.

After Charles Lindbergh's solo transatlantic flight in 1927, enthusiasm for aviation became so inflamed that aircraft production in the U.S. grew by a factor of six, and passenger ticket sales went from 5,800 in 1926 to 417,000 in 1930! Now investment money began to flow into the industry, and major airlines were born. As will be the case with early space travel, it was uncomfortable to fly, but enough people were willing to do so that the market became permanent.

While technology improvements have increased range, speed, and capacity and have helped the industry expand, two factors more than any other caused it to gel: financial assistance and public enthusiasm. Government assistance was critical in compensating for the absence of a passenger market in the U.S. and high operating costs in Europe. Passenger excitement created the mass market needed to break out of the government's cocoon.

Would similar circumstances jump-start a space travel industry? Peter Diamandis believes that his X PRIZE (see Chapter 6) and others like it will impel the public to wild enthusiasm when it's won. If the winning team touches people's hearts the way Lindbergh did, achieving something never before done, with his stick-to-itiveness and courage, then perhaps it could happen. The experience of the Mars Pathfinder Web site deluge in 1998 (800 million hits in about two months) demonstrates that when we enjoy dramatic successes, people get excited. Any stunning achievement in space that shows that ordinary people can participate will have the same effect. Water on Mars or the Moon, sure; an ocean on Europa, yes. But show people that they themselves can go into space and the floodgates will open.

As far as financial assistance is concerned, that's trickier in today's world. The political climate is not favorable to new government subsidies in most countries. In the U.S., Congress, which holds the purse strings, is so fickle that to depend on its largesse and consistency is folly, as a beleaguered NASA will attest.

Governments aren't likely to bankroll entrepreneurs in any case, though they may offer support in the form of tax breaks or other incentives. Venture capitalists require such staggering returns on their investments that their mass participation is unlikely unless the cost of access to space can be brought down drastically or the public evinces such enthusiasm that a huge market is guaranteed. And bankers, well, forget them until conditions are far more stable and the industry is more mature.

All this analysis is not meant to interject a note of gloom. It's just that we ought to be aware that, while some similarities between the aviation/rail and the space industries may exist either now or later, significant differences mean that the space industry may have to invent entirely new ways of bootstrapping itself.

So when you hear people drawing analogies between Earth-based transportation industries and space, remember that there are as many differences as there are similarities.

We're already exploring, and we have an existing satellite industry, but what else could we do in space? Here are a few possibilities:

- *Tourism.* From circling the earth for an hour or two to staying at an orbital hotel to spending time on the Moon, people want to experience space for themselves. We might use space-based facilities for business meetings, hospitals, and sports events as well.

- *Resource use.* We know that space is rich in natural resources like minerals and energy—far richer than the earth. So we could go to avail ourselves of those: clean undiluted energy from the sun, metals, propellants, water.

- *Settlement.* We might set up permanent habitats on Mars or the Moon.

WHAT WE NEED TO DO TO GET THERE

Space is expensive because there's nowhere to go. Even if there were destinations, you have to take your habitat with you, and you must spend fuel to carry fuel. You have to carry your fuel because there are no gas stations in space. If we were able to surmount the fuel problem, either by using some new form of propelling spacecraft or by making fuel as we go, that situation would change and we could carry more farther, which would reduce costs. Once we establish initial settlements with the ability to make their own air, water, and other essentials with indigenous materials, the habitat issue will become less prohibitive. It's chicken and egg, this dilemma, this question of cost. Some day some out-of-the-box thinker will come along and use both the egg and the chicken as fuel or food, but until then, we're stuck with the proverbial question.

Lack of infrastructure holds back development of commercial markets. Infrastructure includes spaceports, traffic control systems, transportation, communications, power, all-weather capabilities. For human involvement, we need habitation, sanitation, agriculture, water, and life support (air, temperature control, radiation protection).

It's not that the Boeings and the Lockheed Martins and the Arianespaces aren't getting human-made stuff off the planet. But they're not opening space up as a place to work and live, and they probably won't until others have taken the initial risks and blazed the way. That's why this book isn't about them. It's about the little guys, the risk-takers, the innovators.

TECHNICAL CONSIDERATIONS FOR GETTING TO SPACE

To get into space, you have to be able to break free of gravity, which requires a lot of power. In order to muster that power, you need to expend a great deal of fuel, at least with today's technology. And in order to get any substantial cargo up there, you also need a lot of power.

Satellites are lofted using rockets that can carry enough fuel to get them out of the gravity well. That's all they need. Once they're up there, the launch vehicle isn't needed any more, so it's discarded and burns up in the atmosphere. You can't do that with passengers. You need a vehicle that not only gets them up there, but can bring them back as well. You need something with life support and the ability to steer finely and to be reused many, many times in order to justify its cost.

One of the issues hotly debated among rocket scientists is whether you can construct a reusable vehicle that's all in one piece—single-stage-to-orbit (SSTO). Most launch vehicles comprise two to three parts (stages), which are dropped to shed weight as the craft climbs. You initially lift with the power of all of them, then drop each spent one as it becomes dead weight, and make it easier for the remaining part(s) to climb. It doesn't sound elegant, shedding this stuff all over the place, but it does make it easier to get past the gravity well. With only one stage, you've got to hang onto the spent parts, and the craft remains heavy. Many rocket designers argue that such a craft isn't possible because getting enough thrust to break through the gravity well and enough mass structure to hold together are iffy propositions. If not, we'll need to travel on multi-stage rockets and either recover the spent parts or build them so cheaply that losing them doesn't matter.

It takes a lot less effort to get from earth orbit to places in the solar system than it does to reach earth orbit in the first place. As science fiction writer Robert A. Heinlein said, once you've reached orbit you're halfway to anywhere in the solar system. The rest is just a matter of steering, with a few course corrections here and there.

— 1 —

Space Tourism? Where'd They Come Up with an Idea Like That?

Until recently, space tourism was regarded as science fiction. However, today space tourism is not only becoming technically feasible, it's on the verge of becoming available, first to the well-heeled, and then to the middle class. But space tourism isn't an end in itself. Supporters see it as a way of building a transportation system that will help us use space-based resources, explore the inner solar system, and eventually settle the Moon and Mars. Read on to find out how this about-to-be industry has leapt off the page and into the real world.

Most people don't admit what they'd really want to do if they could take a space vacation. They'll tell you they want to see the Earth from space or do somersaults in zero gravity. The unabashed ones will tell you the truth: They want to have sex in zero gravity. The thought of floating about without friction or restriction of movement is intoxicating for many. The possibilities seem endless.

So let's talk about that first. What is sex in zero gravity really like? Is it better than sex on Earth? Everyone thinks so, but if you can believe the media, NASA, and the Russians, no one really knows. (Yeah, right. They know. You can't tell me that there's no 100-mile-high club.)

Speculation about sex in space abounds. Real accounts of it—well, try to find one. There are urban legends that circulate every few years, but they're always debunked. NASA denies having conducted sex experiments in space, as does the Russian Space Agency, except on nonhuman mammals like rats to see what happens to their reproductive systems. Which makes sense. There's really no reason to test various methods. People will be more than willing to try themselves.

Harvey Wichman, professor of psychology at Claremont College in Southern California, describes having sex in space this way (though not having gone, he doesn't know from personal experience how well this will work): "To the best of our knowledge, it hasn't happened yet. But let's start with how organisms do it in water on Earth. Very large animals usually do it as a threesome or a foursome. There is an assistant to give a push in the right place at the right time. The nice thing about water is it's quite viscous, and you can imagine that if you've got fins, you've got something to push against. When you're weightless inside a space vehicle, you have to find some mechanical way to constrain the participants. It can be worked out, but it's an interesting ergonomic problem. Sex involves very coordinated activities; making them highly rhythmic is not as easy as one would think. The people who go to space are going to want to do that. If you don't design the facility to make sex easy, they're going to be disappointed."

Some suggestions for sex aids in space include four-legged shorts, a big "sock" that contains the partners, and beds mounted on walls equipped with Velcro-attached sheets and blankets.

If it turns out that sex in zero gravity is as good as everyone thinks, there's the impetus for a space tourism industry right there. Taking tourists to space will not only be economically feasible, but wildly profitable. Everything based on sex is. But no one's talking about it, and despite valiant efforts, no one's yet figured out a way to get from point A (Earth) to point B (a thriving space tourism industry).

It's little known, but there's actually a space tourism industry on Earth already. You can ride in a plane that ascends and descends in such a way that you experience twenty or thirty seconds of zero gravity at a go. Or you can take a jaunt on a MiG 25 Russian fighter plane to 85,000 feet—the edge of space—and see the curve of the Earth. (Space traveler Dennis Tito took one before deciding to purchase a ticket to the *Mir*, a trip that didn't materialize; however, Tito's trip to *Space Station Alpha* did.) However, both of these experiences require time and trouble to prepare for because they're only available in Russia.

But while these types of rides hint at what it's like to be in space, that's all they really do. You can't orbit the Earth, you can't experience sustained weightlessness, and you can't visit the Moon or other solar system bodies. A lot of the reason has to do with international politics.

What Is Space Tourism?

What, besides sex, could you do in space? Patrick Collins, a British economist working on space tourism in Japan, describes some of the possibilities:

- *Look at the Earth.* Author Frank White has published a book of astronaut and cosmonaut interviews called *The Overview Effect* in which space travelers tell of the transforming experience of viewing the Earth from space. Many consider it a spiritual experience that changed their life. They also report that observing the Earth and its features does not grow stale. You can see cities, topographical landmarks, weather, and even fires.

- *Observe the sky.* Undistorted by the atmosphere, the cosmos is stunningly beautiful and easier to see than from anywhere on Earth. And the pictures you can take!

- *Engage in low-gravity sports.* Just imagine flying, let alone gymnastics where you can't hurt yourself by falling and ball games where there's no resistance and the ball can travel almost indefinitely.

- *Observe low-gravity phenomena.* Collins tantalizes you with his list of things to watch: liquids, electrical and magnetic effects, ballistics, animal and plant behavior. He fails to mention watching other people, which also has got to be entertaining.

- *Swim in low gravity or artificial gravity.* If you swim in low gravity, you could also propel yourself out of the water and fly!
- *Walk in space!* This one has to be a show-stopper. It will have to be safe, but imagine what it would be like to walk on top of your hotel or with absolutely nothing surrounding you at all.
- *Hang out in low-gravity gardens.* Growth is likely to be lush, and giant plants will abound.
- *Immerse yourself in simulated exotic worlds* (something you can also do on Earth, though without the reduced gravity).

Harvey Wichman expands on Collins' insights. "When you talk to astronauts, they speak of looking back on our Earth as an experience that changes your life. We do not do that and come back the same person. It is a very spiritual kind of thing, like standing on top of a mountain—a little bit of an epiphany. Astronauts also tell you that looking at the star field from outside of our atmosphere is an awesome experience. Just standing and looking at all the stars shining brightly in the middle of daylight is hard to comprehend. We can't take pictures of this view because cameras don't see the stars. They adjust to the bright sun and miss the stars. The only thing you can really do is paint pictures of what it is like up there, and that too is a wonderful experience. So space vehicles give you two vistas: one of the star field, and the other of Earth."

HISTORY

Even though science fiction writers have accepted space travel and tourism as a given for over a hundred years, it wasn't until the 1950s that someone made a genuine commercial move in that direction. New York's Hayden Planetarium announced that they would accept reservations for trips to the Moon. At least 100,000 people took them up on their offer. In the late 1960s, people got so excited about Stanley Kubrick's movie *2001: A Space Odyssey* that Pan American Airlines emulated the Hayden's example. They accepted tens of thousands of reservations to fly to the Moon.

Then in the mid-1980s, the serious work began. There was a travel company called Society Expeditions founded by a man named T.C. Schwartz. Schwartz, a specialist in adventure travel, and the first to take tourists to Antarctica (a plane still lies on the sea bottom there after the ice gave way—the tourists were all rescued), decided that he should be in the space tourism business. He hired Alaskan-born Colette Bevis to direct a space travel project, and together they coined the term "space tourism."

Schwartz and Bevis began with a market study to see what the economic and technical prospects for a real space tourism industry would be. Part of the purpose of the study was to promote the concept of space tourism. At the same time, they spokc to the U.S. Department of Transportation, the U.S. military, the office of President Ronald Reagan, and congressional staff asking for legislation supporting the development of space commerce.

Schwartz and Bevis were astonished when the announcement of their study at the Press Club in Washington, D.C. made front-page news internationally. Yes, they had signed up people, but to study the market, not to take actual reservations, since there was no vehicle or infrastructure to support such trips. Within the first three months after their announcement, 225 people signed up for the study. They did have to contribute some money—$7,000 to be precise. Five thousand dollars (refundable) of that went into an escrow account with a U.S. bank, with $2,000 and the interest it accrued to be used to operate the study. Participants could withdraw their $5,000 at any time. Participation included priorities for access to the first actual trips, should they occur. Some people took out second mortgages so they could participate. Quite a few of the subjects were internationally famous and/or worked for Fortune 500 companies.

At the same time, Schwartz and Bevis notified their participants that they were seeking ways to create spacecraft for tourism purposes. They put together the first private, that is, nonaerospace company consulting team, which included many people still involved in space tourism: Maxwell Hunter, a rocket scientist who worked across from Wernher von Braun on all the great rockets, George Mueller, a past NASA administrator and current design lead for the entrepreneurial Kistler Aerospace, and astronauts Buzz Aldrin and Byron Lichtenberg. The company was Pacific American Launch Systems, and the vehicle was the Phoenix (see page 39). Pacific American was to provide Society Expeditions with launch services over five years starting in 1992. The service would be a twelve-hour flight in polar orbit for twenty passengers. The vehicle would take off and land vertically.

The first possibility they looked at was a "box" that fit into the space shuttle cargo bay and would carry passengers. Eventually they dropped the idea because it was too difficult to eject in case of emergency. (This idea was later developed by Robert Citron and reborn as SpaceHab. SpaceHab built the module on spec, then persuaded NASA to use it—not your normal business practice. The company now makes annual revenues of close to $100 million.) Then they put out a call for proposals to nonmainstream aerospace companies (not the Lockheed Martins and the Boeings). They selected a design from entrepreneur Gary Hudson—the ill-fated Phoenix. After another press conference, Schwartz and Bevis heard from people in twenty-two countries who wanted to supply launch pads and develop ships. Bevis and Schwartz were as surprised as everyone else when funding began to appear to build the craft and rocket engineers began knocking on their door. They published the results of their wildly successful study.

Then *Challenger* blew up. (The space shuttle *Challenger* exploded in January 1986, killing all seven people aboard.) Bevis spent two intense weeks doing press interviews in support of NASA, stating that all ventures carry risk. She and her sixty-person team worked around the clock to keep the concept of private space travel alive. Who knows what would have happened if they hadn't been contacted by their military liaison, General Abramson, who told them they had to stop further development? The U.S. was just not set up for private spacecraft to be launched in the way that private airliners are. If the USSR or another power saw such craft on their radar screens, they might assume hostile intentions, and a war could start. And so the space tourism industry ground to a screeching halt.

But in the late 1990s, following the breakup of the USSR and its Eastern European bloc, worries about nuclear war diminished and things started to pick up steam again. Other U.S. companies began to follow Society Expeditions' model of taking reservations and putting deposits into escrow accounts. More than that, with the new access to Russian space technology, they were able to offer real space experiences, like zero gravity and edge-of-space flights on which people could experience weightlessness for a few seconds at a time (the former) and see the blackness of space and the curve of the Earth (the latter) (see page 114—you can take one of these flights today!).

Today space tourism consists of these thrill rides and an earthly industry based on themed rides and simulations, visits to NASA facilities, and attendance at space museums, camps, and fairs. Nonastronaut and mission specialist space passengers have been few: a U.S. Senator (Jake Garn) and Congressman (Bill Nelson); a Japanese journalist; and a British chemist. Wealthy California businessman Dennis Tito makes the fifth (he would have visited the Russian space station *Mir*, but it was deorbited before he could get there). An industry it isn't. Much work is being done to help it become one, from market studies to development of vehicles to design of space hotels ... as you will see.

— 2 —

Tom Rogers: Moving Space Tourism From Page Zero To Page One

Tom Rogers has seen it all, from before travel agent T.C. Schwartz's first effort to get space tourism going, to NASA's first official study on the subject in the late nineties, to civilian Dennis Tito's vigorous attempts to fly into space. During Rogers' watch, the giggle factor associated with space tourism has plagued him and other advocates less and less. It's been an excruciatingly slow process, but the times, they finally are a-changing.

The bulk of this interview was conducted in Las Vegas. Rogers and other space tourism experts were meeting secretly with Robert Bigelow of Bigelow Aerospace to help "Mr. B." brainstorm about possible directions for his company. I tried to get into the meeting but was politely rebuffed, so Rogers and I talked comfortably in one of Bigelow's hotels on a bright post-rainy day.

Whenever the subject of space tourism arises, everyone says, "Oh, you must talk to Tom Rogers." Indeed, Tom Rogers is the granddaddy of space tourism in the United States. A physicist and communications engineer who's been associated with space, government, and government contracting since the early 1950s,

Rogers is chief scientist of the Space Transportation Association (STA), a staid (but getting less so) organization whose purpose is to represent the interests of those engaged in developing, building, and operating space transportation vehicles and systems. Its board of directors is full of names of retired military officers and representatives of major aerospace companies; there are also a couple of space entrepreneurs. But it is the Space Travel and Tourism Division (ST and TD) of the STA that is Rogers' pride and joy, and *that* organization is governed and advised by not only the established space industry but also players in the travel and tourism industry, small space and rocket entrepreneurs, ex-astronauts, and financial people. In other words, the ST and TD is a bit radical, and Rogers likes it that way.

What Rogers is particularly proud of is a joint NASA-STA study conducted in 1997 and 1998 on the prospects for a space tourism industry and recommendations to speed its creation and growth. This historic joint venture marked the first time that NASA officially showed interest in fostering space tourism, and indeed, in not acting threatened by the idea. The study (see the "What the NASA/Space Transportation Association Space Tourism Study Found" sidebar) concluded that yes, there is a market for a large space tourism business and that we should, in essence, go for it. With uncharacteristic fervor, the study's government and non-government authors concluded the executive summary of part one with the words, "carpe diem" ("seize the day").

To understand how we got to this point, Rogers takes us back to the mid-sixties. "The first time I began to think about, talk about space tourism, was in 1966. I was in the Office of the Secretary of Defense. My position there was to be in charge of all the research and development that dealt with the command and control of nuclear strike forces—the Atlas, the Titan, the B-52s, the Polaris. I had struck up a professional acquaintanceship with a man by the name of Herman Kahn. Herman worked for the Rand Corporation, and Herman wrote an extraordinary book back in the early sixties, *On Thermonuclear War*, or a title to that effect. He began to think about nuclear war, not in terms of panic and so on but in a professional sense. We had just put up the first eight satellites of the space segment of the first global satellite communications system.

Herman and I were out late one night having a drink, and we agreed on the following two things: one, we would never live to see the end of the Cold War. The second thing we said was when that day came, all of the technology that was beginning to be developed under the Department of Defense and NASA would clearly and quickly be turned to getting American public up into space, and getting the government out of the business, doing what America does. Wrong!"

Later on, when Rogers was working on a space station study for the now-defunct Office of Technology Assessment,

> I got a call, and I can't recall how it came about, from a man by the name of T.C. Schwartz. T.C. was in the adventure travel business in Seattle. This was '83 or '84, and he said, "Do you ever get out to the West Coast?" I ended up in Seattle some months later talking to him and Bob Citron (later a co-founder of SpaceHab). And they had an idea. Why couldn't they put people from the general public up on the shuttle and charge them fares and make a business out of it? T.C. was in the adventure tourism business, so he was used to dealing with all sorts of hazardous, unlikely things, and people with a lot of money and the desire to do a very unusual thing, even with risk associated with it. So we talked and talked, and I said, "How much are you going to charge?" He said, "Well, given what it's going to cost, a few million dollars a ticket." I said, "Forget it." I said, "T.C., the space shuttle is a very costly public asset. I can tell you right now that the friendly federal government will not allow just millionaires to go to space." T.C. finally agreed. We can't just put millionaires up on the shuttle. So then they decided, well, if we can't use the shuttle, we ought to build our own vehicle. They turned to Gary Hudson (space entrepreneur who came up with the Phoenix, a spacecraft modeled on a design by an engineer named Phil Bono) to see if maybe they could build a vehicle. Eventually they found themselves completely over their heads.

They didn't have enough money. They didn't know how tough it was.

It didn't work out.

WHAT THE NASA/ SPACE TRANSPORTATION ASSOCIATION SPACE TOURISM STUDY FOUND

The first space tourism study by NASA, conducted in 1996 and 1997 and released in 1998, afforded the first official sanction to the idea of space tourism in the U.S., and found the following:

After spending hundreds of billions of dollars in public funds on human space flight, the government is still the only customer for human space flight goods and services.

Serious national attention should now be given to creating space travel and tourism businesses.

The general public is more accepting of the idea of public space travel than aerospace engineers are.

Though obstacles exist, the aerospace industry and the travel and tourism industry can work together and with the federal government to overcome them and create a viable, very large general public space travel and tourism business over the next decade (by 2008).

The development path might start with enlarged terrestrial space-related travel and tourism (e.g., simulations, space theme parks, space launch facilities), then move into increasingly higher altitude adventure tourist suborbital flights, short-duration orbital trips, and end up with longer-duration orbital trips and stays at low Earth orbit destinations.

The cost of a shuttle trip for six people stands at $400 million. Turnaround time is half a year, and the chances of a fatal accident are one percent. These characteristics can be reduced by a factor of ten with the next generation of technology, and another factor of ten in the following generation. [That's still $4 million a trip, or $666,000 per person.]

The problem of space sickness needs to be solved, either through medication or the provision of artificial gravity.

What is being discussed in the report is "nothing less than a fundamental challenge to our views of and participation in extra-Earth activities."

You can find Volume 1 of the study on the Space Transportation Association's Web site, located at www.spacetransportation.org. Look for the NASA-STA Space Travel and Tourism Study.

These were the days after Apollo and before *Challenger*. Rogers tells about the struggles that were going on within NASA at the time, which go some way toward explaining how we got stuck.

> By that time, in 1984, the Apollo program was well over, and the momentum carried on for quite a while for two reasons. The first reason was that here was a federal office that had done what it was asked to do. They did it within time, within budget, in the public eye, with panache, in a vital national security activity, to upstage the Soviet Union. And so, they said, "We did what you wanted to do, general public, how about letting us do what we want to do for a while?" And they said, "Fine, what do you want to do?" "Start sending probes out around the solar system." And so that got underway. And there were other things going on at the same time. The whole science fiction world had grown up. We didn't know at that time what was science fiction and what was science fact regarding the solar system. What was possible to do and what was impossible to do. And the other thing was that there was a considerable feeling on the part of many in the scientific community that there

could very well be life forms elsewhere there. And so for the better part of ten years or more, say all of the eighties and for some time thereafter, these two things carried on. And on.

I'd asked many times during the conduct of the (space station) study, looking ahead I could not see any possibility of anything happening but that they'd dribble on. And I asked many people what could happen in the U.S. or in the world that could reinvigorate the civil space program, and nobody could think of anything. So what happened was that momentum carried through to the time of the *Challenger* disaster, and that just scared the hell out of everybody and saddened everybody, and everybody drew the wagons in a circle and said, "Let NASA figure out what to do, press on." And so they did. And that carried us through to about the beginning of this decade [the nineties]. And then it began to become clear that the space station was in very deep difficulty. By the way, the space station's been in difficulty since the beginning, since Reagan said let's go for it in '84. And now over the past several years, the Congress has been thinner and thinner lipped. In the meantime NASA started with Tom Paine, the National Commission on Space and Space Exploration Initiative, talking about humans to Mars. The elder Bush asked for authorization and approval to start human Mars activities, and it was rejected for the combination of two circumstances. One, no one could say we're going to save the world by going to Mars, nobody could say we're going to find gold. But it also talked in today's terms about costing $500 billion. That was a lot of money then. That was the total roughly for all that had been spent on the civil space program for some thirty to thirty-five years.

There was a commission under Norm Augustine [retired Chairman of Lockheed Martin] in 1989 or 1990 that was asked in general terms to take a look downstream about what the civil space program should be. I went to the planned last meeting of the advisory committee; it was held at George Washington University. One of the things they were talking about was making basic research the focal point of the entire program, and the second thing they said, you could expect to get 10 percent per year real growth. Soon thereafter I was asked to talk to the commission, and I might just as well have taken a bath. It was a complete waste of time. And then when the thing was all over I sat back and listened to what they had to say. One of the things they had to say made a great impression upon me. One of the difficulties that NASA faces now is that it receives its authorization and appropriations via the House HUD/VA and Independent Offices Subcommittee, so it must compete with the veterans and housing and so on. Some thought that that's wrong. But one commission member, a former head of House Appropriations Committee, said to the contrary. If what we're doing in the space program isn't at least as important to the general public as housing our poor and taking care of our disabled soldiers, then we shouldn't be doing it. And silence broke out. And I knew at that time that there was the first whiff of change.

Since then I have been working hard, as I see it with the resources available to me, to try to make some deep, deep institutional changes in the civil space area. And to repeat, I had grown up, not as a space scientist or as a space engineer, what I was interested in was the utility of space. What could we do there that would be of economic, social, political value? If you look at the space business entirely, there are two different ways that you can carve it: military and civil. When you look at civil, you can carve it into information and people. The information areas are communications, navigation, remote sensing, weather forecasting, climate. In the human space flight area, we have spent hundreds of billions of dollars, and whatever else you can say about it, you must say that so far there is nothing going on in space involving people with other than the federal government as a

> customer. And more and more as time is going on people have said we shouldn't accept this anymore.

Although science is a worthy goal, it isn't enough in Rogers' view. Back in 1989 after President Bush the elder's speech urging NASA to mount a humans to Mars mission, a national commission advised that the focus of the civil space program should be science. Rogers characterizes this, along with the committee's recommendation that NASA's budget effectively be increased 10 percent per year [which would have given NASA another $100 billion in a decade], as "two of the most ill-advised suggestions ever made to our government." The only way to build a real public constituency for the U.S. civil space program is to give people what they want, and what they want is personal involvement, not the chance to watch a few technicians bounce around in zero gravity. And an effective public constituency for space is something we don't have. Rogers is adamant that "expressions of general public interest in space do not demonstrate the existence of an effective public constituency for large public spending on many of the kinds of civil space activities chosen by our government." Half a billion hits on the Mars Pathfinder Web site does not translate into enthusiasm for spending one's hard-earned money to do the Pathfinder mission in the first place. According to Rogers, the government rationalizes that what they want to do in space is what the general public should expect to pay them to do.

The debate between robotic and human exploration of space has raged for forty years and continues today. Robotic missions, while limited to what machinery can do, are less costly and safer. NASA likes them because it's charged with protecting human safety, and no one risks his or her life on a robotic mission. Human missions allow for more on-the-spot reaction to changing conditions, more in-depth exploration of the kind that can be done by human hands, eyes, and legs—but they're far more complex, dangerous, and expensive. Robotic missions may possess some entertainment value, but essentially the machines are wherever they are to do work. Space tourism for humans—a step beyond exploration because of its mass nature—carries great entertainment value and promises the personal experience of a lifetime. What's more, nearly everyone can do it, and money can be made in the process. (Which is not to say that robotic mining and sample return can't make money, but Rogers sees tourism as the greatest potential wealth-creating opportunity. How much mining and sample return would one really need to do if humans were never to get off the planet and start building habitats elsewhere in the solar system?) Rogers sees the opening of space to the general public as a great gesture of egalitarianism, one appropriate to what he considers the world's greatest democracy (the U.S.). It would also help build that large constituency for the government's own space program, not to mention jump-start the infrastructure for a variety of other activities in space, such as harnessing solar energy to be beamed to Earth via satellites, disposing of nuclear waste in space, beefing up global positioning satellites so that their signals are more robust and secure, and a variety of other projects.

> Now what happened to me about 1990 was this: There was a foundation, the U.S. Space Foundation, and it has an annual meeting in Colorado Springs. I knew the creators of it, and I used to get out there about every year. In 1990, I was making a talk about these things. But the room didn't have many people in it, and a lot of them didn't stay because of the obvious. What they were doing they'd been doing for two generations. It was going fairly well, there was a lot of money, and I was rattling the cage. There was another man at the meeting. His name was Dan Graham. Dan was a retired two- or three-star general, former head of the Defense Intelligence Agency. He and some of his people had come to the conclusion in the early 1980s that the U.S.

should have a space-based defense system to protect against Soviet incoming nuclear warheads on ballistic missiles. They had come to the conclusion that one of the things that had to be done was to get the cost of the system down by reducing the cost of space transportation. And they started about 1990 a space transportation association. When I was talking in Colorado Springs, Dan Graham was in the audience. Later he searched me out, and we chatted. And a few months after that I got a call from him: "How would you like to become president of STA?"

And so Rogers did.

The other thing that happened is that after the OTA study, I was walking down Pennsylvania Avenue one day, and I ran into Dr. James Fletcher. Jim had been a NASA administrator, and I had met Jim during the Atlas days when he worked for Thomson Ramo Wooldridge. He said, "What are you doing?" I said, "I'm doing this OTA study." He said, "What do you know about external tanks?" I said, "What's an external tank?" There was a professor by the name of James Carroll out at the University of California at La Jolla. He had a study contract from NASA that said when the shuttle goes up, its external fuel tank goes up almost as far as the shuttle, and it's let go and destroyed. Suppose you put them in orbit, what could you do with them? So Jim was looking at all this. We became professional friends, and about 1984 to 1987, I started the External Tanks Corporation. Lots of smart people worked on it, but it ended because we just couldn't move NASA at the time to do anything.

I knew the civil space program was in very deep trouble, not for any one programmatic reason, but institutionally. By that time, Gorbachev had come and gone and Yeltsin was in, and we were starting to talk about reducing missiles, silos, and we're still doing that. But here you have another great Cold War institution, NASA, which continues to do today what they did yesterday, because they know how to do it very, very, very well. And they have very smart people, they're nice people, they're hard-working people, and they've grown up to have this extraordinarily large constituency based in Florida and Texas and California. And then Dan Goldin came in. I heard somebody say to me last night, he said, "He's NASA's Gorbachev." But he can't get it to the point where he can get it changed much in any important way. By now people know that big changes have got to come. But if you go to war with NASA, it's very difficult. Because, you see, in the Department of Defense, you talk, you argue, you cajole, and finally if it doesn't work, you kill your opponent before he kills you. You can't kill people at NASA. These are nice guys. It's a very, very difficult thing to know what to do.

I started to give talks about space tourism because I believe that it could be a large new market. I started doing this about the end of 1980. My role in life is to deal with things on "page zero." A lot of ideas can get onto page zero. I try to get a move from page zero to page one, where people start to take them seriously, discover that their self-interests are involved, that there are opportunities and challenges and so on. By the time it gets to page two I'm out of the picture, because by that time a whole new group of people move in. So I'm trying to move space tourism between zero and one.

One day I get a letter from a man by the name of Patrick Collins. Patrick was teaching at the University of London Imperial College with David Ashford (of Bristol Spaceplanes), and he said, "If you're ever over in the U.K., we'd like to meet with you and talk about space tourism." We did, and we came to two conclusions. Number one was that in the United States, I was working uphill because any time that anybody said anything about doing things in civil space, the question sooner or later would be asked, "Well what does NASA think?" If NASA didn't like it you were dead in the water, and that's

> the way it was. But number two, we said in the U.K. and in Europe in general, there wasn't a NASA. And that they could get papers published in peer-reviewed professional journals. So they would start to do that.
>
> One day around 1990, I got a call from Patrick. "Tom," he said. "I have had a chance to take a year's sabbatical," and he said, "I'm going to go to work for a Professor Nagatomo in Japan to work on solar power satellites. What do you think?" I said, "Patrick, three things. One, I'm always for sabbaticals. Two, I'm also for solar power satellites. Three, when you get over there, talk to them about space tourism." All of a sudden I start getting papers and calls. The Japanese caught fire. They said essentially, "We looked at what's going on in the human space flight area in the Soviet Union, good luck, what we want to do is build a business out of it." And they put together a special issue of the Japanese Rocket Society, and I got them to translate it and put it in English, and I put together a paper on space tourism in Japan and its significance to the United States. I started to bring it around Washington. I brought it to the President's science advisor. He said, "We're going to get caught behind the power curve here. Someone else is doing it." And that really began the movement.
>
> I think the Japanese have got a better chance to develop space tourism than we do. The only fully reusable vehicle in the world right now is Japanese. It's an R&D model. It's 3 meters high, and it only flew 3 or 4 meters.

Rogers thinks the Japanese are more motivated than the Americans.

> They ought to be. Patrick Collins observes that Japan is in deep, deep economic trouble. He says, "You know, it's sort of like NASA. What they were doing was just so great through about 1985, all the economists were writing books about how everybody ought to be like the

Japanese." And, he said, "What they didn't do was to appreciate that they were getting into great inefficiencies. The market was pulling away from them. Other people were learning how to build their automobiles at less cost, the labor costs were lower, and they couldn't bring themselves to change. What they've got to do is innovate. So one of these days it's going to completely flip. And all the people that are trying to improve the old things which can't be improved are not going to go fast enough, and it's going to be the new things like solar power satellites and space tourism."

So going back to the Space Transportation Association, how are we going to get the costs down? We started out with a great big plus. The Department of Defense with McDonnell Douglas was working on the so-called DC-X, the first fully reusable vehicle. [This was a prototype, never meant to actually reach orbit.] And it was going along very well. And you don't have to be very, very smart to figure out that instead of throwing the vehicle away, you bring it back and refurbish it. And the second thing is that it would make many, many trips. Every trip you learn something, and you improve it and improve it and improve it. And not only that, but with all the capacity that you've got, you can operate the thing every month, every week, every day, just like an airline flying it. You don't have all these thousands, maybe ten thousand people NASA has on the shuttle. You can really get the technological and operational costs down. But in the Department of Defense, DC-X was really not a mainstream thing for them. Dan Graham, the head of High Frontier, and myself began to get a little nervous about what might happen to the DC-X program. So we searched out Dan Goldin at NASA and we sat down with him, and we said, "Hey, Dan, we don't know each other, but here's what's going on. The train is leaving the station and NASA's not on it." In due course, this led to the X-33 program and the X-34 program. When the press conference announcing the award of the X-34 program to

Lockheed Martin took place in Los Angeles, you had Dan Goldin and the Vice President of the U.S. and Norm Augustine all standing up and saying, "Fully reusable vehicles are good things." And the next thing you know, we've got four or five entrepreneurial companies building up because now it's legitimized.

About that time, about five years ago, I came to the conclusion that we would never get the costs way down unless we had much bigger markets for the new, fully reusable vehicles to serve. As long as you only had one or two or three vehicles, there's no way you can ever get it down. So I started to talk with some people over in NASA about what new things might be done. By that time there's the first feeling over there that Rogers is not mad altogether, maybe we'll just talk to him. To make a long story short, we looked at a lot of things that might become potential markets, and we settled on space tourism. And for the first time in the history of the world, NASA and an independent private sector group decided to conduct a cooperative study with each paying its own way, doing its own thing. It took us two years to do it. Volume One came out in 1998, and Volume Two just came the following year. And it tells you pretty much what's possible.

So now what we're trying to do at STA is to get more and more people involved. We held a conference, the first national conference on space tourism, in Washington in 2000. And we are thinking about doing a few more things. A second space tourism conference was held in Germany, in Bremen. One of the things that happened at about that time is that Richard Branson, who runs Virgin Airlines, said he was interested, he'd been talking with Gary [Hudson]. So we began to see some cracks in the ice. And I think now we're beginning to talk about moving from page zero to page one. We haven't gotten there yet. But I would think short of unpredictable disasters .... All the market studies that have been made tell us that there is a tremendous potential market.

Now let me say something right away. One, we're all inclined to hear what we want to hear. The second thing is that all the market studies bar one have been conducted by space interests—Japan, Germany—and there's bound to be a certain amount of bias. But in the midst of all this about three years ago, there was a major company—at least in Florida—that made the following observation. The tourism business, which is very important to Florida, is continuing to grow, but not at quite the rate, and so they said, "What could be another important tourist business?" And they said, "Suppose people could go to space." They had a professional group [Yankelovich] make some calls and had some focus groups, and Volume One of our study tells you what they found out. Of the total number of people asked, here's the fraction that said they wanted to go: one-third. And it was one-third of 130 million people, which they described as the traveling public, they would stay overnight. And a fraction thereof—7 to 8 percent—would pay $100,000 or more. If this fraction of 130 million would pay this much, how much is the total? If you add it all up it's a trillion dollars. So, to annualize it, say they want to go up every twenty or thirty years, it's bigger than the satellite communications business [about $81 billion worldwide in 2000]. It's tens of billions of dollars. Now even if you multiply those numbers by one over ten, you're still talking about billions and billions of dollars per year. So we feel quite comfortable about the possibilities, with one exception: If a large number of people get killed, then that's up to the gods. We made a discovery during the conduct of the study. And this is in Volume Two. There is a large space tourism business. There's already a large terrestrial space tourism business: people who visit an air and space museum, Cape Canaveral, etc.

> People say, “Why do you want to do this?” The zero-order reason for me is that here we are, one of the largest and in my view the greatest democracies in the world. It’s just not right for the United States of America to have going into space only people who work for the friendly federal government. We’ve got to be out there leading, opening up space to the general public. I’d like to see the United States working with any other country that’s interested. And then I’d like to see us, and this is getting ahead of the game, but here’s a way of starting to pay for what we’re doing in the civil space program. At the beginning of satellite communications, there was no private sector; it was all government money. The government was passing out money. We were sure of the cost. We hoped it was an investment. You can’t tell the latter till you see whether or not money comes back. After less than ten years, money started coming back in the satellite communications area. But in the human space flight area, we’re sure it’s a cost. We don’t know whether it’s an investment or not, but we’re beginning to see that maybe within the next five or ten years, some of the money will come back. But I’m not going to predict. I’m always wrong.

Would Rogers, a near eighty-year-old at this point, go if he could? “Oh sure.” But he gets cagey when asked why and what he’d do there. “It’s none of your business. Now when I say that in space circles, everybody goes into shock because everything that goes on in civil space is planned over years to excruciating detail. Everybody knows, you’re going to flip the switch, you’re going to watch him flip the switch, you’re going to record that he flipped the switch, you’re going to get ready for the next switch.” Then he relents. “I would like to do it because of my acquaintanceship with Michael Collins [the then astronaut who circled the Moon while Neil Armstrong and Buzz Aldrin walked on it]. He used to say to me, ‘Tom, windows. There’s no way you can describe it, looking down at the Earth rotating.’ But I would go first of all because to some extent, I would feel that I owe it to all the people who've asked, 'Would you go?' And I say, 'Sure, I'd go.'"

After the initial interview, Rogers suffered a heart attack. He now doubts that he would be able to go. But he still would if he could.

The Author's Opinion

Rogers is lucky to have been in on space tourism from the beginning. From T.C. Schwartz's efforts in the 1980s—small and ultimately unsuccessful, but important because they planted a seed—to the groundbreaking late-nineties study on space tourism by NASA and the Space Transportation Association, Rogers has seen and had input into it all. No wonder he garners so much respect within the industry.

His most important points are related to public opinion. First, the only way to build a constituency for space is to give the public personal involvement. Not vicarious involvement, but the chance to go themselves. Not watching "technicians bounce around," but themselves getting to bounce around. Second, expressions of public interest do not translate into a mandate for large governmental spending on space à la NASA. Opinion polls evidencing support are fine, "gee whiz" sentiments during robotic Mars missions are great, but neither means that NASA is spending money the way people would prefer if they had their 'druthers.

As we have seen in examining space industry precursors such as aviation and railroads, government's leading the way is often critical in getting new transportation industries off the ground. But there comes a time when watching others grows stale, and doing becomes the order of the day. Freight trains aren't nearly as interesting as passenger trains; airmail service doesn't move people the way jumping on a jet themselves does. If space is going to be all about science, or all about robotic exploration, there will come a time when most people will yawn and tune out, and *that* will consign space to the junk heap.

— 3 —

Harvey Wichman: You Can't Throw Your Socks on the Floor in a Spacecraft

Being in space is a lot more complicated than just floating around. In this chapter, space psychology professor and ergonomic specialist Harvey Wichman explains why messy eaters are personae non grata in space, and why you'd better watch your behind when you swim down the hall in a spacecraft.

Much of this interview was conducted by a Canadian TV crew that was producing a program on space tourism. Dr. Wichman and I met them at John Spencer's (founder of the Space Tourism Society) house in West Los Angeles. As we sat outdoors surrounded by bougainvillea, first Dr. Wichman was interviewed, then I. We both ended up on the cutting room floor.

Living in Zero Gravity

We've all seen films of astronauts floating weightless in space, turning somersaults, swimming through doorways. Wouldn't it be great to do that? Such a feeling of liberation. You could actually fly! I don't know about you, but I can't wait.

However, living day to day in zero gravity isn't all fun and games. There are logistical issues that a lot of people don't think of, all of which have to be designed for and thought through. You may be surprised by some of them. Harvey Wichman guides us through the familiar and not-so-familiar:

> Weightlessness turns out to be *much* more trouble than you would ever think. Trouble in things like brushing your teeth, washing your face, trying to go to the bathroom. You must not have any loose fluids around. You can't have a toilet with water in it.
>
> We design showers for the ISS [International Space Station, now known as *Space Station Alpha*]. Whether they'll ever use them I don't know. And the reason is that water in space does not act like water on Earth or in 1 g behaves. A good way to think of what water in space acts like is think of an automobile that you have really waxed very carefully, and now it rains on it. And what you find is that all the water beads up into globs. That is the way water behaves in space. On Earth, gravity spreads that water out, gravity spreads it over your body when it lands on you with a shower. But when there is no gravity the surface tension of the water pulls it into beads. So, if you fill your hands with water in space, you really can take your hand away and your other hand away, and you just have this big thing up—it looks like Jell-O. And you can push it onto your hand. Now you want to wash your hand, and it just breaks through your fingers in big clots like Jell-O and they start drifting around.
>
> I'll tell you how you design a shower. First of all, water is a precious commodity in space so you design the shower the way you would for a motor home or a small boat. You have a sprayer that you wet yourself with, then you hang it up, then you soap up, and then you take the sprayer and wash yourself off. Here's where the first problem comes. When you pull the trigger on the sprayer and the water rushes out one end, this has become now a water rocket. And you are holding it so

> you and it will now begin bouncing off the walls and this rocket engine shoots you around and you hit the wall. So you've got to have a way of securing the person in there. So what we designed in the shower we did at Rockwell was a big rubber donut on the bottom of the shower that you stuck your toes under, and that would allow you to hold yourself in position while you used your little water rocket to wash yourself. Even at that though, the water is bouncing off the wall back and forth against you and the wall, you and the wall. So to make the water go down, we add a big squirrel-cage fan at the bottom of the shower drawing air into the top. The problem with that is, even room temperature air, when it rushes over you at high speed, evaporates water and it chills you, so you have to run it through a heater first and warm it. And then as this air rushes downward, the bouncing water globules get entrained in it and slowly work their way down and go down to the bottom. Then when you're finished with your shower, we designed a big squeegee that just went all the way around the shower and dried it off, because you've got to dry that all off because you might then have globules of water floating around and getting into your radios and burning them out, short-circuiting them, this type of thing. Because nothing will fall down. It's going to drift away in the air. Generally speaking the best way to deal with cleaning your body in space is to have a nice warm moist towel and some soap. Wash yourself off with that. Have two soapings and then another towel to be kind of a rinse towel and then a dry towel.

What about going to the bathroom?—the question everyone asks. Space toilets use flowing air rather than water to move waste through the system. On the shuttle, solid wastes are compacted and stored on the ship until landing. Wastewater is vented out into space.

There are some minor physiological *differences* between males and females to be concerned about. But for all practical purposes, for short durations, let's say a week or two, they are irrelevant.

Bringing kids with you on a *space* trip might be more challenging than you realize. Wichman designed a simulation of a passenger compartment for kids. If you think about it, we know zip about kids in space. "Most of the space flight simulations have occurred in the military or quasimilitary organizations like the ESA or NASA."

Wichman designed his facility for two *adult* crew members, two adult passengers, and two children between twelve and fourteen.

> Once the children get very much older than fourteen, the parents will probably leave it to the children themselves to go to space, and also they're beginning to be about the size of adults and have adult characteristics socially.
>
> Managing liquids in weightlessness is simply a monumental problem for earthlings. Toilets cannot use water, spitting out toothpaste and rinse water in a basin is not acceptable, nor is pouring from one container into another during cooking. ... Finally, urine collection and vomiting will require special training for space flyers. Vomitus is very corrosive and must not be drawn into apparatus via air currents. Its odor is very repulsive and thus it must not get into filters. The simple "barf bags" provided in seat-back pockets on airliners are not adequate for space flight. Fortunately, bags are now marketed which are very satisfactory for either space flight or terrestrial flight. Training in the management of stomach upsets will be required for space flight but is not required for airliner flight.
>
> Eating in space is not very much different from eating on Earth. The original space foods that were eaten by Apollo astronauts where you squeeze a tube of a whole meal into your mouth, [are] not [used] any more. Pretty much the meal that people first use in space when civilians go will be very much like microwaved

food. What will people drink? They'll drink pretty much what they drink on Earth, but it will be drunk through a straw. A straw with a little valve on it because if you just had a regular straw and you sucked on it, when you stopped sucking, there would still be fluid moving up the straw, and it has momentum and it will just keep right on going and pretty soon you've got a little squiggly Kool-Aid or something out here, and it starts to drift away in the air stream. And you mustn't let that short out your radio so you've got to get hold of it.

You have to be careful with hot liquid. We're used to using our lips to run over the edge of a coffee cup or teacup and the lip is a little tougher than the tongue and it first tests the temperature. You could scald yourself by yanking fluid up into your mouth if it were too hot, so that has to be carefully checked out ahead of time.

When you heat a microwave oven for food in space, with a little bit of careful planning you can have your peas in a sauce. And the sauce kind of holds things together, and the surface tension of the liquid in the sauce keeps it in contact with the compartment that the peas are in on your tray. It's surprising. You can eat a large number of foods that way.

One of the things people have to be trained to do in space, since you cannot just walk because walking requires that gravity pull you down against the floor for you to push against. You translate. That's the term used for moving about in space. So to translate you have to push yourself off from something. The most frequent thing you push against is the wall, so you have to be against a wall or have leverage against a piece of apparatus that is solidly attached to the space station to push yourself away from. Once you push, you're underway and you can't stop until you get to the opposite end. This means that before pushing off to translate, a passenger has to look around and make sure that somebody else hasn't already pushed off and is translating because once they are en route and you are en route, if you end up on a collision course, you're going to collide. And if you're not paying attention, you may bang heads or something like that. It can be a serious issue. So people have to be trained to translate, and there are new etiquettes that go with translating.

If I walk down the aisle of an *airliner* to the bathroom and somebody is coming from the restroom, to pass in the aisle it's a narrow aisle, we each turn to make way for the other. I look at the person, the person looks at me. We smile to indicate that we have no hostile intention, or maybe we nod, and then we pass. And we have momentary eye contact, but not too long, because that's very difficult to do to people. Then we go on our merry way. And that's the only way you can pass in an airliner or on Earth unless you happen to be walking on your hands. In space though, one person could be oriented upright, that is, head toward this end, and the other person, head toward this end. So, you could pass somebody and push off from the wall, this person is working over here, head this way, you're working over here, head down this way, because there is no up and down, and you go by fanny to face. It turns out this has been a real problem for cosmonauts who spend a lot of time in space. I mean that's rude. If you were the one translating, you have elected to pass by someone else, you owe it to that person to reorient yourself before you push off, so that you pass by face to face. Who would even think of this unless you have knowledge from prior space flights like what we have gotten from NASA flights and from life on Russian spacecraft.

When I've interviewed astronauts I've asked them, "What are the kinds of things that really bug people?" What really irritates them? It's when somebody's a sloppy eater. The next time you're in a restaurant look at the floor. If it's a carpeted floor, you'll see the carpet is stained. We do drop food. But on earth it falls down. If a pea rolls off the edge of your fork, it falls on the ground and you hope nobody notices. In space, if a pea comes

> off your fork it goes drifting by and your neighbor is sitting and eating, and here comes your food now, and they don't like it. They've got to reach up with their napkin and capture your peas or carrots or whatever it is in their napkin or your Kool-Aid or Tang or whatever it is that got out of your straw that you weren't careful about. That is a real irritant.

Being in space is going to be different enough from being on Earth that you'll need to think about it carefully before you go. Advises Wichman,

> It requires a little training ahead of time. You've got to be meticulous. When you have people crammed together this tightly you have to really look out for everyone else. You throw your socks on the floor, they go to the floor and they bounce up and drift away in the air, and somebody else working someplace, here comes one of your socks ... no no no. That's no fun. And people are going to get irritated with you very quickly. So people have to be very polite, very courteous. Everything you do, you have to think, "What is the effect that I will have on other people?" And that is easy enough for adults, but not too easy, and way more difficult for children. Children tend to be forgiven and excused in our society for all kinds of nonsense. The children are going to have to be very good about what they do with their things. Can children be taught to do this? Will children find that the environment calls forth this kind of meticulous behavior from them? I think so. And I think that will be part of the fun of it for the kids.
>
> Will people with children be able to tolerate it? My guess is yes, but you see this is why we need these simulations to be run now. You don't have to be in space, in weightlessness, to do a study of the effects of confinement on people. And so now is the time to be doing it. If it turns out to be fun, fine. If it turns out that the children drive the parents crazy, now's the time to find out

> about it before somebody pays $50,000 to take the children to space.

Wichman also explains that air circulation will be a big issue. "Since convection currents do not occur in weightless environments, air circulation that seems natural on Earth does not take place in spacecraft. Warm air does not rise. Thus the failure of a small room fan in a cozy stateroom can be fatal. Any air that is to be moved in weightlessness must be moved mechanically. This is the major cause of the loudness in current spacecraft (about 72 db in the shuttle—like driving on a freeway at 60 mph with the windows down). Quieting spacecraft while assuring that there will be no stagnant air will be a serious challenge."

CONFINEMENT

Weightlessness won't be the only challenge for passengers and tourists. Spaceships and hotel rooms are going to be small. Really small. And if you go on long flights, like to Mars, that's going to become a major stressor. Wichman describes a test he's conducted that examines people's susceptibility to claustrophobia. "You take a 36-inch diameter rubberized ball and you put the person in it and zip 'em up with a little air-breathing apparatus so that they don't suffocate. And you tell them we'll come back and get you sometime—we'll let you know when—and walk away. Now people who could tolerate this would pass the test. Most of the people can't tolerate that." In many cases, that's true even if they know the crowding lowers the price of their ticket because more people can be squeezed into a small space.

Wichman doesn't think these challenges will be show-stoppers.

> Given the constraints that people are going to have on them in the early days of space flight, it's reasonable to ask whether people would really pay lots of money to be subjected to these kinds of difficulties. Difficulty going to the bathroom, separating your urine from your feces, and having to drink everything through a straw. Having to be very

careful about what you eat. You mustn't have things that can come off and go drifting away. I think the answer is yes.

Why do I think the answer is yes? Because I've run a couple simulations of this with people who are volunteers who responded to an ad in the newspaper that said, "Help make space history. Volunteer to participate in a forty-eight-hour space flight simulation." And the kind of people who volunteered were people who were interested in space. Teachers, engineers, all kinds of people. The big question was, and I was asked to do this by McDonnell Douglas after we designed a passenger compartment for their Delta Clipper vehicle. They wanted to know, now that you've designed this compartment, will people really be able to tolerate being there. So, we built a mockup of the passenger compartment in my laboratory, and we subjected people to it for forty-eight hours. And we simulated as carefully as we could a flight into space. Now the one thing we couldn't simulate is weightlessness. The only way to produce weightlessness on the Earth is to fly through a parabolic arc. NASA has an airplane that does this. They pull it up into a steep climb and then they push it over and go into a dive, and during that period while the airplane is going over into the dive, everybody experiences weightlessness. But that only lasts about thirty seconds. And that's the longest you can simulate weightlessness, so our people had no weightlessness.

What we did simulate was the confinement. And we simulated the interior noise. At the present time, spacecraft are noisy inside. And they are noisy inside because on Earth, in a gravity environment, the air that is closest to the Earth is compressed by the air above it. So it is denser. So, as I inhale, the cool air that I bring into my body is warmed to the temperature of my body. And I exhale and as soon as it comes out, it rises like a bubble. And fresh cool air flows into this area for me to inhale the next time. These are called convection currents. In space, that doesn't happen. Air is just as dense at the top

> of the room as it is at the bottom of the room because it is weightless. And the consequence of this is that when you breathe air in and then breathe it out it just goes floating out, and when you breathe it back in you bring it back in again, and then you breathe it out. And, in fact, a critical issue in the design of passenger compartments is you mustn't have any nooks and cozy corners where somebody could get stale air accumulating and suffocate. Any air that is going to be moved through a space vehicle has to be moved mechanically. So, there are motors and fans everywhere in space vehicles directing, moving this air in a nice circuit so that it goes through the filters and absorbs the carbon dioxide out of the air, adds a little bit of oxygen into the air, and then also absorbs all the odors from people's breath and all this kind of thing. Cooking odors, whatever, have to be mechanically or chemically taken out of the air in a filter system. The consequence of this is that the vehicles are noisy inside. About the same loudness as driving down the freeway in your automobile at 60 miles an hour with the windows open.

The experiment was designed as a space vacation. The volunteers looked forward to it; they expected to have fun. "And they did," says Wichman. "They had so much fun that the people took each other's names and addresses and planned reunions. You have to have seen the look on the faces of those people when we opened up the airlock and they came out of the simulator after forty-eight hours. They were exuberant. They just had a wonderful time."

For one thing, you can design facilities in ways that minimize the confinement problem. So far this hasn't been done for real because people who tolerate small spaces have been deliberately selected. But when you do design small facilities, you can turn disadvantage to advantage if you're clever. Wichman explains:

> People love cuddly cozy spaces. I once did a study of dormitory rooms in which students had built lofts. The lofts could be just a bunk bed type of arrangement or it

could be that the bed was put up high and then that left extra floor space available for students. And I remember having the student eagerly take me into his dormitory and show me his study space. And here it was, tucked under his bed, and it was a little tiny crammed cubicle that was sort of like a navigator's station on a World War II bomber. Just this little cubby. He loved it! If an architect had designed that and tried to sell it to someone they probably wouldn't have bought it. But he designed it, he built it, and this was his place and he could study because he was under his roommate's bed. And so he could stay up late at night and have the light down below, and the roommate could be up in the dark. A very clever arrangement.

TRAINING

Habitats for orbital flights are expected to be about 5 by 10 meters—16 by 40 feet—about the size of a Cessna airplane. But these flights will be similar enough to those of the Concorde that existing regulations will cover them. There will have to be some training before you go, and simulation will be part of it, Wichman is certain. "When we actually begin flying to space, the primary purpose of a simulation is going to be to train the people for living in that particular space vehicle. Stop and think about it for a moment. We do train people who fly in airplanes. It takes about five minutes. Flight attendants give a little lecture while the plane is being pushed back and taxied out to the runway. They tell you how to operate the escape doors, they tell you where they are, they tell you how to use a flotation cushion or to find a flotation vest beneath your seat and how to operate it. They don't tell you how to use the toilet because toilets on airliners work just like toilets and bathrooms at home, so that's no problem. When going to space flight, we're going to have to tell people how to work things and how to make things happen and how not to bang into each other when they're floating around and all of that. That is going to require much more than the five minutes that is originally there."

Later on, orbital habitats will be much larger. The passenger compartments in the ships that take the people to and from the hotels will probably be about the same size as those for the first orbital flights, but they will carry many more passengers. The space hotels will have much more room and possibly have areas in which g forces are experienced, such as Bigelow Aerospace's plan for a 0.4 g spinning space hotel. Says Wichman,

> It will probably be necessary for passengers to be certified for space flights. Instead of a three- to five-minute spiel by flight attendants while taxiing out for takeoff, space passengers will need something like a twenty-four-hour indoctrination in which they learn emergency procedures, group behavior dynamics, how to use a space toilet, liquid management, and space flight norms and mores. This will also give people a preflight opportunity to test how well they respond to the cramped quarters that will be typical of the first orbital flights. Just as air crews are now trained in high fidelity flight simulators, space passengers will need to be trained in simulators and then given certificates good for some fixed period of time after which they must be renewed.
>
> Simulator training may also be required because passengers on space flights may need to assist flight attendants with chores and other functional activities. In fact, it is often the case on adventure tours of one sort or another that part of the satisfaction in the trip is derived from actively participating in a meaningful way.
>
> This brings us back to an issue that relates to NASA's selection of astronauts. How does anybody know they are going to be satisfied or capable of tolerating being confined in a spacecraft? One way to do it is to pay $50,000 and go up and fly, and then if you get up there and go berserk or something—I don't know what they do. Wrap you up and stow you, I guess. The way to find

> out is to purchase a simulated space flight on Earth for a couple hundred dollars, and spend a couple days in this simulator and see if it gets to you or not. If it does not, you're probably going to be all right, but it's a way for people to self-select themselves out. As opposed to some governmental agency or the airline deciding who may or may not fly. So, in all likelihood, when space flight does begin, what will happen is people will purchase a simulated flight, during which they get their training. And they'll get a certificate that will be good for some period of time—let's say for the sake of discussion a year. At any time during that year you can fly on this particular spacecraft because you've learned how to use its toilet and its microwave oven and so on. And you know how to operate it there. At the end of the year if you haven't flown, or even if you have flown, you may have to renew your certificate, or maybe every time you fly you get a year from there, that renewal on your certificate. In that way space flight will be different from just running down to the airport and hopping on a flight to someplace. Training will be more involved than the training that takes place every time an airplane flies.

Work is already being done on a training facility. When it secures funding, a proposed National Spaceflight Training Center in Houston will provide education and astronaut training for adults and college students. Trainees will participate in space shuttle and Space Station missions in very detailed, extremely accurate simulators. The Center will also offer preflight training for actual suborbital and orbital flights once they occur. In the meantime, you'll be able to participate in its "Sub-Orbital Astronaut Program," which includes a week of training, touring the Johnson Space Center, seeing IMAX films, and a drawing for a free ticket for a suborbital flight.

No one is quite sure how much training will cost, but you can bet it won't approach the price tag for training an astronaut: $1 million.

PASSENGER BEHAVIOR

From Harvey Wichman:

Another primary way in which space flight differs from airline flight is in the duration of the nights. Because space nights last more than a full day, the opportunity for full-body cleansing and changing of clothes will be required. Airlines don't have to deal with such issues. The problems of water management make this a serious issue. Certainly showers will come late in the evolution of space flight. Once again, simulator training for whole-body cleaning with moist towels will be important.

Airlines with flights that seldom exceed twelve hours in length feel no obligation to provide passengers with facilities in which to engage in sexual behavior. Nevertheless, it does take place and generally falls within the category of nuisance-type passenger misbehavior. However, with the longer duration of space flights many passengers will be disappointed if they don't have the opportunity to engage in sexual behavior while in orbit. Normally, in our society, we simply do not discuss such things publicly. But the management of body fluids in weightlessness where they could be ingested by inhalation is a very serious matter that must be discussed from the beginning. Arranging the necessary restraints to enable a couple to engage in coitus during weightlessness is a significant ergonomic challenge. No less difficult will be devising ways to be sure that all associated body fluids are properly restrained. Awkward and difficult as this may seem, it may be necessary for some simulator training to assure that this activity does not endanger either those engaging in the behavior or others on board the craft.

I am tempted to say that airlines have no obligation to teach people how to behave except in emergencies. However, there currently is much public and airline

> industry interest in what is sometimes referred to as passenger rage or passenger misbehavior. At the same time there is much concern expressed in the media that airlines are not living up to their service obligations. Regardless of where the fault lies, it is too frequently the case that disorderly behavior occurs either among passengers or between passengers and crew members such that an airliner makes an unscheduled stop to remove one or more passengers. This is not possible in orbital flight so actions must be taken beforehand to dramatically reduce the likelihood of such behavioral problems. Recent interviews that I have conducted with airline managers and flight attendants make it clear that alcohol ingestion is frequently associated with in-flight misbehavior. Probably the use of alcohol should be banned on space flights.
>
> There is a significant difference between orbital flight and airline flight in the motivations of some passengers. Airline transportation is often for the purpose of getting from one place to another. The first orbital space flights will be for purposes of vacation and adventure. It is important that passengers be well behaved for safety's sake but also in order for everyone to have a good time. In my laboratory we have demonstrated, in forty-eight-hour civilian space flight simulations, that negative interpersonal interactions among passengers can be dramatically reduced by simply giving two hours of pretraining before a simulated space flight.

All of this suggests that the cursory preflight training for emergencies conducted by airlines as airliners are preparing for takeoff will be inadequate for space flights. Astronauts and cosmonauts are normally socially prepared for space flights by careful screening and long periods of working together before a flight. They also occupy the role of employees, not paying passengers.

SAFETY

Space travel is dangerous. Despite forty years of space flight, we have little experience on which to base design for ordinary passenger travel. And in order for space tourism to attract investment and consumers, every aspect of it will have to be safe. It takes a long time to prove that something is safe. Trial after trial without mishap doesn't necessarily do the trick. You have to test the systems repeatedly under as many conditions as possible. We've had almost 100 years to prove the safety of aircraft, a history that has involved hundreds of thousands of commercial, military, and private flights and some deaths. Now we can predict statistically that flying is safe within a certain margin.

Facilities will have to be guarded from mechanical failure. In addition, adequate numbers of rescue facilities, vehicles, and personnel will have to be available.

In addition to mechanical safety, anyone who goes into space will face the issue of possible harm from other people. The stakes will be far higher than they are on Earth because of the air situation. An intoxicated, unstable, or angry guest or even staff member might be able to inflict damage that would cause the ship or facility to lose life support. In a sense, this situation resembles that of air travel, so could similar security measures suffice?

In May 2000, the Associated Press reported that Japanese airlines had begun to warn passengers that they will be forcibly restrained if they become unruly, which includes threatening others or refusing to heed orders from the crew. The policy came in response to a more than doubling of incidents of rowdy behavior by passengers in 1999 (330 incidents that included drunkenness and sexual harassment). In one case, an offending passenger was held in his seat with adhesive tape. The intoxicated man had been shouting and pushing passengers and crew. Another man hit a flight attendant in the face after she asked him to stop groping a female passenger.

Space facilities will also have to be safe from a health standpoint. That means that we need further biomedical research. Germs will travel quickly in confined space. We all know that your chances of catching a cold rise dramatically when you're on a long flight. How will we prevent epidemics? What do we need to know about taking medications in space? Will people have to be pre-screened for health problems? What kind of training should crews have, and what procedures will be doable and not doable in space?

We'll also have to guard against collision with debris and other vehicles. Theoretically the answer is to get rid of all the debris, but that isn't possible. Therefore, we'll have to know what the odds are of collision, the risk of damage, and appropriate preventive and remedial measures.

Interestingly, many space advocates, although rabid about possibilities for the future, wouldn't go right now even if they could. Patrick Collins cites the safety issue as one of the main reasons he wouldn't go, at least to *Mir*, which is now a moot issue. (He must have changed his mind. In 1989, he applied to fly to the Russian space station, and came in fifth out of 4,000 applicants.) So much for the image of space advocates as fearless daredevils. In all fairness, there's also the time issue, not to mention the cost.

On Earth, adventure travel is increasingly beset by lawsuits from tourists who suffer mishaps like falling off a horse or rafting disasters. This trend means more risk for travel agents, tour operators, and attractions, with the possibility of costly penalties and should be noted in discussions of the economics and feasibility of space tourism.

FACILITIES

If you were an architect or engineer designing a space hotel, you'd have to think about many issues that designers on Earth can ignore. Interior space will be at a premium, but you'll need to impart a feeling of expanse to fight claustrophobia. Lack of fresh air in space will result in more than sick building syndrome; good circulation will be critical. You'll need to design zero gravity areas so that people can move, sit, and sleep easily and without getting in each other's way.

Patrick Collins argues that regardless of these requirements, we could build space hotels today with existing technology. Life support, attitude control, navigation, and communication can all be based on those we've used before. In fact, we've had the technology since the space station *Skylab* in 1973 and 1974. Actually, what we'll need for tourists is simpler than what we need for a scientific research station, though guests will demand higher levels of comfort and safety.

In "Space Hotels: The Cruise Ship Analogy" (see Bibliography), Collins and Fawkes suggest a space hotel model based on cruise ships, which have become larger and grander every few years. This analogy has also been suggested by numerous science fiction writers and space activists, including visionary author Arthur C. Clarke, space architect John Spencer, and activist Jason Klassi. The analogy is an apt one: both space hotels and cruise ships comprise self-contained worlds that must supply entertainment as well as life's essentials. Of course, it will take a while before orbital hotels reach the level of comfort and luxury enjoyed by cruise ship passengers. The following suggestions are based on the assumption of a mature rather than early stage space tourism industry.

Fawkes and Collins suggest the following activities and facilities:

- *Observatory lounge.* Everyone is going to want to look at the Earth and the stars. Therefore, the authors suggest that orbital hotels feature as many windows as possible, with telescopes and binoculars for magnification. Imagine a black, sugar-dusted expanse that fills most of your field of view! It would be hypnotizing. The authors even propose that there be some viewing areas that give the impression of being outside, a challenge for the engineers who design such facilities. You could almost feel that you were outside without having to deal with the life-support and safety issues involved in going outside for real.

- *Promenade deck.* This walking area would hug the perimeter of the hotel and provide a place for exercise, observing, and socializing. A place to see and be seen.

- *Range of cabins.* As on cruise ships, cabins will probably run the gamut from low-end inside rooms without views to large suites with external windows. Each cabin will include "en suite" bath and toilet facilities.

- *Beauty salons and spas.* Guests will not only desire the usual hair-styling/nail/facial treatments but will no doubt be motivated to seek new zero-gravity treatments, especially those that help counter the face swelling that occurs in zero gravity.

- *Gym.* Many earthly hotels already feature gyms. It will be even more important to keep fit in space, and guests are sure to want to try new sports and exercises not possible on Earth.

- *Video and games arcade.* A standard feature wherever kids are found.

- *Cinemas.* One of the issues the authors raise is that of excluding films that might alarm passengers. In the same way that cruise ships don't show *Titanic* and airlines don't schedule films that depict airplane crashes, space hotel managers may decline the opportunity to show movies about hostile space invaders and space disasters. Maybe.

- *Satellite TV.* This service will keep guests in touch with the latest news, sports, and other events back home.

- *Telecommunications facilities.* This amenity goes without saying. Most people don't want to be out of touch any more, so hotels will provide e-mail, phone, fax, and videophones. Optional, of course, for those who want to get away from everything.

- *Formal dinners.* Meeting the "captain" and senior crew may continue to be a highlight for passengers as it is on cruise ships. (Hotels will be a hybrid between ships and buildings, and are likely to feature some kind of captain.)

- *Variety of restaurants.* Resorts don't survive these days unless they offer a variety of food and eating environments. And since space guests are likely to come from many places around the world, hoteliers will want to cater to diverse tastes. The authors also suggest that restaurants will offer a major franchise opportunity for space hotel operators.

- *Bars.* Cruise ships make quite a bit of money on alcohol, which, unlike meals, is not included in the price of a cruise. Imaginative bartenders and liquor companies are bound to come up with new drinks and business opportunities.

- *Entertainment.* A staple of cruise ship life is live entertainment, including shows and lectures, which takes place in theaters or club-like environments. On space hotels, there will undoubtedly be sports displays as well, such as zero-gravity gymnastics, which may take place in special arenas.

- *Swimming pools.* Zero-gravity and artificial gravity environments will necessitate and provide opportunities for new types of pool designs based on unconventional shapes. Author Arthur C. Clarke has suggested a spherical pool; Collins and others are working on ideas for artificial gravity pools in donut shapes.

- *Hot tubs.* Imagine sitting in a hot tub gazing out at the universe. What greater luxury and treat I cannot imagine.

- *Shops.* Because of space limitations and the necessity of importing merchandise from Earth, wares will have to be carefully selected, although in some cases, shoppers will be able to purchase them in space and have them delivered from Earth-based warehouses. Nevertheless, travelers report that shopping is, along with eating, their favorite activity, so there will be plenty to buy. If items are somewhat unique and different from what they can find on Earth ("Made in zero g," for example), guests may find them particularly desirable. Eventually, as humans take up residence in space, a larger and larger proportion of items will be manufactured there, facilitating importation.

- *Art auctions.* On cruise ships, art auctions are run by third parties, adding excitement for guests and revenue for the operators. As with shopping, not all items would be carried on board, but those that would would be extra valuable, with a space hotel added to their provenance (history of ownership and residence).

- *Casinos.* Gambling is a major revenue source for cruise ship operators. Regulation may be an issue, but it's unlikely given cruise ship precedent and the expansion of legal gambling on Earth, particularly in the U.S.

- *Photo and video gallery.* If people love taking vacation videos and pictures now, they're going to be fanatic about doing so in space, which may be a once-in-a-lifetime trip for them. The official hotel photographer could take pictures and videos of the guests, but they will no doubt want to do so themselves as well. Head-mounted video cameras will record guests' activities almost through their own eyes. For those who don't own such devices, rentals and instruction will no doubt be available.

- *Spacewalking.* Even though safety will be a major issue relating to this activity, someone will work out practical ways for guests to walk outside the hotel. Qualified instructors will eventually lead small groups who have completed training, as they do with scuba diving on Caribbean cruises.

- *Bridge.* Guests will be able to visit the hotel's (ship's) bridge, which houses the main controls and systems.

- *Transportation.* On cruise ships, all passengers embark and disembark together. Space hotels will not follow this model. Transportation to the hotel will originate at a variety of spots on Earth, with each following its own schedule. However, economics will demand that significant numbers of guests arrive and depart on each flight, so guests will not be trickling in several at a time as they do in earthly hotels. In some cases, passengers and cargo will fly in and out together. This means that arrivals and departures and loading and unloading will have to be carefully coordinated.

The trip to the hotel will be a short one. The spacecraft will be launched to orbit, which takes only about five minutes. After that, it takes but a few hours of zero-gravity flight before the ship docks. Upon disembarking, passengers will travel along a zero-gravity access tube, holding onto a cable or railing as they propel themselves along. Collins and colleagues who run the Space Future Web site (www.spacefuture.com) advise that the way to move in zero gravity is to think of your center of mass, just behind your belly button. If you push against something, the line of the push should go through your belly button. If it doesn't, you will start to rotate around your belly button.

A number of architects and companies have tackled the problem of space hotel planning and design. Fortunately, they don't have to start from scratch. The basic principles of space habitation have been well understood since *Skylab* safely housed people in the early 1970s. Unfortunately, they have no paying clients ... yet.

Bigelow Aerospace has dived into the orbital hotel business feet first. Robert Bigelow, who's already built a fortune of about three-quarters of a billion dollars on the hotel business, wants to create the "someplace to go"—the destinations necessary for lowering the cost of space transportation. He thinks that the rocket entrepreneurs might be better able to attract investment for their vehicles under such a scenario. He's planning spinning hotels that can accommodate 150 people in an artificial gravity environment. Each hotel would spin around a half-mile axis. Bigelow, who's investing half a billion dollars of his own money in the venture over fifteen years, doesn't plan on building and launching the hotels alone. Rather, he sees his company as part of an eventual consortium that would invest the billions of dollars necessary to do that. On the way, Bigelow may work with aerospace companies like Boeing to build an industrial space station, where artificial gravity mechanisms can be perfected.

Space Island Group in West Covina, California is just one of the companies planning to build an orbital hotel from used space shuttle fuel tanks, which are currently discarded and left to burn up in the earth's atmosphere. (Some of the others have fallen by the wayside, including Hilton, which was interested in collaborating with Space Island but changed its mind.) Each tank is as large as a 747 jet airliner; a dozen of them, joined together, could accommodate from 400 to 600 people. Gene Myers, president of Space Island Group, estimates that the first hotel would cost about $10 billion and the second $5 billion.

The Hawaii-based luxury resort architectural firm Wimberly, Allison, Tong & Goo, known for its elaborate, fantasy-like designs, is one of the most visible and highly credentialed companies working on space design. Its design features both zero-gravity areas and areas with gravity.

Hans-Jurgen Rombaut of the Rotterdam Academy of Architecture has come up with a design for a lunar hotel that uses indigenous building materials and exploits the moon's low gravity and lack of wind. The slender fantasy-like structure, with its suspended teardrop-shaped "habitation capsules" and 500-foot tall slanting towers, would not be possible on Earth. Each capsule, meant to look like a small spaceship, will feature its own freshwater supply and a rubbish disposal unit. Guests will help prevent muscle deterioration during their stay by climbing to the restaurants on top of the two towers, also accessible by elevator. Within the towers, visitors will be able to play low-gravity games like indoor mountaineering and flying.

The architect deals with extreme temperatures and cosmic ray bombardment by shielding the structure with a 50-centimeter-thick hull. The hull's two outer layers are composed of moon rock. In the middle is a 35-centimeter layer of water secured between glass panes. The water absorbs energetic particles and helps mediate the temperature; the rock provides extra protection. Windows are made by hollowing out the moonrock layer, leaving the water and glass in place.

CLOTHING

Why do all space aliens seem to wear shiny silver suits? You'd think that extraterrestrial fashion never changes and never varies from one planet to another. Surely our analogs out there in the galaxy have adopted new styles by now.

For us, silver suits may or may not eventually be hot numbers at space resorts, but one thing is certain: Some functional clothing will be needed for those living and working in space. Such apparel will be necessary in several situations:

- In case of emergency, where pressure and/or life support is lost
- To go outside on places like the Moon
- To accommodate bodily changes that occur in zero gravity

In unpressurized environments above 63,000 feet, people need to wear something that supplies oxygen and that maintains a pressure capable of keeping body fluids liquid. (The air pressure at high altitudes is not sufficient for keeping such fluids from boiling.) Inside pressurized spacecraft, hotels, and space stations, such clothing is only necessary as a precautionary measure. But outside, there are environmental hazards that need to be mitigated. Suits for "outdoor wear" need to operate in a vacuum and protect against radiation, impact from micrometeoroids, and extreme hot and cold. On the Moon, suits will also have to offer protection from jagged rocks while remaining flexible enough to allow for easy movement. Today's space suits are unable to provide maximum protection against radiation, which means that until that changes, tourists will only be able to go outside during periods of low radiation intensity and limit their exposure as much as possible.

Suits for outside will have to contain life support, such as cooling and ventilation, urine-collection devices, and perhaps biomedical sensors and drinkable water. They will also need to be fitted with communications devices. Designers face trade-offs among complexity, cost, flexibility, and weight. It isn't easy to make a suit that is airtight but allows the wearer to breathe and perspire. And making suits to fit all body sizes and shapes is a major challenge.

A spacesuit tear can be fatal. A person can survive for about twenty-two minutes with a one-eighth inch hole in the suit, so both U.S. and Russian spacesuits have backup oxygen packs to keep them pressurized as long as possible. After that period of time, the pressure will drop to the point where the person will suffer permanent brain damage, then death. A larger tear can be fatal almost immediately.

Suits must be designed for the environment in which they operate. Environments with up and down directions offer markedly different challenges from that of zero gravity. For example, suits in gravity-bound environments will have weight. Pumps that keep the oxygen flowing through the suit will face higher loads than those in zero-gravity suits, and pressures may be too high to allow cooling sublimators (pumps that force heat out of the suit) to work.

In zero gravity, clothing should include lots of pockets and anchoring devices for carrying things around. It will also need to allow extra room for changes in body shape during weightlessness. In any space environment, clothing will need to be easy to maintain, comfortable, and tear-resistant.

The Russians have gone a long way toward creating complete space wardrobes. The Russian Space Agency's Institute for Biomedical Problems (IBMP), which designs the clothing in association with Kentavr-Science, Ltd., insists on clothing that is not only biomedically sound, but also comfortable, practical, and emotionally satisfying. Yes, that's right. Cosmonauts may choose clothing of any color or colors they like. After all, they have to look at their surroundings all day and night with no relief.

Cosmonauts may choose from twenty-one different items, including short and long cotton underwear (lace bras and panties for women), outerwear, socks (shoes are used mostly for workouts), warming suits made of synthetic materials for cold weather, a changeable suit made of cotton and lavsan, an operator's suit with built-in springs to work the back and other muscles, and a light suit that comprises a polo-type shirt, shorts, and socks. Cosmonauts also get accessory belts for their tools and whatnot.

The Author's Opinion

I'm amazed that air travelers put up with what they do. Frequent fliers know the routines: You buckle up when you're sitting, even if the seatbelt sign is off; you put your coffee cup on a little tray so that the flight attendant can pour without spilling; you get out of the way as quickly as possible when you're blocking the aisle; and so on.

Space travelers will need to become familiar with many more procedures than that. Just learning how to eat will be a challenge, as will not crashing into people and making sure that your habitual actions don't result in possessions being flung about. What will happen when people have had too much to drink and forget not to stick their rear ends in people's faces or let their clothing float about and block the air vents? And the issue of children in zero gravity—well, there will have to be specific artificial gravity areas for their sleeping and eating if we are not to have to exclude them from space travel categorically.

People on airliners occasionally have bouts of claustrophobia or become ill. Doubtless the same will happen on space flights and in space hotels. Wichman's idea of preflight simulation seems a good one, if time consuming. We can learn much from such trials and may derive solutions from our experience with them.

I wonder how space crews will deal with unruly passengers and guests. Airliners can make unscheduled stops and oust them. You won't be able to do that in space. Will crews be equipped with tranquilizer guns, or even tablets? I doubt that—it smacks of overstepping one's authority.

Obviously, a lot of careful planning must be done in designing vehicles and facilities, in establishing procedures and training for tourists and staff. One problem is that you can't simulate zero gravity on Earth, though you can make extra gravity with a centrifuge. It will take a long time for these issues to be worked out. Right now, all we have is *Space Station Alpha* and the very circumscribed histories of *Mir*, Skylab, and other space projects. How will we prove our designs and concepts? Through long and arduous experience. And that means heightened risk for the first tourists, probably for quite some time.

But I'm optimistic. People learn and adapt to new ways until they become second nature. One day all these concerns will seem quaint.

— 4 —

Space and the Body: Are We Robust Enough to Venture Out?

It's not talked about that much, but the effects of space on the body are a major consideration in whether and where we'll get to go in space. Much remains unknown. Can you take aspirin in space? What happens if a child is conceived and carried in space? If you remain on Mars for a year or three, can you return to Earth's gravity safely? This chapter explains what we know about the physiological effects of space ... and raises serious questions about what we don't know.

Physiological Effects of Space

One of the major issues for tourism, and even more for long-duration flights, is that of how space affects the body. So far, we have lots of questions and few answers. One reason is that few people have spent much time in space at one go. A few astronauts and cosmonauts have lived on the Russian space station *Mir* for months at a time; that's *all* we have to go on. And since we can't simulate the space environment on Earth, we can't study the problem easily. Early flights sent up dogs and monkeys and a few humans for brief periods, which gave us some information about the effects of space on mammals, but because flight lengths were so short, the information is inconclusive for evaluating longer stays.

One of the reasons we know so little is that NASA has specifically avoided situations it can't control. Harvey Wichman, a Claremont College psychology professor who has planned living quarters for *Space Station Alpha* (see Chapter 3), explains. "When we were designing facilities for living on the ISS [International Space Station, now called *Space Station Alpha*], we were designing those for highly selected astronauts. NASA's way of dealing with a lot of the difficulties of living in a space environment was to not try to solve those environmental problems when they could be avoided, because often they would be expensive or added to the weight of things and they were trying to keep the weight down. So their technique for dealing with it was to select astronauts who weren't bothered by those things."

The fact is that we know little or nothing about how people with acute or chronic medical conditions react in space, since astronauts and cosmonauts have been carefully screened for health problems. (Remember what happened to the fourth astronaut in *Apollo 13*? He was kicked off the mission because he'd been exposed to mumps, which he'd never had.) Nor do we know much about people of various ages: babies, children, and people over forty-five—John Glenn notwithstanding—and nothing about how pregnant women fare in space.

ZERO GRAVITY

Most advocates agree that you'll have to have artificial gravity on orbiting hotels for a variety of reasons. First, zero gravity makes people sick, at least in the beginning.

Second, we know from the Russians' experience with space station *Mir* that long-duration exposure to zero gravity results in bone and muscle loss as well as muscle atrophy. Astronauts' experiences involve having to be carried off in a stretcher and sometimes facing recovery times of months. The remedy, at least two hours a day of aerobic exercise, is not a palatable one. While tourists who visit orbiting hotels for a week or two won't have to worry, staff who stay for a month or two will. What if they don't follow the regimen strictly? Will anyone be liable for damages?

Some other physiological changes that occur in zero gravity are immediately apparent. Because there's nothing pulling downward, the blood that accumulates in the legs, feet, and lower torso spreads more evenly throughout the body, swelling places like the face, that aren't used to such opulence. You can feel all stuffed up in the sinuses, and your head may pound. The heart pounds faster as it tries to push out the extra blood that comes into it. Because there's fluid where there normally isn't, the body thinks it has a surplus and expels as much as two pints of water over a couple of days. As a result, you can become dehydrated and anemic because the thickening of the blood signals the body to stop producing red blood cells.

We also know the following about long-duration exposure to zero gravity:

- When returning to 1 g after a long flight, say, two weeks, people experience difficulty maintaining stability of the head when moving from one place to another. They find it hard to focus and keep their balance.

- Muscle volume decreases 20 to 30 percent after long-duration flights, especially in the first four months. Tendons and ligaments also progressively deteriorate. Recovery takes thirty to sixty days. On Earth, muscles and bones are kept in shape by working against gravity. In space they atrophy—there's nothing to push against that has any weight, hence no resistance. As a result, it's easy to injure yourself.

- Muscle strength falls by 10 to 40 percent immediately after long-duration flights. It takes thirty days to recover.

- Bone loss increases by up to 2 percent per month, and we don't know where the plateau is. Not knowing the plateau or endpoint is a real problem. All other physiological systems stabilize at an endpoint, but not knowing the endpoint for calcium can lead to the creation of medical emergencies. Countermeasures are being developed. The best of these are assistive suits that squeeze the body and prevent calcium loss. Such suits also help propel the wearer, which is useful on Mars and the Moon. Other countermeasures being researched include artificial gravity beds.

- Protein metabolism and synthesis are decreased in long-duration zero g.
- The immune system becomes depressed. Several functions of white blood cells are decreased, making for an increased potential for infections from a sealed cabin, microgravity, and floating bacteria. However, increased rates of disease in astronauts have not been noted.

According to Dr. John Charles, a NASA physiologist, we have no useful data on g levels other than 1 or 0, which means that what will happen to us on Mars, during a long stay on the Moon, or under artificial gravity, are mysteries. We do know that countermeasures to zero gravity aren't terribly effective, although you can wear a "g-suit" that compresses the lower body (it's tight in the hips and loose in the ankles) to help sustain blood pressure, and you can ingest salt tablets and water to replace lost fluid. Space travelers will need to return in a reclined position in order to increase tolerance to g forces.

G TRANSITIONS

There's something else we know for sure: Changes in gravity stress the body. Every time you ascend from a celestial object or descend to it, the g level, that is, gravitational force, changes. In addition, if you change from artificial gravity to zero gravity or vice versa, that's a g transition. The shuttle gets to 3 g's on ascent and 1.2 to 1.9 on descent, making at least two g transitions. On a trip to Mars you'd have 0 g during travel, 3 to 5 g's on deceleration, and 0.3 g's on the planet—at least a four-g transition. In order to prepare for such a trip, passengers would need to experience these levels as part of their preflight training.

LARGE G FORCES YOU CAN EXPERIENCE ON EARTH

The highest g forces on today's roller coasters are 6.5 g's on the Drier Looping in Germany and the Moonsault Scramble in Japan. Colossus at Six Flags Magic Mountain in Valencia, California gets up to 3.2; Golden Loop at Gold Reef City in Johannesburg, South Africa, reaches 5 g's. G forces on amusement park rides are not sustained as they would be during space flight, however. Jet fighters maneuver at 6 g's.

G transitions worry physicians the most. They cause the following:

- *Inner ear changes and space sickness.* The inner ear helps maintain balance and orientation. When the fluid and small bones don't detect the pull of gravity, as they normally do, you can become disoriented, though some people adapt after about thirty to forty-eight hours. The reason many people who travel on the space shuttle are sick, some physicians say, is that NASA staff asks them not to take medication before take-off so that they can collect data for research. Precautionary medications can be used before and during fight to control symptoms, however, and training in biofeedback and control of sudden head movements can be helpful.

  When the dizziness subsides, you may still having trouble sensing where your hands and feet are. Dr. Kent Miller, who advises Buzz Aldrin's Share Space Foundation (see Chapter 6), says that it is estimated that 60–70 percent of all first-time space travelers will be sick for twenty-four to thirty-six hours. That may comprise a significant problem for recreational short space travel.

- *Neurosensory problems,* such as difficulty focusing on a target like a display panel.

- *Changes in blood pressure.* Some astronauts have had trouble maintaining blood pressure levels when standing up after landing. This interruption in the body's normal functioning essentially disables the autonomic nervous system (the part of the nervous system that regulates involuntary action, as that of the heart and intestines). Heart rate can reach as high as 170, though it falls quickly. (Normal at rest heart rate is about 60 or 70; 170 is the peak of the aerobic training range for a twenty-year-old.) Physicians tell us that the greatest cardiovascular stress occurs right after landing, which means that if you have to evacuate in a hurry, you could have problems functioning afterward.

- *Difficulty in maintaining postural stability,* that is, the ability to stand upright, after a week or two (with repeated flights, the system learns how to adapt).

Genyo Mitarai, a physician at Chukyo University in Japan specializing in aerospace and equilibrium physiology, addresses the issue of g forces during acceleration and deceleration, processes different from those one encounters in an aircraft. He says that as far as we know, "the maximum endurable level of acceleration depends on the direction acting on the human body. Four g acting from head to foot will cause blackout, while 2 g is the maximum in the reverse direction."

But maybe new designs will mitigate some of the g stress. Psychologist Wichman offers this optimistic prediction: "It seems unlikely that acceleration loads in future space flights will exceed 2 g's, equal to what is experienced in a 60-degree banked turn in an airliner. New rockets being designed for carrying freight to space are not expected to exceed 2 g's on the way to orbit. Part of the reason for this is that objects built for use in orbit can be designed structurally for zero g loads. Thus there is a great advantage in not having to make things sturdier than necessary just to be able to survive the eight- or ten-minute flight to orbit. Passengers will benefit from this factor. Orienting passengers during acceleration on departure, and deceleration on re-entry so that g forces are experienced from front to back across the long axis of the body will be sufficient to manage such loads for almost anyone. The high g loads experienced by astronauts and cosmonauts riding in spacecraft derived from ballistic missiles will become a thing of the past."

Radiation

One of the other critical health considerations for tourists and space travelers is that of radiation, which comes in two flavors: sporadic bursts, caused by solar flares (solar particle events), and cosmic rays, which are constant emissions from other galaxies not dangerous on short missions. Solar flares were a major worry for Apollo astronauts and mission planners. They are immediately deadly to the unprotected. To some extent you can guard against the periodic flare-ups by building a shielded refuge in the middle of your spacecraft or structure, but it would have to be quickly accessible to do any good, especially because people can't move quickly in microgravity. So far, no one has come up with a way to protect the body against cosmic rays or to repair the damage they cause, though chemical radioprotectants (drugs) and some modified clothing have been investigated. On Earth and in near space, we're shielded from them by a magnetic zone, something neither Mars, the Moon, nor deep space has. There is also surrounding radiation from the Van Allen belts, contained in a specific region of space called the South Atlantic Anomaly. These belts trap radiation that is highly dangerous for space travelers, but they also form a shield that protects those who journey in low orbit below them.

Cosmic rays, particles composed primarily of ionized particles, travel at near-light speed. They ram themselves deep into the body, causing a nuclear reaction in the tissues—one that causes massive damage to our DNA and puts us at great risk for cancer. (The National Research Council has demonstrated an increased lifetime cancer risk for voyagers to Mars of 40 percent and has published standards and recommendations for long-term space flight accordingly. Interestingly, this level so exceeds OSHA standards that it is currently illegal for NASA to send people to Mars!) They also cause mutations in the microorganisms that live on our skin, in our mouth and intestines, and in the atmosphere of the ship or facility. Certain mutations could change these bacteria into deadly attackers.

Radiation limits are affected by factors such as the type of tissue exposed. The tissue most resistant to radiation is brain tissue because it's highly specialized and doesn't reproduce. However, other types of tissue require extra protection, which can be incorporated into clothing such as lead shirts and pants. Health effects are separated into early and delayed effects. Dose is a useful measure for evaluating acute effects and the risk of delayed effects. There is no safe dose of ionizing radiation, but guidelines for reasonable exposures (as low as reasonably achievable, the "ALARA" principle) to astronauts have been set and differ according to age, type of tissue, sex, exposure time, and mission.

HEALTH ISSUES ON LONG JOURNEYS

NASA has analyzed various missions conducted between 1988 and 1995 and has come up with following statistics, which can be used to envision what health issues might arise on a long journey. In 1996, NASA and a nonprofit consortium of twelve university laboratories, known as the National Space Biomedical Research Institute, found that of the 279 people who participated in space missions in those years, all but three suffered some sort of illness during the trip, presenting 175 biomedical risks. Four of the risks were classified as Type I: grave, very likely to occur, and without protective countermeasures. Some of these risks resulted from zero gravity. The team concluded that the chance of breaking a bone on a three-year mission to Mars is about 20 to 30 percent, and that an acute medical crisis is likely during such a trip. Part of the basis for this estimate is experience with long-duration missions carried out in close quarters like submarines, where there is a 6 percent chance of an accident requiring emergency care in a year.

The mental health situation on a Mars mission could be especially tricky because the ability to abort the flight and/or mission is extremely limited due to the tiny and infrequent launch windows. Once the crew makes it to the planet, at least there will be resources waiting for them. Because the circadian rhythm during the journey is disturbed, however, sleep quality can be low, and people can become touchy and easily confused. They also can exhibit poor judgment. Other mental health issues might arise as the result of the distance from Earth, monotony, isolation, and confinement. In such an environment, mental health might be precarious, and people could turn on each other. We know that such an outcome is possible from studying missions in similar environments on Earth: More than 10 percent of participants develop serious adaptation problems, and up to 3 percent experience psychiatric disorders like depression. Because of the communications delay between Earth and the spacecraft or Mars itself, real-time psychiatric assessment and treatment wouldn't be possible without a specialist on board.

On the planet itself, one possible hazard is exposure to unfamiliar toxins. The mere knowing that one can't breathe the atmosphere will increase not only physical but also psychological stress.

Returning to Earth will be difficult, too. *Mir* astronaut Andrew Thomas, who spent 141 days on *Mir* in 1998, said that when he returned to gravity, he was dizzy, everything he picked up felt like it weighed a ton, and he had trouble walking. Normal balance skills took weeks to return.

A DIVERGENCE OF OPINION ON SPACE STRESS

There are those who think short trips to space would not stress the system unduly. In a 1999 paper presented at the 2nd International Symposium on Space Tourism in Bremen ("Space Activities, Space Tourism and Economic Growth"—you can find this and all of Collins' papers at www.spacefuture.com), Patrick Collins, an English economist living in Japan (see Chapter 5), thinks that anyone in sufficiently good health to ride a scheduled airline flight could make a short trip to and from space without ill effects and without training. If John Glenn could do it at age seventy-seven, then most other people can, too.

Dr. Melchor Antunano, an aerospace physician from the FAA (Federal Aviation Administration), disagrees. He says that heart patients can fly safely in a pressurized aircraft, but that microgravity poses a significant hazard for them. Because the blood rises suddenly from the lower to the upper body, extra stress is put on the heart. He's also concerned about the effects of recirculated air in such an environment, stating that upper respiratory infections would be easily spread and pose a great risk for passengers with lung and bronchial problems.

In addition, he worries about exposure to very low or no pressure, to temperature and humidity extremes, and possibly to decreased efficacy of antibiotics and other medications, as well as increased virulence of microorganisms. In fact, we don't know at all how medications—even aspirin—will affect us in space, and it will take years to find out. Rapid decompression in space could result in sudden death due to the evaporation of all body fluids, which would occur if passengers and crew weren't wearing pressure suits.

But some health experts feel that aerospace medicine is far too conservative. Part of the reason for the lack of risk-taking is the connection between aerospace and the military. Most of our experience observing health effects on the body has come from military endeavors, especially the flights of military test pilots, but also those of astronauts, who still dominate the U.S. astronaut corps. In addition, there's little experience with women's health issues in aerospace and the military, and physicians, most of whom hail from the Air Force, tend to be overcautious in the absence of real data. This lack is a serious problem and carries implications not just for tourists but for all women who wish to live and work in space.

X PRIZE founder Peter Diamandis (see Chapter 6), who is a physician by training, doesn't think our lack of knowledge is a valid reason not to go. "The human body is an amazing machine that can basically adapt itself to anything," he says. "One of the interesting things I point out to people is all of the changes that occur when you go into space are the body adapting itself to maximize itself for the environment. They're natural. They're what you would expect to happen. It's the body offloading fluids, you don't need this extra muscle mass, you don't need this extra bone mass, so it's the body just adapting itself to the environment. The body is dynamic. It's when you become a skier or a jogger your bones get more dense and your muscles become stronger because you're adapting to the stress you're putting on them. You go into space and it removes the calcium, it removes the muscle because you don't need it. There's extra maintenance, energy cost to maintain that that you don't need in space. So it's the body just being very true to the environment."

Surgery and Medical Treatment

If there were a medical crisis that required surgery on board a zero-gravity ship or facility, both patient and doctor would face enormous challenges. Because there's nothing to hold it together, blood becomes a spray, a collection of droplets that float out all over the room. Tissue density changes, which means that response to normal surgical practices is unpredictable. Surgical instruments weigh nothing in the doctor's hands. We don't know how anesthesia will behave in space, and we don't know how wounds will heal.

Dr. John Charles, a NASA physiologist, says that there's a good chance such crises would occur on flights to Mars. He puts the risk of illness on a flight at one per three missions. There's also a good chance of one intensive care unit admission per flight. How would the crew deal with that? Only so much medical equipment can be taken on these first flights.

One possible way of dealing with these issues is telemedicine, that is, consultation with remote specialists and/or robotic operation of instruments by them. However, we lack experience in telemedicine, especially in emergency situations. The fact that communication is delayed for minutes each way between Mars and Earth also limits how much we can rely on telemedicine, though delays between Earth and the Moon are on the order of just a few seconds and therefore probably tolerable for consultation purposes. Operating instruments remotely is not an option, at least in surgical situations, even with a delay of just four seconds.

Harvey Wichman addresses the issues of pregnancy and birth:

> We have no experience with this at all. The only experience we have in space that is relevant to that issue right now is the Russians trying to hatch quail in space. And they're just having a dickens of a time because the number of abnormal quail hatched from a quail egg that doesn't experience gravity is outrageous. An occasional quail doesn't seem to have much in the way of obvious problems, but it's a serious issue, giving birth in space. A time is going to come when we will have spinning space stations that will produce an artificial gravity. Then all of these points won't matter. But until that happens, they will. Simulations have been run on Earth of a woman giving birth to a baby as it would be done in space, where her body is encased in a plastic bag which has arms that go in so the doctor can manipulate the baby as it comes out. All the tools necessary are velcroed inside the sterile chamber, which is there for the purpose of managing the body fluids from the woman during the birth process.

In fact, we don't know if conception can occur in microgravity or in a high-radiation environment. We don't even know if it can occur in low gravity, as on Mars or the Moon. We don't know how such environments will affect sperm production.

Dr. M. Theresa Verklan, a member of the Space Nursing Society, cites other important unknowns in respect to space childbirth and development:

- How are growth and development affected by microgravity? Will muscles waste?

- How will carbohydrate metabolism be different from that on Earth, and what will its effect be? Bones may not develop strongly. Will the body structures be able to support gravity after landing?

- Will the left ventricle take over dominance from the right ventricle, as happens in a normal baby?
- Babies' heads are sensitive to flow and pressure changes—will they have cerebral hemorrhages?
- How will we deal with babies' defecation?
- The uterine muscle contracts throughout pregnancy, helping tone the fetus. How will it perform in microgravity?
- Where do you put the waste from surgery?
- What will childbirth be like in zero g?

THE AUTHOR'S OPINION

You can find great differences of opinion about the risks of going into space. It is true that practitioners of aerospace medicine tend to be conservative. On the other hand, some enthusiasts may minimize the risks. But here on Earth there is still no agreement about the effect of microwaves and high power lines on human health, so how can we expect consensus on space medicine? Only time and experience will tell.

Zero gravity will be a large part of the attraction of space vacations. Tumbling and cavorting to start with, all kinds of sports—many of which haven't yet been invented, the aforementioned sexual activities, and just simply watching how things work differently from the way we're used to. Therefore, facility designs will probably incorporate both zero-gravity and normal or reduced-gravity areas. Such places will be expensive at first, being more complex to build than simply locking a few modules together. But as with all technologies, the price should come down as the learning curve flattens.

The people who risked their lives to explore Earth knew they might die, as in fact many did. The people who volunteer to go into space in the early days of tourism and development will have to realize that the same is true for them. Plenty will consider the opportunity worth the cost. Nevertheless, it will be critical to inform them of the risks, and there will have to be liability waivers that limit the exposure of carriers and insurers. The issue of such waivers will be complex because of the need for insurers, passengers, and carriers to agree on their specifics.

But I think it can happen. I think the lure of space will be so irresistible that a lot of people will want to go regardless of the risks. There are experiences to be had, prizes to be won, and images to be created.

— 5 —

Patrick Collins: Would *You* Pay $25,000 for a Space Vacation?

We've looked at some of the concerns of being in space. But whether anyone gets there at all will depend on far more earthly factors, like economics. Is there a market for space tourism? If so, how big is it? This chapter looks at some of the answers space tourism researchers have found, and raises a few questions along the way.

Although I had heard Dr. Collins speak at conferences, we first met and talked via e-mail. We finally got to visit in person at a conference in Los Angeles, where all was made clear.

Assessing demand for a new product or service can be tricky. When you ask potential customers if they'll purchase your offering, they might say, "Sure," and then never go near it once it's actually on the market. The way you phrase the questions has a lot to do with the answers as well.

In the case of space tourism, you're talking not only about a brand new service but one outside the realm of anyone's experience. Even if you describe in detail what a space visit would be like, the idea remains very hypothetical. You can say that riding into space may feel like going on a roller coaster, you may liken space sickness to seasickness, you may describe weightlessness as akin to swimming, but in fact, the actual experience will be different from any picture you paint. And no matter how detailed you get, some of your respondents are going to twist what you say, infusing the picture with images from *Star Trek* and other science fiction and passing everything through their own internal filter. So, how can you get responses you can count on?

The unfortunate answer is that you can't, but you have to try anyway. A few valiant researchers have conducted space tourism market surveys, and their results have been encouraging.

MARKET SURVEYS

In 1993, the first major space tourism survey was conducted in Japan by Dr. Patrick Collins, while a Science & Technology Agency Fellow based at the National Aerospace Laboratory in Tokyo. The findings were encouraging: 70 to 80 percent of those surveyed said they'd like to visit space. Collins conducted another survey in North America in 1995 and found that almost as high a percentage of people there wanted to visit space: 60 percent. A 2001 survey by Harris Interactive, sponsored by Space Adventures, found that 51 percent of those surveyed in the U.S. and Canada want to travel into space. Other surveys didn't come up quite as rosy. A 1994 Berlin airport survey found that only 43 percent of those surveyed wanted to go to space, while a 1999 U.K. survey showed that barely 35 percent of those surveyed wanted to go.

People who did want to go were willing to spend a lot of money to do so. Seventy percent of the Japanese surveyed would pay up to three months' salary. Sixty percent of the Britons would do the same thing, as would almost 46 percent of North Americans and Germans. But three months' salary isn't really enough, or it won't be for a long time. Price tags of $25,000 and up are being proposed—a year's salary for the Americans at the time of the survey. Twenty percent of Japanese and a tenth of the others would pay a year's salary, which would meet or exceed that ticket price.

Most people surveyed wanted to stay between a few days and a week. Those who didn't want to go most often cited safety as the reason, followed by preferring to dream, being unable to swing the expense, and feeling that the whole idea wasn't realistic. Some were simply disinterested or felt they were too old.

Right now, these surveys are all we have to go on. Ron Jones and Buzz Aldrin (see Chapter 6) are proposing to do another one—this time a very comprehensive study involving a larger sample, but based on Collins' work.

MIDDLE-AGED AMERICAN MEN AS A POSSIBLE MARKET FOR SPACE TOURISM

A September 1998 article in *American Demographics* describes how "boy boomers" want more and more adventure toys like motorcycles and snowmobiles as they age. In fact, they are the core constituent of the market. In 1980, the average age of the 5.7 million registered motorcyclists in the U.S. was twenty-six. Now, almost 60 percent of motorcycle owners are between the ages of thirty-five and sixty-four. Sales of motorcycles rose from a low of 280,000 in 1991 to 501,000 in 1999 and were projected to reach 644,000 in 2000. (The 1999 and 2000 figures are from a November 2000 article in *Dealernews*.) And these "boys" have money. They're going for fully equipped Harleys at a price between $18,000 and $28,000. More than half of these earn over $50,000 a year; about a fourth make more than $75,000 annually. One of the drivers of this trend is middle-agers' attempt to recapture their youth. Might these fellows be a good target market for space tourism?

HOW ACCURATE ARE THE SURVEYS?

There are so many things you can't account for in market surveys. For example, there's the "jaded" factor. Some people have surmised that once space tourism becomes widespread, it might lose its glamour and become less desirable. Right now, the rarity of travel into space contributes to its appeal. Will this continue, or will travelers need to be offered newer and more flamboyant attractions to hold their interest? There's no way to tell whether such a possibility is reflected in the survey results, and in fact, it would be difficult for respondents to know how they'd feel under such circumstances.

Another factor that may influence results is the widespread perception that traveling to space is stressful, and so only unusually fit people can go. Much of this feeling is attributable to the fact that most space travelers have been carefully selected people. It's hard to say whether the idea of space sickness informs people's thinking, since it's not well known that it occurs. People may be concerned about other sorts of discomfort, like going to the bathroom in space. If people think doing so will be messy and uncomfortable, some may not want to go.

Of course it's one thing to say that you'd go and another to follow through. You have to have the time and money. You've got to be in reasonably good health. You have to live in a society that allows freedom of movement, and you need to be conditioned to accept leisure travel as a desirable activity. Another critical consideration is the state of the economy. Barrett (see Appendix B) mentions various factors: education, life stage, and fashion.

Purveyors of space tourism services can influence few of these factors. The one over which they have the most control is price. They can't raise people's incomes, but they can work to lower costs. They can also give space tourism as positive an image as possible, whether it be one of glamour, excitement, relaxation, or wonder. But they'll need very good public relations because there will be accidents, deaths, injuries, and other negative events that will tarnish the image. Because space travel will be essentially a discretionary activity, they'll need to work hard.

What to Do About Price?

A partial solution to the money problem is that of a lottery, as suggested by moonwalker Buzz Aldrin. A lottery would allow at least a few people who couldn't normally afford a trip to go.

Lotteries are extremely popular in the U.S. and the U.K. In 1997, 79 percent of the U.K. population played the main weekly draw of National Lottery for a total of £3.2 billion, about $4.8 billion. In the U.S., lotteries took in about $38 billion in 2000. (In comparison, the U.S. personal computer industry totaled about $16.7 billion in 2000.) However, these figures represent people hoping to win *money*. Parting with one's money in order to win a space trip may prove to be an entirely different story.

Aldrin and his nonprofit Share Space Foundation want space travel to be an egalitarian phenomenon, particularly in light of the high ticket prices. As part of their mission to establish a credible foundation for the space tourism industry, they're trying to come up with a way to enable at least 5 percent of tourists to be chosen at random. He admits that it's very difficult to get a national lottery going. In the U.S., too many states are opposed to the idea. For one thing, a national lottery would take away from their own revenue generated by statewide lotteries. So Aldrin has been looking for corporate sponsors to underwrite the would-be tourists instead.

SUPPLY AND DEMAND

Getting from here to there is the challenge. If the surveys reflect reality well, then there is a substantial potential market out there. But the real market—the one that exists today—is a different animal. Currently, fifty expendable rockets are built each year; they launch fifty payloads. According to Collins, if reusable launch vehicles were developed that could fly once a week, there wouldn't be enough business for them. Even if there were only one reusable vehicle, it could launch all fifty payloads and meet the entire annual demand for launch services, so why build more? Conclusion: The demand for expendable launch vehicles would drop to zero, and manufacturers would be out of a market. This situation accounts for the disinterest of manufacturers in developing reusable launch vehicles. In addition, the government space agencies that have supported the manufacturers want to protect them: They effectively subsidize the companies and have largely, though not totally, resisted funding the development of such vehicles. NASA allocates about 2 percent of its budget for this purpose; the European and Japanese space agencies earmark less than 0.2 percent.

If manufacturers made fifty reusable vehicles each year, in five years their fleets could handle 12,500 flights! Therefore, for rocket makers to have as much work as they do now, demand would have to be fifty times greater than it is today, and the market would have to grow extremely quickly in order to keep each year's new production busy. Collins calls this the 50:50 problem. The number of satellites launched today and in the foreseeable future couldn't possibly meet that requirement. Therefore, Collins argues, a new market needs to be developed. In his opinion, passenger flight is the only market capable of generating that kind of traffic.

Tourism probably will start with short trips lasting a few hours, then grow into longer holidays requiring orbital accommodation. Then people will need accommodations. The demand for them will emerge gradually. At first, these facilities will be "hostels," then larger and more sophisticated hotels. Eventually, the cost of the transportation component of a space trip will fall and more money will be spent on accommodations, though both will become less expensive as demand grows.

But the growth of the industry will depend on the robustness of its market, and even though people say now that they'd spend this or that, will they think so in ten, twenty, or thirty years? Despite survey answers, no one's even sure that given the chance, such a market would exist today, and we certainly don't know whether it will exist in the future. One factor that might make the difference between success and failure is the development of a different kind of propulsion—lasers, for instance, which may make space transportation a whole lot cheaper. Since no such alternative exists today, we have to come up with business plan-type numbers based on today's realities. That is what Patrick Collins and the Japanese Rocket Society have attempted to do.

If economic conditions do change substantially—for example, as a result of global warming—what might the result be? If the world's weather becomes more extreme as it has in the late 1990s and early 2000s, more money will be spent on new infrastructure and cleanup, by governments and individuals. Or, consider the decline in fossil fuel supplies—same outcome; there's little that affects economies in as major a way as energy prices. In either case, people will have less disposable income to spend on pleasures like orbital trips and space holidays, no matter how much they might want to go into space at least once in their lives. Then what?

THE COST OF GLOBAL WARMING

If our weather becomes more extreme, world economies could be affected profoundly. Two Public Interest Research Group (PIRG) studies found the following:

Worldwide, the economic loss related to natural catastrophes in the 1990s was 7.3 times the comparable cost in the 1960s, adjusted to present value.

Damage by Hurricane Andrew in August 1992 cost more than $30 billion—more than twice NASA's annual budget—and left a quarter of a million people homeless. (The whole U.S. VCR market is about $2 billion.)

In 2000, weather-related natural disasters around the world caused $31.4 billion in economic damages and the loss of 8,851 lives. Weather-related natural disasters around the world in 1998 caused $89.5 billion in economic damages and the loss of nearly 44,000 lives.

Economic losses due to extreme weather in the United States amounted to $204.3 billion in the1990s; worldwide losses were $646.2 billion.

Insured losses worldwide due to weather hovered between $1 billion and $5 billion per year from 1960 through 1988, with the lower figure being the most common. Since then, they've reached as high as $25 billion, with all but three years exceeding $5 billion, and two exceeding $10 billion. (The entire computer data storage market is just over $40 billion.) U.S. insured losses from weather-related catastrophes from 1990 to 2000 totaled nearly $94 billion. (The size of the entire global satellite industry in 1998 was about $66 billion.)

Source: "Flirting with Disaster: Global Warming and the Rising Costs of Extreme Weather," October 27, 1999 and April 17, 2001. See www.pirg.org.

ECONOMIC BENEFITS OF SPACE TOURISM

If people end up adopting a more frugal lifestyle for the reasons just stated, what will happen to the economies of the world? Some economists argue that widespread adoption of a low-consumption lifestyle will stunt economic growth. Patrick Collins thinks that the space industry could reverse this trend, bringing unparalleled prosperity to the world. In fact, he goes so far as to say that the space industry is acting irresponsibly by ignoring the opportunity to develop space tourism services. Collins thinks that low-cost access to space will provide companies with new opportunities for diversification, and the pressure of competition in existing industries will ease. At the same time, there will be less incentive for companies to encourage wasteful practices. He thinks combining a lower-consumption lifestyle with a large and growing commercial space travel business will lead to unlimited economic demand and high employment. Space tourism will earn an economic return on the huge investment governments have made in developing space technologies that haven't reached commercial markets. By creating new industries, space tourism will also help counteract today's deflationary tendencies, which, he argues, are caused by overcapacity in older industries.

Collins also believes that space tourism will be socially beneficial by raising standards of living all over the world. As he sees it, we'll have more resources, less despair, and less competition among countries, and people will enjoy more career choices.

Collins cites Adam Smith, who favored the division of labor for its ability to promote economic growth. Smith maintained that the larger the group of people cooperating economically, the wider the range of activities they perform. Steelmaking wasn't developed in a society of only 100,000 people, nor was electricity generation developed in a society of only a million or computers in a society of 10 million. A wider range of activities is necessary in a larger population if economic growth is not to lead to unemployment. As the size of the population increases, industries operate on a larger scale and need a smaller proportion of the population to produce ever greater output. Those who leave the older industries as they restructure are re-employed in new industries. Continual innovation is needed in parallel to generate a growing range of industries. This innovation (read "space tourism industry and other space development") is what Collins thinks we need in order to propel our economic engines.

Collins asserts that space passenger travel isn't available today not because it's difficult or expensive to develop—at least by comparison with current government budgets for civilian space activities—but because government space agencies aren't trying to develop it. He says that since governments assume that passenger space travel will be a private business activity, it isn't their responsibility. (This is true today, though in the past—see the Introduction on the development of the airline industry—it was not, at least not as completely.) Collins believes that they have the wrong idea, and if they continue to ignore the demand for passenger travel, their budgets will continue to be cut due to declining popular support because their activities will be of interest primarily to themselves.

COSTS

While it's difficult to estimate how much it will cost to develop the infrastructure for even the most basic of space holidays, someone's got to try. Patrick Collins and the Japanese Rocket Society have come up with a plan that involves building 52 fifty-passenger craft over a ten-year period. They estimate the development cost at $12 billion, and the cost of manufacturing each vehicle beyond the prototypes, of which there will be four, at about $700 million. When you consider that the sale of each vehicle will recoup some of the costs, Collins puts the net development, certification, and manufacture cost at $36 billion.

To put these numbers in perspective, let's look at costs associated with developing certain kinds of airplanes. While such vehicles are not strictly analogous to space tourism craft, they're the best comparison we've got. We can't use the space shuttle as a model because 1) it's a government project involving layers and layers of "pork" that wouldn't be present in a commercial venture; 2) it carries only seven people; 3) it doesn't have to make a profit, and couldn't at a cost of $10,000 per pound carried; and 4) it flies only three times a year rather than once a day.

THE MAGIC NUMBER: $12 BILLION

Consider the following statistics about aircraft development budgets:

- The Boeing 777 cost somewhere between $5.5 billion and $12 billion to develop.
- The Boeing 747, developed in the 1960s, may have cost on the order of several billions of dollars. The company admits that they spent $750 million to develop it. Cost overruns were horrendous, and some people connected with the project assert that they brought the bill well up into the billions. In today's dollars that would be about $3.8 billion.
- The new Airbus A3XX, which will carry between 470 and 650 passengers, will cost $12 billion to develop, according to the company, though rival Boeing claims the tab will be closer to $20 billion.
- The Boeing 707, which carried about 140 people, cost twice as much to develop as predicted, nearly $400 million in the 1950s ($2.4 billion in today's terms).
- The Concorde, which travels at twice the speed of sound and carries 110 people, cost $12 billion to develop—in the 1960s. In today's dollars, that would be about $55 billion!

> Half of that cost was for the propulsion system, which was based on already existing technology. (The Concorde is powered by modified Rolls Royce Olympus engines, which were originally developed for U.K. transonic jet fighters with an operational life of 2,000 hours. The engine had to be modified so that it achieved a 60,000-hour operational life and an acceptable noise level. Even though it was based on existing technology, it still cost $6 billion ($36 billion in today's terms)!

Looking at these examples (why *do* so many of these craft cost $12 billion to develop?), one could assume that the development of space vehicles on the order of the Japanese *Kankoh-maru,* a fully reusable one-stage vertical take-off and landing launch vehicle for passenger service, could cost substantially more—orders of magnitude more. No one has ever developed a feasible reusable passenger space vehicle before. Who knows what challenges might arise? New technology costs a lot more to develop than systems based on existing technology. Engineering is about experience, and the more experience you have, the lower the cost. Each of the cited planes ran into huge cost overruns. If that happens when the technology is known and the industry is proven, what could happen with something brand new?

The Concorde was about a lot of new technology, even though its propulsion system was based on technology already in use. The fact that the Concorde flies higher than other passenger planes meant that a lot of additional work was necessary to determine whether it was safe for passengers to fly so high. The 707, 747, 777, and A38x all fly at the same height—a standard one—and are subject to the same certification regulations. No new certification regimen was required for them. No wonder the Concorde cost about four times what the 747 did! Therefore, it's not illogical to assume that the *Kankoh-maru* will cost at least four times as much to develop as the 777, Concorde, or A3XX.

Collins isn't worried. He feels that we're already most of the way toward developing the necessary technologies. He's not daunted by the high stress on the vehicle resulting from the fight against Earth's gravity well, for example. He feels that modern materials cope with the heat just fine. Pressure isn't an issue either, he believes, being close to that experienced by jet fighters on a routine basis. (Jets maneuver at 6 g's, while the maximum acceleration of the *Kankoh-maru* would be 3 g's.)

COMPARISON WITH THE AVIATION INDUSTRY

Comparing timelines, contributing factors, and costs of space tourism with those of the airline industry is a favorite pursuit of those working to develop the former. "It took sixty years for the airline industry to mature and for air travel to be considered safe ...." "You can turn around an airplane in X amount of time; therefore you should eventually be able to turn around a spacecraft in the same amount of time ...." "The 777/747/Airbus 340 cost X amount of dollars to develop; therefore, a spacecraft that can carry X number of people should cost Y amount of dollars to develop ...." Even though the scale of space tourism activities may be smaller than the airline industry, analysts assume that similar operating concepts will be applied to these new services.

How analogous is the aviation industry? First of all, let's set one thing straight. The American and European aviation industries grew out of two governmental activities: war and airmail. They did not evolve from private enterprise acting on its own. World War I jump-started the industry, with both technical improvements and numbers of planes increasing exponentially. After the war, leftover planes in the U.S. became toys. You could buy one for a song. You didn't need a license for yourself or certification for your plane in order to fly; you just flew, and lots of people did. They gave rides, dusted crops, mapped terrain, took commuters from place to place, and ran bootlegged booze. What you didn't do was open a commercial airline with published routes and schedules. U.S. railroads worked just fine for getting people and goods from place to place. They were accessible, reliable, and safe, whereas airfields were hard to get to, and planes were small and not as safe as most could wish.

In Europe, it was different. Railroads had been severely damaged; the door was open for something new. In France, Germany, and Britain, governments came to the aid of startup airlines (though after some delay in Britain due to then Secretary of State for Air Winston Churchill's insistence on self-reliance for private enterprise). They provided handsome subsidies that jump-started what would eventually become Air France, Lufthansa, and British Airways.

Eventually, the U.S. began to catch up. The key was airmail. It took a while to get going. At the beginning, there were no maps, and pilots had to recognize topographical features from the air. You couldn't fly at night. But eventually runways were lit, pilots were required to pass qualifying exams, aircraft had to be inspected, and the system became institutionalized. Who paid for it all? The U.S. government.

In the beginning, U.S. passengers were tag-alongs. Not so in Europe. With government subsidies, European and British airlines were building real passenger airlines. What really got passenger airlines going in the U.S. was the public enthusiasm inspired by Charles Lindbergh's solo flight across the Atlantic in 1927. After Lindbergh, people *wanted* to fly. In 1926, 5,800 people bought air tickets; in 1930, 417,000 tickets were sold in the U.S. In 1926, a thousand planes were built; in 1929, 6,200 were made.

And that's how it started. Commercial aviation was not profitable at the beginning. (In the case of the popular Airbus, which is heavily subsidized by European governments, it still isn't.) Those who tried to run commercial airlines soon discovered that operating costs were too high for the number of passengers the planes could carry. There was also stiff competition from other modes of transportation, especially railroads. The evolution of the industry was slow, and governments helped it along in major ways by offering financial support and by becoming customers. (The militaries of various countries bought lots of planes; in fact the German Luftwaffe and the U.S. Air Force were instrumental in the development of the jet engine/jet plane, as was Charles Whittle in the RAF in Britain.) They also provided infrastructure like airports and traffic control and took responsibility for certification, licensing, and other safety issues. The big question is, if governments hadn't helped as substantially as they did, would there have been a mass commercial aviation industry today? Answer that, and you may have the answer to the commercial space question.

The early aviation and space travel industries are not completely analogous. In the case of air travel, there were people scattered about who needed transportation. In space, that isn't the case. There's no one and nothing out there other than *Space Station Alpha* and some satellites that need sending back and forth. So, which comes first—the transportation or the settlement? In either case, how do you justify the endeavor economically, and where does the money come from? On Earth, it was the need to get mail from one place to another that did it, or it was the need to replace broken rail routes. There are ways to get satellites up and down and to repair them without developing a whole new transportation infrastructure. *Space Station Alpha* can hold only seven people; its existence, which is slated to end by the year 2015, doesn't justify a whole new system, either. So what does?

As for economics, it's interesting to look at some of the numbers. In aviation, basic decisions such as the number of passengers a vehicle could carry made a huge difference to the economics. This is where Patrick Collins' and the Japanese Rocket Society's assumptions become so important. At fifty passengers a craft, how do the numbers look? How do they look with larger capacities?

RETURN ON INVESTMENT

Patrick Collins notes that over the last fifty years, taxpayers around the world have spent almost a trillion dollars for civilian space activities, with about half going for human space flight. These expenditures, in his opinion, have resulted in little return on investment: a few tens of millions of dollars per year from human space flight. If the trillion dollars had been invested in commercial endeavors, it would have fueled a major industry: Several hundred billion dollars a year would have been generated, more than ten million people would have had jobs, and annual profits would have amounted to tens of billions of dollars.

At current spending levels, over the next thirty years government space agencies will spend $750 billion, which will maintain employment for its current level of half a million people worldwide. By contrast, $25 billion invested by businesses typically generates annual sales turnover of $25 billion per year (though the investment-to-revenue ratio varies with each industry). The profit margin might be about 10 percent, which allows the businesses to repay investors perhaps twice their initial investment, say $50 billion over twenty years, while creating 500,000 permanent jobs. Over thirty years, $750 billion could create permanent jobs for 15 million people or more. Collins invites us to contrast these two outcomes. The $1 trillion taxpayers have paid for civil space activities would have generated a space business comparable to civil aviation, with annual turnover of hundreds of billions of dollars if this were commercial investment.

Such a proposal requires a shift in the way we think about government. Traditionally, governments did not try to make money. Rather, they spent money on things the private sector couldn't, and when those projects were far enough along, then and only then did private industry assume responsibility for them. In recent years, more people have started to suggest fundamental changes in the purpose and behavior of governments. So far, those changes haven't happened to any major extent.

Telecommunications and broadcasting satellites generate $20 billion per year in revenues. Governments spend $25 billion per year on civilian space activities, which bring in no appreciable revenue.

PRICES

Economists have estimated demand and outlook for space tourism from the market surveys mentioned previously. Much of their focus has been on prices, i.e., how many people will travel for what price and how much capacity will be needed to support them.

In 1985, Society Expeditions offered short orbital flights for $50,000 each, though as it turned out, they weren't able to provide such flights in the end. They received several hundred deposits of $5,000. (If 300 people had signed on, they would have taken in $15 million.) The company estimated that demand would reach about 5,000 per year at that price (a total of $250 million), and 30,000–40,000 per year at $25,000 (a total of $1 billion). At half a million dollars, they predicted a demand of just 100 passengers a year (a total of $50 million). However, these tickets would take people into orbit for only a few hours. No hotels or long stays were involved in the assumptions, although there would be a long orientation period on the ground of seven to ten days.

In a 1997 article ("Space Tourism—How Soon Will It Happen?"), David Ashford of Bristol Spaceplanes Limited predicted that once orbital tourism is established and viable, a vacation at an orbital hotel will run about $10,000. He thinks at that level, the number of tourists would be at least a million per year. (Looking at the figures in the Japanese and North American surveys, Ashford first came up with fifteen million per year but revised the figure downward to allow for excessive optimism among respondents.) Ashford based his price on an assumption of reusable launch vehicles being as mature as airliners are today—a state that, in the case of the aviation industry, has taken nearly a century. In that case, vehicles would be able to make several flights a day and be in service for twenty years, just like today's commercial airliners.

Patrick Collins thinks that world demand could be six times greater than that in Japan, resulting in a $12 billion per year market at a price per trip between $12,000 and $24,000. Figures prepared by him and the Japanese Rocket Society indicate a profit margin of 33 percent—well above that of the airlines. Such figures are based upon a per-passenger per flight price of about $30,000 and fifty-two vehicles flying almost once a day.

> Certifying a new vehicle involves more than 1,000 test flights and more than three years in order to collect extensive statistics. The cost of certifying a small business jet can run in excess of $100 million.

The JRS study concluded that the development of *Kankohmaru* and the production and operation of four test vehicles through 1,200 test flights to achieve certification for passenger carrying will take ten years. Once the service begins, eight vehicles will be made per year with annual passenger numbers growing by 100,000 per year, reaching a million in ten years of operation. In thirty years, there could be five to ten million tourists traveling to low-Earth orbit per year.

CAPACITY

If demand ends up resembling that which the surveys predict, hotel rooms as well as vehicles will be needed. Collins and his colleague Kohki Isozaki estimate that if a few hundred thousand people traveled into space per year and stayed for two or three days, the orbital population would comprise several thousand guests and a staff of one thousand. When a million people visit orbit each year, there will need to be accommodations for 10,000 people. Market research shows that most people who would like to go into space want to stay for two to three days or longer. Therefore, if five to ten million passengers traveled per year, 30,000 to 80,000 guests would have to be accommodated simultaneously. An 80 percent occupancy rate would require capacity for 35,000 to 100,000 guests. Hotels also would need to provide rooms for staff, who would not be commuting to and from Earth every day. If you figure, as Collins and Isozaki do, that there would be one staff member for every two to three guests, that's an additional 10,000 to 40,000 people. Since staff would work in shifts of probably two to three months at a time, the total number could run from 20,000 to 80,000, so we're talking about a capacity of between 45,000 and

140,000. Depending on the size of the hotel, these numbers could mean up to about 250 hotels!

Collins predicts that thirty years from now there will be 100 hotels or more in various orbits (high-inclination orbits for economical access from high latitudes and to give guests views of much of the earth; equatorial orbit—the cheapest to reach—for those more interested in zero-gravity activities than in the range of views of Earth; polar orbit to give views of the whole earth; and highly elliptical orbits to give views of the distant earth). There probably will be at least one propellant service station in each of the main orbits, and there probably will be a regular supply of water from the moon and comets. Hotels in lunar orbit and some on the surface are also likely. These tourist activities will represent a turnover of $100 billion per year, but even that will be a few percent of the revenue from civil aviation, which is projected to reach several trillion dollars by then.

COMPARABLE TRAVEL ON EARTH

Looking at what people spend on Earth helps put these prices in perspective and helps evaluate demand for space tourism. The $10,000 to $250,000 price compares favorably with extravagant cruises, adventure tourism in Antarctica, and a return flight on the Concorde. However, it is important to emphasize that at first, space tourism will be a "roughing it" experience, not a luxury one, so comparing its price with that of deluxe travel on Earth may be only partially valid. There may or may not be a significant overlap between luxury travelers and the first space tourists. On the other hand, it's not valid to say that only adventure travelers will want to go into space at the beginning; many people with high incomes like to do unusual things, to challenge themselves, and to be first on the block, so even if they're used to being pampered, many will go just for the experience.

SOME TYPICAL VACATION COSTS (PER PERSON RATES EXCEPT WHERE NOTED)

- A six-month cruise in the master suite of the QE 2: more than $350,000
- A two-week Kenya and Tanzania family safari: about $5,600
- A one-week crewed yacht charter in the British Virgin Islands on a 50-foot sailboat: about $6,700 for six people
- A twelve-day Alaska Iditerod adventure: $8,400
- A three-week Audubon Society tour of islands around South America: $28,000
- Climb up Mount Everest: $35,000 to $65,000
- Fifteen-day cruise to Antarctica: $7,500 to $13,000, excluding transportation to South America
- Round-trip fare on the Concorde, Paris or London to the East Coast of the U.S.: $9,000
- Trip around the world on the Concorde: $62,000
- Safari in Africa by private luxury jet: $37,800

The sidebar "Some Typical Vacation Costs" shows that some people will spend lots of money for vacations and/or treks. Luxury travel market size information is hard to come by, but we do know that the 5,000 luxury travel agencies who belong to industry group Virtuoso generate more than $2.5 billion annually in travel sales. (Virtuoso's agencies are located in North and South America.)

According to the *New York Times* in May 2001, the Travel Industry Association of America reports that half of all American adults, or ninety-eight million people, took adventure travel vacations in the last five years. Thirty-one million of them engaged in "hard-adventure" activities like mountain climbing, sky diving, and spelunking. These travelers are predominantly young, single, and employed. They spend $465 per trip, a far cry from the $10,000 or $25,000 ticket price forecast for space tourism. On the other hand, such travelers tend to have high household incomes. Forty-eight percent of them earn $50,000 or more per household, compared to just 39 percent of the American population as a whole. Eighty-two percent of them have attended college, compared with 70 percent of the total population.

Space Adventures, the company that provides today's rudimentary space tourism, reports that since 1994, nearly 2,000 fare-paying, "regular" people have flown at the airbase they work with in Russia. This includes flights in MiG-21, MiG-23, MiG-29, Su-30, L-39 and the "Edge of Space" MiG-25 aircraft. Approximately 40 percent flew in the MiG-25 to 82,000 feet (25 km) at Mach 2.5.

Cruise lines accommodate seven million passengers per year. The cruise business is projected to reach $50 billion over five years. Cruises are expensive and time-consuming, so they may be somewhat comparable to space vacations. On the other hand, the safety issues are far less troublesome.

As of 2000, the company's zero-gravity flights in Russia had been available to the public for two years. They have flown five flights per year with ten passengers per flight, or about fifty people per year. They anticipate twelve flights with 120 passengers in the coming year.

The income from these flights is, roughly and conservatively, over a period of seven years:

800 MiG-25 flights	At a price of $12,600	$10,080,000
1,200 other MiG flights	At a price of $6,725	$8,070,000
50 zero-gravity flights	At a price of $5,400	$270,000
Total (2,000 passengers)		$18,420,000

These prices also include a hotel stay, ground transportation, and a variety of nonflight items. They don't include airfare, and they don't include extras like a ride in a centrifuge (an additional $1,150), a video ($750), and meals other than breakfast. But we can say that there is roughly a space-based tourism industry now approximating $2.5 million plus per year, and that there's a demand from about 300 people per year.

That's not much. It's far less than the earthly American adventure travel industry, which serves 6.2 million people per year.

One of the reasons may be the difficulty, at least for Americans, in getting to the tour site, and the quality of that environment (Russia can be really cold). There also may be post-Cold War residual feelings of unease. If the flights were transferred to the U.S., there might be more takers. But they'd have to be better publicized.

WHEN WILL SPACE TOURISM HAPPEN?

When you build planes, whether for air or space, you've got to fly them repeatedly before they're certified as air- or spaceworthy for passenger use. This process takes a long time! Collins mentions a figure of 1,200 test flights and explains that it will take ten years before that's completed! The Concorde, which flies under less extreme conditions, took at least four or five years to test. This reality, which is seldom mentioned by space activists, carries serious implications for the industry and means that mass space tourism is not just around the corner.

Collins reports that space tourism is considered feasible within fifteen years by senior NASA staff, though few people know this fact. He adds that in 1998, leading aerospace organizations formally admitted that passenger space travel is feasible, likely to be economically profitable, and will become the major commercial activity in space. The 1998 joint report by NASA and the U.S. Space Transportation Association cited by Tom Rogers (Chapter 2) acknowledged that space tourism was likely to start soon and could grow into the largest activity in space. In July of 1998, the American Institute of Aeronautics and Astronautics published a report that said that public space travel should be viewed as the next large new area of commercial space activity. In September of that year, Keidanren, the largest economic and business organization in Japan, published a pamphlet called "Space in Japan." That work identified the only activity considered to have promise for commercialization of space activities as the Japanese Rocket Society's work on space tourism. And in late 1998, Dan Goldin referred favorably in three speeches to the promise of space tourism. We don't know yet how Goldin's replacement, Sean O'Keefe, will feel about such issues.

Harvey Wichman has this to say: "My prediction is that in about ten years a vehicle should be ready to carry people to space inexpensively enough so that at least some fairly wealthy people will be able to get there. Ten years, fifteen years, something like that. A hundred years from now there will be large space stations that are spinning, with full earth gravity in some locations and less in other locations. At first, space travel will consist of suborbital flights that just break into space and come right back down. There's a big difference between getting up to space and going into orbit. The latter requires that not only do you go up, but you then accelerate to 17,500 miles an hour to keep yourself in low earth orbit. That takes a huge expenditure of energy."

Space property rights advocate Alan Wasser (see Chapter 18) offers this analysis:

> It could go a number of ways. The most obvious is, I think, very unlikely, which is that we have another kind of Kennedy moon race with the government funding and doing it and sending astronauts to Mars. It's possible but not really likely unless totally new developments come along to change the current situation. A lot of people still focus on that simply because it's the way we did it last time and everybody always fights the last war. I just don't see the taxpayers being willing to pick up that tab. There was a special situation with the Cold War and Kennedy and the assassination and other things that were going on behind the scenes. I just don't see them being repeated.

Which gets to the possibility of its being a commercial venture, which I think is far more likely. Somewhere along the line, some circumstances will produce an incentive for private financiers to shoulder the costs. At the moment, just what that would be doesn't exist, but I figure there's a lot more chance of that coming along than that the taxpayers will do it. So, the key to me is identifying an incentive. Private industry could easily raise money to do it. The problem is that there's no reason they should. If somebody invented or discovered a product for which going to space would make it available and profitable, that would be fine. One of the recent hopes was that helium 3 in a particular isotope useful for fusion power would be the economic driver. It's available on the Moon and hard to get on Earth. The problem is that fusion power isn't turning out to be much of an economic driver for anything. There have been lots of items like that, and there continues to be because the need for an economic incentive is so obvious. I think the day the economic incentive is accepted and found, we could be within ten years and maybe even less to the human settlements on the Moon and humans on Mars. It won't take that long to do it once the reason exists. The problem is the reason could take fifty years at the rate we're going. Or it could come tomorrow.

THE AUTHOR'S OPINION

The *Challenger* disaster in 1986 is still shaping ordinary citizens' image of space travel. The explosion was rare among space flights, but notable and frightening because ordinary people died. Had all the travelers been astronauts or cosmonauts, people would have grieved, they would have been horrified, but they wouldn't have identified with the victims as they did and do. It will take a lot of positive experience to push that memory to the back of potential tourists' minds.

Surveys are fine—and they're necessary—but they tell you only so much, and only at the time they're taken. People's attitudes change; that's what keeps Roper and Gallup and all the other survey companies in business. The space tourism market surveys that have been taken are encouraging. There is demand, even at high prices. However, my feeling is that what will really determine the market will be word of mouth after the first flights. We all know that word of mouth can make or break a feature film, a car, or a tourist destination. If people go to space and love it, a tourism market will develop. If they complain about the experience or the price, the industry will die. Many people have to fly to get from one place to another. No one has to take a space vacation.

I'm also concerned about prices. It's one thing to say you'd spend a month's salary on a trip, but what if you don't want to go by yourself? Then it's two or four months' salary. And that's a difficult proposition for most people.

Another driver of the market will be health issues, which we know from Chapter 4 are unresolved. As a result, younger people may be the most enthusiastic, but the least able to afford the prices. Middle-aged people may have the money but might be more circumspect about the effect on their bodies. We don't know yet.

As far as the economics are concerned, I think Patrick Collins may be overly optimistic about costs. Considering the overruns on the commercial planes mentioned previously (we're not talking about government projects here), I think developing vehicles for mass tourism will take many tens of billions of dollars—maybe more than $100 billion. And that will mean either higher ticket prices or a need for the vehicles to carry a lot of passengers.

On the other hand, if some new propulsion technology, such as lasers, comes along, costs could drop significantly. Scientists are working on new types of propulsion. We don't have the final story yet.

I do agree with Collins that there hasn't been much economic return from the various space programs. There could have been, but governments resist the idea of making a profit. They've paved the way, which is great, but they've monopolized space, which is troubling. We may all be on this Earth together, as Kim Stanley

Robinson says (see Introduction), but if governments have to be the ones to get the public into space, it never will happen. That isn't their charter. Therefore, it boils down to governments either yielding their monopolies or admitting that the public is never going to get to space and facing the consequences of that admission.

As far as when space tourism is going to happen, I think it will happen, but not as soon as lots of people like to believe, and only when someone lights on a way to make it profitable quickly (which might appease venture capitalists; see Chapter 7). We've heard lots of rosy predictions, but none has materialized. So, I'm cautious about saying when space tourism will happen, or whether it will at all, much as I'd love it to. Perhaps I should set aside the royalties from the sale of this book so that in twenty or thirty years' time I can pay for a ticket.

— 6 —

Peter Diamandis, Buzz Aldrin, and Ron Jones: How in Tarnation Are We Going to Get Off the Earth?

Many people don't realize that we couldn't go to the Moon today if we wanted to—we don't have any ships that can reach it. NASA threw away all of the Saturn V rockets we used to get to the Moon in the sixties and seventies. (Can you believe that?) Nor do we have any vehicles capable of carrying tourists into orbit since the shuttle carries only six or seven people—not an economically viable number. Where will such vehicles come from? This chapter looks at a couple of the possibilities.

I first spoke with Peter Diamandis at a space conference in Houston. As usual in Houston, it was incredibly muggy outside, but we were nicely cocooned in an air-conditioned restaurant. Space activists milled around trying to catch a glimpse of Babylon 5 *actor Bruce Boxleitner, who was a scheduled speaker and guest. Diamandis ordered eggs, I got coffee, and the interview took off.*

Ron Jones and I have spoken a number of times, both in person and on the phone. He and Buzz Aldrin have attended every space conference I've ever been to, and each time I've gleaned more interesting and critical information from them. What follows is a mosaic of that information.

"The quickest way to become a millionaire in space is to start as a billionaire." Peter Diamandis smiles at his own joke. He happens to be right. As the Iridium bankruptcy, numerous launch failures, and the experimental X-33 reusable launch vehicle cost overruns illustrate, space can be hard on the pocketbook, and word gets around. "When you go to an individual for money to invest in a space project, if he's not knowledgeable you can probably convince him. If he is knowledgeable, he'll run very fast."

It's tremendously difficult to raise money for space projects, but Diamandis doesn't think we should let a little thing like that stop us. He has some ideas for improving the situation. Like competition. Especially competition. A good old-fashioned race. After all, that's what got us to the Moon back in the days when Americans were happy to contribute their tax dollars to beat the Russians. The U.S.'s status as the world's only superpower today has made us complacent. We're too far ahead, there's no one to beat, so we take our competitive instincts to the stock market and the Internet, and, every four years, the Olympics.

Diamandis thinks we can channel our natural instincts in an even more constructive way: outward, into space. In 1995, over a series of sushi meetings in Colorado, together with astronaut Byron Lichtenberg and friend Colette Bevis, he founded the $10 million X PRIZE. X stands for exploration and the X factor—that which is unknown. It's also the Roman numeral for ten, which is apt because it's a $10 million prize. Designed to stimulate competition and spur the creation of the first ships able to address the market for suborbital tourism, the award will go to the first privately financed company that can launch three people up to sixty-two miles—the official beginning of space—on two consecutive flights in a two-week period. So far, twenty-one teams from five countries have signed up to compete. Mostly they're financed through private individuals. No one has yet won.

Because the prize describes desired performance criteria rather than a particular design, contestants have come up with a wide range of ways to reach the goal. They're proposing rockets that take off vertically and land with the help of parachutes, ships that are launched from the air, ships that are refueled while in flight, ships that are towed to a particular altitude and then let go, craft designed to operate from conventional runways.

So far, more than half the X PRIZE is funded. The money comes from corporate and private sponsorships, an X PRIZE credit card, a sweepstakes, Olympic-type sponsorships, and possibly film and book rights. The business community in the city of St. Louis, where the foundation's headquarters are, put up $1.5 million in return for visibility.

THE X PRIZE AT A GLANCE

- $10 million award.
- To spur creation of first spaceships able to address market for suborbital space tourism.
- Vehicles must be privately financed and constructed, and must be able to fly three people into space (an altitude of 100 kilometers or sixty-two miles) twice within two weeks.
- Object is that vehicles will be able to generate revenue after the competition; costs should be reasonable because vehicles will be able to fly twice in rapid succession for only the cost of fuel and minimal turnaround labor.
- Announced May 1996.
- There are now twenty-one teams from five nations registered to compete.
- Prize describes desired performance criteria rather than a particular design.

- Designs proposed to date include liquid and hybrid rockets with vertical take-off and parachute recovery, air-launched, air-towed, and air-refueled designs, systems designed to operate from conventional runways, hypersonic waveriders.

"What I'm trying to do is spark a nationalistic approach again," explains Diamandis. "I really want there to be a competitive fighting between the Japanese and the U.S. and the French and Germans. I want an America's Cup mindset. I want a race. Humans love to compete. It's in our genes. We evolved through competition. And we respond well to competition."

The other thing about competition is that having ten or twenty different vehicles competing with each other will help bring costs down and efficiencies up. Diamandis feels that this sort of competition will yield the best business models and operational vehicles.

His enthusiasm for competition is rooted in precedent. Back in the 1920s and 1930s—the early days of aviation—numerous prizes helped jump-start what we now take for granted: a safe, reliable, economic air transportation industry. One of those prizes—the $25,000 Orteig prize—inspired Charles Lindbergh to solo across the Atlantic. His success demonstrated that long-distance air travel was feasible and broke a psychological barrier behind which other pilots, engineers, investors, and entrepreneurs had been hiding. Today, Lindbergh's grandson Erik, also a pilot, is a trustee of the X PRIZE Foundation.

Because of the publicity they generate, prizes can garner the attention of the "right" people, an especially tall order in today's risk-averse financial community. "There are hundreds of billions of dollars of capital that could flow toward space if it turned the right people on," Diamandis reasons. "And if you have the right billionaire from this city [St. Louis] fund a team only because his rival from another city funded another team, that we have to plug into. And as a benefit of that you'll get all the ancillary economic engines that go on. Right now I'm trying to tap into hundred of millions of years of evolution and competitive spirit. Does that make sense?"

It does. Almost all rich people got to be rich because they competed. Vigorously. Perhaps obsessively. Bill Gates (Microsoft's Chairman and Chief Software Architect), Larry Ellison (Chairman and CEO of Oracle), and Richard Branson (head of the Virgin empire) can't stand to see their rivals get the better of them. They don't mind risking everything in order to win. Sometimes they go down in flames, but there isn't one of them who doesn't think the reward worth the risk. It makes them feel alive.

Do such people actually care about space? Some of them do. Ellison is reputed to be a space enthusiast, and Branson has formed a new company, Virgin Galactic Airways, which he hopes will some day carry tourists into space. However, it doesn't matter all that much whether they care about space or they don't. The point is that they will become involved in the contest for its own sake if its profile is high enough.

The people who really care about space are the X PRIZE contestants, most of whom are engineers and rocket scientists, none of whom is particularly wealthy. And exotic as their work sounds to most people, rocket scientists don't get rich. They're competitive, but in a cerebral way, and most lack the killer instinct. They're far more focused on the dream than on crushing the competition, and as a result, most of them have little business training or experience, or a bent in that direction. This "deficiency" makes them easy targets for critics who charge that they don't live in the real world.

Diamandis admits that not all his contestants are budding business geniuses, but he believes most of them are getting a bad rap on that issue. "I split the teams into three parts," says Diamandis when challenged.

> There are those who are very real—the Kellys [Kelly Space and Technology, from San Bernardino, California], the Pioneers [Pioneer Rocketplane, from Solvang,

> California], and so forth. About a third of the teams have some decent business sense and are building some hardware and are moving forward.
>
> About a third are on the cusp. They have a good idea or they have good people. Don't have any money yet. And about a third probably don't have the business sense and have just a concept but have the ability to potentially attract the right people or the right partners. So it's fair to say that a large number of the X PRIZE teams have a long way to go, and obviously we do attract first and foremost the would-be rocket scientists and not the businessmen.

In fact, in the six years since the prize was announced, almost none of the teams has actually built and flown hardware. The exceptions include U.K.-based Starchaser Industries, California's Scaled Composites, and the Canadian Da Vinci Project. Contestant David Ashford of Bristol Spaceplanes, who is an aeronautical engineer, explains why: Nobody has any money. "There's an enormous credibility problem," he says, "because you go to your bank manager and say, 'Please can I borrow $50 million to build a little space plane.' You give your presentation showing what a fantastic return on investment can be had. And he looks at you and says, 'Well, that's very impressive, Dave, but if it's as good as you say, why isn't NASA doing it?' You've got an enormous perception problem. The space tourism scene now is that you've got a handful of enthusiastic startup companies, and a hundred or so reasonably knowledgeable and enthusiastic individuals. NASA doesn't want to know. None of the big government space agencies wants to know. As of now, there's not one big name linked to the idea of space tourism. There's not one big corporation really committed to it or one major public figure."

This lack is particularly notable because according to Ashford's estimations, an orbital vehicle—not a mere suborbital one like the X PRIZE craft, but one that can go all the way to orbit—could be built for a mere $1 billion, the cost of one shuttle launch. Of course, that's still a lot of money—more than any group of venture capitalists is going to lay out—but it's not that much for a large corporation like Boeing or Lockheed Martin to swing, assuming they could expect a return on their money, or government backing. Ashford bases his estimate on a 1993 feasibility study of a fully reusable space "cab" he conducted for the European Space Agency, in which he concluded that such a vehicle could be developed for $2 billion utilizing designs from the early to mid-sixties and using no new technology. He's since halved his estimate. The study was broadly endorsed by the British government. In 2001, NASA claimed that it needed $4.5 billion for its Space Launch Initiative (SLI) to study how to make a similar vehicle.

Ashford cites a comment by an unnamed X PRIZE contestant who told him he could make a little orbital vehicle for even less—$140 million. Ashford believes that such a feat, at least a prototype where the cost depends heavily on who does the work, might be possible given a small team of "ultra-competent" people. Ashford qualifies his statement. "I'm talking about a very restricted situation, a one-off prototype. When it comes to production and operations, the scope for cost saving is nothing like as great," he explains.

Ashford happily reports that there is less of a perception problem than there used to be (Our interview took place right before Dennis Tito's successful flight to *Space Station Alpha*, so the situation might be even better now), and that doors are opening. But it's very slow. He thinks the X PRIZE has had a lot to do with the change. Because of it, there's now so much activity in the private sector that there's more credibility for people who are pushing the idea of reusable launch vehicles and space tourism. Up to now, says Ashford, people in the financial sector have been unaware of what is possible because of NASA's very effective public relations machine.

And yet, the X-15, an experimental rocket plane that last flew in 1968 (Neil Armstrong, the first man to walk on the Moon, was one of its test pilots), would win the X PRIZE—if you could squeeze two more people into it—were it operational today. It could fly to space and back—just up and down, but still to space, just as an X PRIZE vehicle is supposed to do. The X-15 is to this day the only fully reusable vehicle to have been to space and back.

Ashford believes that the team to win the X PRIZE will be the first one of those with a "sensible" design (some are a bit fanciful, he thinks) to get the money. If he's correct, then winning is much more about business than technology. "In hindsight," says Ashford, "the X PRIZE was too ambitious. It should have been a stepping stone set of prizes. It should have started off with a single-seater that will go to fifty kilometers, which would be a lot easier to do. And then somebody would have done it. You can't actually do the X PRIZE job for the prize money. My estimate is £50 million ($75 million). That's if you pay in cash. Obviously when you do something like this, you get deals; people give you bits and you stick a logo on a fin. I could probably have done a single-seater that goes to fifty kilometers for $10 million." And that money you can get from investors because the prize money will cover it.

Will insurance be prohibitively expensive? No, says Ashford. Even though safety is a big concern, contestants won't be able to fly unless the plane is deemed reasonably safe by the FAA (or the CAA in England). And that safety rate will probably be one fatal accident per 1,000 flights, preferably per 10,000. In that case, insurance premiums shouldn't be exorbitant. "What's a life worth? Divide that by 10,000, plus a big profit for the insurance company," he says.

The lack of contestant business acumen doesn't particularly worry Diamandis, who himself has struggled to make various businesses work. It's all right to attract the rocket scientists

> because we have a registration process that basically throws out the quacks that are obvious—you know the people who are using anti-gravity drive or UFO technology. But we have to be careful not to throw out the proverbial Charles Lindbergh who comes with a different solution because Lindbergh was considered the quack of his day.

> You know Lindbergh was turned down. He went to go buy the airplane he wanted and he was told, "There's no way we'd let you fly our airplane—you'll kill yourself." He was turned down by many people in his effort to go after it because he was not considered a serious player. So, we have to be careful not to disallow the registration of someone who has a different approach that could work or who could find the right angle. That's exactly what we need. We don't need the traditional approach. Traditional doesn't work.

Of that, Diamandis is sure. "You know, large corporations and industry never make anything happen. It was Apple Computer that led off the computer revolution and IBM that came in there with the muscle to make it happen. So, it will be these entrepreneurial X PRIZE teams and it may be small governments or small regions that enable it. And then you'll have the Lockheed Martins and the Boeings come in and the U.S. government come in, but not to initiate."

One of the reasons that large corporations don't make things happen is that they're risk-averse, largely because they have to satisfy their stockholders. "I was on a panel speaking with the presidents of Boeing, Lockheed, and Martin Marietta in the early days," says Diamandis, "and I asked them a question. I said, 'Why aren't you building space tourism vehicles? The market's there, it's going to be there.' And their answer was, 'We don't build a vehicle until the customer is there to guarantee us a profit. We didn't build the triple-7 at Boeing until United and American and Delta had placed enough orders. We didn't build the space shuttle until the U.S. government placed a firm, fixed-price order. We don't build anything until someone else takes the risk.' It's going to be the entrepreneurs who build these first vehicles, get a business going, start flying hundreds of people per year, and prove that there's a market. And then you'll have the larger vehicles come in afterwards."

It's not just financial risk that worries the likes of Lockheed and Boeing. People are horrified at the thought of injury and death these days, much more so than they were in the past. Or at least that seems to be the case. "Part of it is that failure is so visible.

When two brothers in Dayton, Ohio, or I should say Kitty Hawk, crashed their first few airplanes, no one knew about it. Now [with] the Internet, everybody knows about it," says Diamandis. But there's more. "The question now is who is to be blamed. For me there are two major issues: one is trial lawyers—a litigious society, that someone is at fault, someone must be punished, and the second is that it's so visible that the press just makes a tremendous amount out of it."

The question of claims against the contestants is an important one. If they are not protected from liability claims, they may not be willing to experiment. And potential investors will say, "I'm not going to take a risk on this dangerous project if there's a chance that my money will be eaten up by litigation." Diamandis requires all teams to obtain whatever insurance their government demands and to comply with (or obtain waivers exempting them from compliance with) all regulations that pertain to the flight of their vehicle, whether they be international, national, regional, or local.

With all the hoopla over risk, one might wonder who actually would fly in these untested vehicles. It turns out that there's a bit of a fudge factor. Even though you have to be able to fly three people to space, contestants are required to fly only one person on each of the two flights. The other two seats can be occupied in any fashion the teams choose, including a dummy payload, that is, the weight and volume equivalent of two people. If only one person goes up, it will be a test pilot, a class of people used to risking their lives.

When Lindbergh flew back in 1927, a lot of things were different. There were fewer laws with which to comply, for example.

> In the early days of aviation there wasn't any regulation and people were allowed to try and fail. [And die.] And to try and fail is the most important part of innovation. Today most people are spending their money on lawyers, on paper designs, there are very few people building hardware because there's so much anticipated

> hurdle that you just can't imagine what the FAA is like. Here is an organization whose purpose in life is to prevent risk. You go to them and say, "This is safe," and they say, "Prove it." And you do your best to prove it and after you do that, they say, "Well, that's good, but what about these other ten things—improve those." By the time you're years into the process, they go, "That thing you improved in the beginning, the rules have changed now—go improve that in a different way."
>
> You can never prove that something is 100 percent safe. It's impossible to do. Consequently, there is an unachievable expectation when you compare space vehicles to airliners today because airliners had a period of fifty years, sixty years without this kind of bureaucratic oversight to experiment and learn. I don't think any of the vehicles that were designed back in the '30s, '40s, and '50s could ever come into existence today given the FAA. So it's a big problem. And it's never been about the technology. That's the biggest issue. It's never been about designing new forms of propulsion or new materials. It's all been about financing and regulatory. It is doomed for failure if people try and apply the FAA to space. It will not work. Do not pass GO, do not collect a dollar.

But Diamandis isn't a whiner. "Are there solutions? Of course." One of them is to regulate passengers as well as vehicles. Diamandis and economist Patrick Collins (see Chapter 5) have proposed a concept called "accredited passengers." Because it is so difficult to prove that a spaceship is safe enough to carry any and all members of the general public, why not create a special class of passenger for those qualified to take undue risks? Citing the precedent of the U.S. Securities and Exchange Commission, which protects the general public by regulating who can invest in startup companies (they call their special class "accredited investors"), Diamandis and Collins would like to see the FAA approve the idea of "accredited passengers." Those who would qualify would have to prove to the FAA that they understand the risk and have had sufficient training. They would then be exempt from certain regulations that apply to scheduled aviation services. Such a move would allow the industry to get a toehold, reason Diamandis and Collins.

Going overseas is another possible solution. "I've spent some time speaking to folks in Australia saying that Australia is probably ideal. Some government somewhere needs to take the lead in saying, 'We want to make ourselves the haven for space flight and we're going to provide infrastructure, the regulatory freedom, the insurance support,' and someplace like Australia would be ideal."

But in a way Diamandis wishes it would be the U.S. He worries about American competitiveness. He finds it ironic that Russian facilities and technologies are used by American tour companies—one such is Space Adventures, which offers zero-gravity and high-altitude flights from Russia; that there's a partnership between American company SpaceHab and Energia regarding the Russian portion of *Space Station Alpha*; and that the way to get tourists into space for U.S.-backed MirCorp was to use a Russian space station as the destination. Why, he asks, can't Americans do space using their own technology and facilities?

Maybe they will. If Australia becomes the seat of a thriving space industry, the U.S. won't be able to stand it, Diamandis believes. "I think the seed for change is going to come out of greed or fear," he says. "In other words, if Australia starts to do it and industry starts to flock there, then and only then will the rules change to allow it to come here. And greed in a positive sense, people seeing so much capital." And so we're back to the race idea once more.

Diamandis can come across as flashy. He looks like a movie star. His voluminous resumé makes him look either astonishingly brilliant or unable to commit to one thing. He's a licensed pilot, a medical doctor, founder of several companies and one of the founders of the International Space University. You might expect a frenetic manner from such a doer, but Diamandis speaks slowly and carefully. He discovered space when he was ten years old and has made it the core of his life ever since.

> There was a moment when I was in fifth grade when the Apollo program was going on and we were having a discussion in class. I was hearing a presentation about planets. And that's the moment that I crystallized as the defining moment. What it really was, who knows? But from that moment on, space has been front and center for me. And I think it's okay to say it's important just because.
>
> Space is important to me on two levels. One, heartfelt and spiritually, it is my mission in life. I don't feel the need to explain why. I just know that it's what I was born to do. It's got [what] I want to call almost spiritual overtones in that regard. I believe that my purpose on the planet is to help carry humanity to the stars. And that's what I do each and every day, and that's what I love.
>
> Now I can rationalize it many different ways. But I think first and foremost we are explorers as humans. I think we have become the pre-eminent species on this planet because we are explorers and that space is just part of that inner drive, that exploration gene, I call it. And we need to explore space, we need to explore the cosmos because we will become a better species and a better humanity as a result of that. I think space is the ultimate destiny of humanity and is critical because we are too much at risk on this planet. We're too much at risk for disasters, both natural and man-made, and we need to spread the eggs out of one basket, so to speak. You know all of the discussion of the benefits that space can bring to Earth—that's not the reason I'm doing it. It's not for those. The early European explorers going to Europe were rationalizing all the benefits that the new world would bring to Europe, but the final result, the real reward for that was the creation of what we have here. Not for the benefit of Europe, but the fact that this was meant to exist and would exist and for the benefit of

> who we are now. So for the thousands of generations ahead of us who will be born around this and other star systems, that's the reason we do this. That's the reason space is important, because it is.

Like so many other space enthusiasts, Diamandis wanted to be an astronaut, so he became a doctor. "I grew up in a family of doctors, and ever since I've been a kid I've always wanted to be an astronaut, and so I went to MIT as an undergraduate. I looked at the statistics and figured out that you either were a test pilot or a physician. Those are your best chances of going up in the astronaut corps."

Surprise! The medical route into astronautdom is a well-kept secret.

"Basically it's an attribute that's a positive thing to have because you're dealing with people in space and physicians can be trained to do other things, but it's hard to train someone who's not a physician to be a physician. Anyway, having said that, I said, 'Well, then that sounds like a good course of action. I'll make my folks happy and I'll get a chance to be an astronaut.' And so I pursued medicine as a career angle and did research in space medicine as part of that."

Diamandis went to MIT and followed a pre-med course, then went to Harvard Medical School. He also became a pilot. The research was interesting, but it led to a change of direction, and ultimately to the X PRIZE.

"The research I was doing was on space station design, artificial gravity, on motion sickness, on countermeasures for long endurance space travel. How do you keep the body in shape on long missions—a fascinating question. But along the way I decided I didn't want to be a government employee. Wouldn't want to be an astronaut that route, so that's when I first started becoming interested in how do you privately fly into space. And that began a long series of roads that led me to the realization that the ships don't exist."

If prizes are so obvious, why did it take so long for someone to come along and set one up? "I think everyone has just expected that these vehicles would be developed by the government or would naturally happen through the normal entrepreneurial process," says Diamandis. Times have changed. Government space development no longer works the way it did in the 1960s, and as business has grown bigger, being an entrepreneur has become a more complex and formal endeavor.

Not everyone agrees that the X PRIZE is a good idea. When Diamandis and the others were working out how to define the prize, they asked, "What's the space equivalent of flying across the Atlantic?" At 100 miles up, re-entry temperatures are too high, so they lowered the height to sixty-two miles. Some observers are skeptical that the goals the X PRIZE has set will really advance the cause of space tourism and getting humans rather than small cargo into space. Flying up to sixty-two miles in a small vehicle and coming right back down doesn't get people into orbit, and it does not provide the capacity necessary to support much of a tourism business. You need large airliner equivalents. The critics are right in a sense. It's true that none of the X PRIZE vehicles is that big, and, in fact, many of the vehicles in the competition are specifically designed to serve markets other than space tourism.

Diamandis dismisses the criticisms. "It's like saying, you know the Apple computer. The Apple IIc was a waste, what we really needed was a Pentium chip. Those vehicles will never exist unless you take the first steps. You need to have an industry."

He believes that suborbital is the way to start. It's easier than orbital, less expensive, and safer. The energy needed for Mach 5 to reach the sixty-two-mile height is one twenty-fifth that to achieve Mach 25, which you need to orbit the earth.

Some agree with David Ashford that the size of the prize is too small. It takes at least that much ($10 million), and up to $80 million, just to develop prototype spacecraft, and that doesn't include liability insurance. Further, there's no guarantee that a particular contestant will win, so they all have to go out and raise money as though the X PRIZE didn't exist. Diamandis and his staff see this spending as a plus, however. Citing the example of the Orteig prize, they point out that competitors raised and spent $400,000—sixteen times the amount of the prize. They also stress that the average America's Cup contestant spends $50 million to $80 million per ship to win a zero-dollar prize. Anyway, you have to spend money to make money, so there's no point in complaining.

Skeptics also point out the impracticality of some of the vehicles. For example, some of them generate sonic booms, a phenomenon that kept the Concorde from flying over the U.S. One of the contestants, Bert Rutan, a well-known and respected aircraft designer, makes his vehicles by hand. Some observers question whether such structures will hold up over time.

Nor do the naysayers stop with technical criticisms. Some of them worry that the whole thing will turn into one big fizzling setback if all the contestants fly and fail. In that case, they might not be able to get funding again.

On the other hand, if someone succeeds, what will happen to the industry? The X PRIZE Foundation believes that multi-billion dollar business opportunities will be created in new markets such as same-day package delivery, space tourism, and rapid point-to-point passenger transport. That's possible, but only if the economics and regulatory environment are right. And there's much involved in both: infrastructure to support all the new flights, environmental concerns, markets, profitability, launch and landing restrictions, competition from other ways of doing things.

A Moonwalker's Approach to Space Tourism

Former astronaut and moonwalker Dr. Buzz Aldrin and colleague Ron Jones have come up with a unique approach to jump-starting a space tourism industry. Their approach ensures not only that a ride into space is available to everyone, but that NASA will be equipped with vehicles for going back to the Moon and exploring Mars. Feeling that space should be an egalitarian proposition, Aldrin founded the nonprofit Share Space Foundation in 1998 to lay a credible foundation for an emerging space tourism industry and to help open up space to everyone. Jones is its first executive director. The founding money came from FINDS—Foundation for the International Non-governmental Development of Space, of which Rick Tumlinson is executive director (see Chapter 16).

The underlying assumption of Aldrin and Jones' strategy is the more people fly, the lower the cost will be due to economies of scale. Lowered costs will drive vehicle development toward reusable, evolutionary systems. Mass travel will force a new approach, promising vehicles capable of routine, reliable airline-type operations in sizes that can address market demand as it grows and changes. Small X PRIZE-type vehicles that can only carry small numbers of people to suborbital altitudes will not fit the bill, nor will any of the too-small vehicles that entrepreneurs are currently working on. One-stage (single-stage-to-orbit, or SSTO) vehicles are not up to the task (see page 27). Aldrin and Jones believe you need two-stage-to-orbit vehicles capable of high flight rates that can eventually accommodate eighty to 100 people per flight.

Aldrin and Jones envision progressively evolving steps. The first passenger vehicle would be sized to replace the shuttle's crew-carrying capability (eight to twelve passengers depending on the orbit) and would use an Atlas-sized reusable rocket as the first stage. The next step, a vehicle that would eventually carry eighty to 100 tourists, would need a reusable first stage with the power of a shuttle solid rocket booster. Such a powerful reusable rocket, if it could be developed in partnership with the private sector, would give NASA the heavy-lift capability it needs to get back to the Moon and on to Mars much more cost effectively than if the agency had to develop it at public expense. The private sector, on the other hand, would use these same large reusable boosters to deploy voluminous orbital hotels in a single launch.

NASA has been unable to justify economically the development of a heavy-lift booster, tying the development of same to politically unsupportable shuttle upgrades or costly humans-to-Mars missions. The Japanese Rocket Society's *Kankoh-maru* (see Chapter 5) is designed to carry only fifty people, half the load Aldrin and Jones think is economically optimal to prove the business viability of space tourism. Also, most of the new entrepreneurial vehicles are designed to carry cargo, not tourists. No one else has devised an integrated strategy that would marry the public's desire to explore space with the private sector's need to make a profit.

Aldrin and Jones' strategy is based on a building-block approach using what we already have—rocket and airplane technology—and giving the private sector what it is always looking for, a new arena in which to make money. They would introduce reusable vehicles immediately. The fastest way to do that is to make an existing throwaway booster reusable. They would start by encasing a proven safe liquid rocket fuel within an airplane that returns to Earth for reuse after it deploys an upper stage with satellite for a high-altitude launch to orbit. The trick is choosing the right sized "flyback" rocket to maximize commercial potential. Serve an existing market, such as satellite launches, then move out. The best-sized booster could fly by itself with an upper stage attached to orbit small satellites. It could fly in a dual configuration about a central core upper stage to put up larger satellites. Or three or more boosters could be clustered about a large core to put up very large payloads. With this "stepping stone" approach, say Aldrin and Jones, the entire global satellite market can be launched.

What about system growth? Aldrin and Jones say once the flyback "workhorse" booster has effectively demonstrated reusability and made improvements in reliability, the next step could be to replace the shuttle's solid rocket boosters, making the vehicle safer in its last years of operation. The new shuttle reusable booster would become the bridge between what we have now and a next-generation, totally reusable, two-stage-to-orbit shuttle replacement down the road.

Aldrin and Jones have their own vehicle on the drawing board. They call it the *Star Booster*. They think the vehicle could do suborbital flights by 2005—if they can get funding and build a convincing business case. *Star Booster*'s first stage consists of an existing rocket (an *Atlas III* or Russian *Zenit*) sheathed in a winged vehicle. The *Star Booster* would detach from its payload a third of the way to space and return to Earth like an airplane. Aldrin and Jones believe that because their vehicle would use existing rockets, it offers a way of achieving reusable space transportation cost effectively, greatly decreasing the burden on taxpayers to fund an entirely new propulsion system. Gradually *Star Booster*, using existing upper stages, could evolve into a family of vehicles suitable for different uses: launching small to large satellites, supplying *Space Station Alpha*, carrying tourists (these vehicles they call *StarBirds*), or launching space hotels for the commercial space tourism industry. Eventually, these vehicles could become Aldrin-invented interplanetary craft called "cyclers," which would take people to the Moon or Mars. In fact, the team believes that their family of vehicles can be configured to meet the needs of a lot of different customers including the military, NASA, tourists, etc., and offers the most versatile utilitarian space architecture put forth to date.

The two innovators feel that the only way to proceed is to get the private sector working in cooperation with government. They're convinced that government help is required to help birth a space tourism industry. Jones says that most private vehicles designed to satisfy only the niche markets like supplying the space station or taking people on suborbital flights are evolutionary dead ends and won't lead to mass public space travel. Jones says that the unfortunate reality is that the government failed the American people when it stopped pursuing advanced propulsion technology twenty-five years ago. As a result, chemical propulsion is still too costly and the private sector just isn't ready, nor does it have the resources to solve our national space launch problems on its own. He says the private sector has embarked on a haphazard shotgun approach to space launch and is learning the hard way that the job is way too expensive to do. In his opinion, even giants Boeing and Lockheed Martin aren't large enough to get such an industry going. Besides, they're not motivated, since fleets of reusable vehicles would cut into their expendable launch vehicle profits. (This is Patrick Collins' view as well. See page 99.) You need multi-industry involvement, really, says Jones. That means getting the travel, tourism, hospitality, and cruise industries involved too.

Aldrin and Jones have ideas about how to build the first hotels cheaply as well. Use the large reusable liquid boosters, wrap them around a core composed of the existing space shuttle external tanks modified with a fully outfitted habitat on top of the stack, and launch. Their approach would allow a huge hotel to be constructed and hoisted in one flight. You could use it in orbit as a commercial space station or resort, or you could put a propulsion system on it and send it to deep space, creating one of Aldrin's proposed cyclers. Cycler vehicles would be permanently stationed in space and continuously cycling between the Moon and Mars. (They would essentially be perpetual motion machines, using the gravity of the planets, that is, cycling, to propel them. Always in motion, one would be going, one coming, and when in the vicinity of the Earth or destination planet, they would release or be intercepted by smaller ferries that take people to and from the surface.) This possibility introduces a totally new approach to the exploration of Mars, an approach that the pair feels NASA has yet to grasp.

In this way, Aldrin and Jones say, a government-commercial sector partnership can develop the infrastructure it needs for NASA to explore Mars while fostering the growth of a new space tourism industry. This idea is quite different from the traditional taxpayer-financed straight-to-Mars approach some others favor.

Building on Patrick Collins' work (see Chapter 5) and that of the Aerospace Corporation (an industry consultant that has conducted studies on space tourism with promising results), Aldrin and Jones have been asking the U.S. Congress to fund a detailed space tourism market research study. What is needed now, they say, is research and analysis to better define the passenger space travel market, an assessment of the types of architectures that can best respond to other national needs like civil Moon or Mars exploration or space power satellites. The last will include real business plans covering an entire industry. With the new administration (George Bush the younger), they are more optimistic than ever that the government will demonstrate some of the wisdom it so aptly showed forty years ago when making wise choices that led to the success of Apollo.

THE AUTHOR'S OPINION

I like the idea of the X PRIZE. I think Peter Diamandis' idea of promoting a competition makes sense. Many people are inspired by trying to outdo each other, and they learn from each other in the process. Nor is the benefit for the winner just the money: it's the reputation and the potential for business. That's good will that you can't buy at any price.

The fact that the X PRIZE winner will have achieved suborbital flight rather than orbital doesn't matter. You have to start somewhere. Make the goal achievable and proceed from there.

I do worry about what will happen if all the contestants fail. That, or deaths, will set the struggle back significantly. Then what? The American public possesses little understanding of the engineering process and less tolerance for risk and danger. Perhaps the Russians are better at such things.

My other concern is that I think the contest itself needs a higher profile. One of the problems is that it's taking years for anything to happen. The public doesn't have that kind of attention span. But until the contestants get the funding they need, there will be no events to be reported on, no edge-of-seat happenings. Another Catch-22.

the [illegible] and the potential for business. That's good [illegible] you can't have it [illegible]

[illegible] than traffic does [illegible] you have to [illegible] Make [illegible] and proceed [illegible]

So [illegible]

[illegible] Once the problem [illegible] its [illegible] to happen. The public doesn't have that kind of attention span, but [illegible] will [illegible] events to be reported on the edge-of-seat happenings [illegible]

— 7 —

Kirby Ikin, Tom Olson, and Paul Contursi: The Bottom Line— Funding and Insurance

In Star Trek, *no one talks about how Star Fleet's ships and crews are funded. In the 24th century, humans have transcended the need for money. But in the 21st century, we still need it in order to make space happen. Where will it come from?*

Kirby Ikin is in the space insurance business and is a member of the board of the National Space Society. The insurance information in this chapter is based on a talk he gave at one of the Society's conferences.

I heard Tom Olson and Paul Contursi speak about their Colony Fund idea for the first time at a Mars Society conference. Excited by their innovative idea, I pounced on them at once. We then spent the better part of an afternoon outside the conference rooms at the University of Colorado at Boulder talking about it.

POSSIBLE WAYS OF FUNDING SPACE DEVELOPMENT

- Venture capital.
- Debt financing (loans).
- Tax-free bonds.
- NASA, military, or other governmental organizations earmark some money for private enterprise, i.e., grants, or become paying customers.
- Bootstrap. Start small, pay off quickly, and expand primarily by using the resources of space itself.
- Tax breaks.
- Raise money through other self-supporting endeavors like merchandising and entertainment, then divert part of the profits to space. Keep the initial business robust so it supports itself and space efforts. Eventually space business contributes more of its own share.
- Space investment bank or investment fund.
- Sponsorships.

Most of us have at least a nodding acquaintance with the way established businesses get their money (borrow, issue stock, rack up sales), but few people understand startup financing. How can a company that isn't selling anything and may not do so for years keep paying the bills, especially when it's developing complicated and expensive technology?

VENTURE CAPITAL

One way is through venture capital, an often misunderstood form of financing. Venture capital is the investment of high-risk money in order to create a rapid rate of growth in the value of a company and then gain a significant and short-term return on the original investment. Translation: These people want insane profits fast, like eight to ten times the value of the initial investment. If you can't give that to them, you'd better find another funding source. They'll eat you alive.

How can venture capitalists (VCs) get away with demanding such outrageous terms? Because unlike banks, they invest with no security whatsoever. But their willingness to forego collateral does not mean they'll put their money just anywhere. Venture capitalists are careful investors who look for a sound business proposition, protection in the form of patent or copyright, an exit strategy that provides a method for returning the investment to them, a strong management team, and investor-partners to help them share the risk. Even with all that, they expect at least a third of their investments to lose money and another third to break even. They rely on the spectacular successes to offset those.

Many of today's space firms find it difficult to fulfill these requirements. For example, a sound business proposition means a real product for a real market, and the company must identify how this market is to be reached. The company must consider risks, competition, and organization. The proposition must show a potential return of between eight and ten times the amount initially invested within three to seven years! You can see why venture capitalists flocked to the dot-com companies for a while.

What's the product in today's space market? What's the market? Let's say that your product is a vehicle for taking tourists into low-Earth orbit. The market is companies that will operate space tours for adventure travelers. The first risk is inability to provide the product at a reasonable cost. Today, that's problem number one. Other risks include safety problems and/or perceived safety problems, inability to secure affordable and dependable insurance, governments that won't let you or your customers fly, lack of suitable spaceports, potential lawsuits from disgruntled tourists, and the important problem of lack of operating experience. What are the comparable risks for investors in dot-com companies? Lack of market, competition, and failure of communications infrastructure and/or hardware. No death, no physical injury, no gravity well. Yes, government regulation could threaten the business; yes, lawsuits could be filed; yes, the company could fail to deliver what it's promised to its customers. But these are normal business risks, and ones that can be managed. In a space venture, you lose a spacecraft and you lose a sizable portion of your investment. If people die or are maimed, you've got not only their immediate problems but a tarnished image for the whole industry. Potential returns? Probably not between eight and ten times the amount of the investment, and certainly not within three to seven years, especially considering how long it takes to certify vehicles in the United States: 1,200 flights.

All investors need exit options, a way to get their money out of the company. If the company is public and the investor owns stock, the exit comes when he or she sells the stock. If the investor is a venture capitalist, the exit is either the sale of the company to another company or its going public and issuing common stock that can be openly traded. No valid exit strategy, no investment. If you are writing a business plan, remember the exit! (Would you want to invest your money in something that gave you no way of or time frame for getting it back?)

Generally speaking, VCs do not play a significant part in the day-to-day management of the business, largely because they are not qualified to do so. What they do do is look for a management team with significant proven business experience. If a key officer already has experience successfully building a company to exit, venture capitalists will be impressed. Most people don't realize that VCs invest as much in people as in ideas. The officers of the company must have an entrepreneurial attitude—an attitude prodded and fanned by the granting of equity options that can make them huge sums of money if the company succeeds—and are not simply looking for comfortable and safe long-term employment.

The importance of a venture consortium is often overlooked and/or misunderstood. It is unlikely that a venture capital firm will invest in a start-up on its own. Normally, VCs expect to join with one or two others to create a consortium, which allows them to share the risks. One of the VCs will be the lead investor or cornerstone; this firm will negotiate the terms of the investment with the proprietors and management team and then initiate the necessary legal proceedings to formalize the arrangements.

Different VCs favor different investment sizes. Those who come in early will invest as little as a quarter of a million dollars, while those who come in later expect to invest between eight million and fifteen million dollars.

Another thing that most people don't realize is that early venture capital investments do not entitle the investors to own the entire company, but rather only a third to a half of it. The rest is left open to allow for more investment, which occurs if the investors have confidence that things are perking along, or if new investors find the opportunity attractive.

Let's say two venture capital firms invest a half million dollars apiece at the beginning, for a total investment of a million dollars. And let's say that that million dollars represents about a third of the company's equity. (The other two thirds are held by the founders.) That means that the company will need to grow enough to be worth 100 percent of its equity, or $3 million, and it will need to grow another ten times that to provide the kind of return the VCs want. So, in three years, the company will have to grow so that it's worth $30 million! (One million dollars times three times a growth factor of ten.) These figures are typical and go a long way toward explaining why most businesses do not get venture funding. It isn't easy putting together a credible plan that shows that kind of growth.

As if this who-owns-what-percent-of-the-company business weren't complicated enough, there's another factor to consider: Investments become diluted when other investors put money in. Let's say you invest a million dollars in a company that's worth $3 million for 30 percent ownership. Great, you own a third of the company, but your shares aren't all that valuable yet because nothing much is happening. As the company grows, your shares will become worth more because the company will have developed, but as new investors put money in, your shares will be worth a smaller share of the company because you'll own a smaller percentage.

Confusing? Not really. Think of it this way: You buy a million shares of Shimmering Silver Rocket Company stock at $1 each, which gives you 30 percent ownership of the now $3 million company. The rest of the company is owned by the founders. Now, as the company hires people, develops intellectual property, does publicity and public relations work, and lines up customers, the stock rises in value, say, to $5 a share. This can happen even though the company hasn't yet made a cent in revenue because it has cash, personnel, an image, a potentially profitable product under development, etc. Now your stock is worth $5 million. Not bad, eh? But while your stock was rising, some other investors came along and bought four million shares, which were created by management in order to raise money. (The company is not yet public; these shares go to private investors.) Now there are five million shares outstanding, of which you own one million, or 20 percent. The company is worth $25 million (five million shares valued at $5 each)—about eight times what it was worth at the beginning. You've made a 500 percent profit, but you own less of the entire company than you did before. You've been diluted. (Beware: not all companies grow this impressively. Your equity may be diluted without the compensation of tremendous, or any, profits.) If this bothers you, you can invest more money and increase your ownership, but don't be surprised if the other investors see you and raise you, like in a poker game. (Beginning to see why venture capital investing is not for the faint of heart?)

Part of the moral of this story is that a smaller chunk of a much bigger pie is worth having, so dilution may not be all that terrible. On the other hand, if the company doesn't do as well as hoped (and most don't), this dilution can diminish the motivation of the management team, which is good for no one. In that case, the investors might give management more options—limited by conditions of performance—to give their motivation a shot in the arm.

The earlier the money is invested, the greater the risk. In the beginning, the company is unproven—a mass of potential and carefully supported wishful thinking—so "early stage" or "seed-corn" funds are often quite small. Shortly after this initial investment, what's called a "first round" investment of about $10 million (£3 million to £5 million in the U.K.) occurs. Later investments depend on the needs of the company. These don't necessarily indicate poor performance. In fact, in some cases, significant later sums are invested because the company is showing lots of potential and needs more money to make it happen.

At any time during this process, some other company may want to buy Shimmering Silver Rocket Company. If the amount offered is high enough, that may be a good thing—it's an exit, and investors will get their money out. If it's on the low side and the shareholders think they will be able to do better later, they may refuse the offer. If the company is sold, typically key management will be asked to stay on for two to three years, since it is part of the company's assets and can help provide continuity for a while. However, that's negotiable and may not happen. After that brief transition period, it'll probably toddle off and leave the task to those better suited to running an established company. (Most entrepreneurs do not do well under such conditions.)

On the other hand, if the value of the company has grown attractively, the company may decide to go public, that is, issue stock for public sale. This step, also an exit strategy for investors, will bring the company into the realm of the more familiar for most people. The investors get their money back. The company's stock is traded on the stock market, and additional funds are now raised by borrowing money, issuing bonds and/or more stock, and doing joint ventures with other companies (also an option for companies that are still private). Oh yes, and of course, sales.

Venture funding hasn't exactly stormed the space community, though there has been some in the satellite market. Given the dangers of space travel, the uncertain regulatory environment, the lack of markets, the absence of a place to go, the long wait till exit, and the lack of business experience of most space entrepreneurs, VCs don't find the idea attractive. Chances are that what is needed is investment on the scale of that won by Silicon Valley. That and pursuit of markets by big established companies like Boeing, Lockheed Martin, and maybe Virgin. Once these companies decide that space provides attractive business opportunities, they'll jump in and sweep along the rest of the world with them. Public companies like Boeing and Lockheed Martin won't risk shareholder money that way. The Virgin Group, which is private, will be interesting to watch since its founder, Richard Branson, is known for bold moves.

ANGEL FUNDING

In the early stages, angel funding comprises private investment by wealthy benefactors who would like to get their money out but who realize that that may never happen. The angels can be the company's founders or other individuals. For many space start-ups, this is the only method of funding available since their businesses are just too risky for VCs or the public market. Jim Benson of SpaceDev started his company that way—with his own money (see Chapter 13). Blockbuster author Tom Clancy also is reputed to have given money to the defunct Rotary Rocket Company, which came to an embarrassing end: It was seized for nonpayment of property taxes.

A SPACE INVESTMENT FUND

And then there's another idea. Let's say you or I would like to contribute toward making space happen, but we don't have lots of money to burn. We could buy stock or bonds of public companies that do space work. We can't give money to NASA—there's no mechanism for that—but we can contribute funds to space advocacy organizations. We can volunteer our time and do lobbying and other outreach. Or, we could put our money into a long-horizon fund that invests in space companies, public and private.

Tom Olson and Paul Contursi are Mars Society members who are working on a way for you to do just that. Fired up by Robert Zubrin's Mars Direct plan (see Chapter 8), Olson and Contursi started thinking about ways to break the financial logjam surrounding space. They came up with the idea of the Colony Fund, a pooling of venture investment money contributed by ordinary people around the world who want to make space happen and are willing to wait for their returns. The Fund's goal is to help develop the entrepreneurial space sector including, ultimately, the exploration and settlement of space. It will invest in technologies, missions, products, and services it sees as promising for accomplishing that goal. Each investor will have a personal stake in humanity's space endeavors.

The idea sprung from Olson's head one day after he saw that the Mars Pathfinder Web site got fifty million hits the first day or two of operation, 800 million hits over a period of two months. He started thinking about the implications of those numbers—a silent majority that supports space initiatives but doesn't make itself heard—and he put that together with some deep thought about the American economic system, and boing: the Colony Fund.

> I took a look at that 800 million hits and I thought there might be a fiscal constituency that could be a lot more viable and easier to tap into than trying to turn those numbers into a political constituency. People are so burned out on the political process and the infighting, the internecine warfare that goes on in congressional funding every year. This is nonsense. And so I thought, "What if a venture capital fund could be made available to the individual player?" It's basically selling a dream and saying, "This is a dream not only can you buy into, but down the road you may profit from it." And it creates a very long-term kind of financial commitment so that people involved in building the dream are not subject to the vagaries of government funding and those battles that go on every year for continued appropriation. You know you have a long term funding source so you can just relax and do your job and get the stuff done.

Olson reasons,

> If the Mars Society is really serious about this being a global effort, then why shouldn't everybody get a chance to play and have this be globally involved? What would be the best mechanism to do that? I sat down and that 800 million number kept haunting me. I kept saying maybe we should try to do away with all our old-style traditional thinking and do what Zubrin himself did with Mars Direct on the engineering side. Let's sit down with a blank sheet of paper on the economic side.

> So I've got these two things here: Pathfinder's 800 million hits, and an economic system in this country and by extension the world in which every investment instrument ever created by the human brain is just that. It is an artificial product of human thinking. These don't follow any specific laws of nature like gravity or Einstein's laws. These are artificial constructs that in most cases work very well, and people in Wall Street create new investment vehicles and constructs all the time. So why not take it a step further and invent still a new one?

So far, Olson and Contursi have spent a year and a half doing just that. The idea is this: Create a long-term fund that invests half its money in conservative instruments and the other half in space missions and their products. Start with small missions, bootstrap your way up, and exit in thirty years. Minimum investment: $1,000, possibly decreasing to $100. Capital funding target: $50 billion.

"To me this is the thing that allows the greatest number of people to play, from individuals to corporations to even governments," says Olson.

> It allows everybody to play on a level playing field. They all have a stake in the operation, they all have a stake in its future, and they all have a stake in its outcome. And when you sign in, you sign in for a long-term commitment.
>
> Everybody can profit from it in the long run. Maybe you personally if you're young, or for your grandkids if you're older and you want to buy some shares for their future. Thirty years from now, you take those thousand dollar shares, now they're worth ten thousand, even adjusted for inflation, they're worth ten thousand or more. You can cash those in and literally buy a ticket to go to the colony yourself. At that point, the Colony Fund doesn't have to be a venture capital fund with those kinds of limits any more. At that point people have a choice. They can cash in their shares or they can roll

> them over into a more traditional kind of bank entity—the Colony Fund could turn into the First Bank of Mars. Now they're in the business of giving traditional loans—now they're getting into the land grant and the loan business. Of course, now you have a lot more in that account to play with and to use as more traditional loan capital for actual Mars colonists to go out. You've gone beyond the business of starting a colony. You're now in the business of grub-staking individuals—people who want to build your dreams and realize a new world. Nothing else I've seen presented so far is looking toward that. More near-Earth, short-term profits. They're still thinking on that old paradigm of how do I appease the investors next quarter. What's going to be the rate of return? Even NASA—they did a study on privatization. They set up some rules and said, "Well you assume a $5 billion initial investment and you've got to make a minimum 50 percent rate of return over a ten-year period, and that assumes a rate of taxation of 34 percent and all this other stuff in order to make something viable." No wonder nobody's doing it if you go by their rules. But who says we have to?

Olson explains how the fund will make money—by clever marketing, among other things. "The big events, like the first human mission," he says.

> If the Colony Fund's bankrolling that event, nothing involved in it's going to be free. So, you're going to make money on pay-per-view and on selling soil samples. Here's an example. The Lunacorp people [a company that plans to provide virtual entertainment from the Moon] want to send a probe to the Moon and bring back I think ten kilos of lunar soil. They want to give half of it away and sell the other half at $6,000 a gram. Now what you've done there is, first you've created an immediate short-term value for extraterrestrial soils, at $6,000 a gram you've made $30 million on that. The cool thing

> about that is you've given away the other half to science or whatever, but now you've set a commercial value for it, that's now a charge-off for your taxes. So, they've killed two birds with one stone by doing it that way.
>
> What will Mars soil be worth? Well maybe you can charge $10,000 a gram the first time around. The dirt from the first human mission becomes a novelty item that you can charge a lot of money for because no one's ever done it. A quarter metric ton of Mars soil could pay for the flight.

And that's just for openers. Olson figures that with TV rights, third-party spinoffs, T-shirts, toys, videos, CD-ROMs, DVDs, and so on, there will be plenty of income. But there's more.

> One of the things I thought of was when the first astronauts land on Mars they'll have little side cams—3D digital cameras on the helmets. When they're out there on duty they're recording everything they do and say, and that stuff gets uploaded every night. You take that raw material and our programming mavens in Silicon Valley create that into the ultimate 3D virtual reality arcade game. You strap on the goggles and gloves and you get to go to Mars for an hour. And it won't be a programmer's dream: You're seeing what those people are seeing for real. It's going to look real to you. What would you pay for that? Businessmen in Japan pay $1,000 a pop for green fees to golf. What would they pay for something like this? These are the kinds of marketing opportunities that you can exploit short-term and long-term that are also, as the Fund is going to be investing in them, going to make a profit; that brings profits back into the Fund.
>
> There are a lot of marketing opportunities right here on Earth before you even set foot on Mars that will make profits for the Fund. Then the idea is to find resources that will make a colony viable; as you develop, you get your product back. That product could be knowledge, soil, anything else

> that Earth needs that we can get back cheaply and sell for a lot of money. [Robert] Zubrin mentions deuterium, for example. There's five times as much deuterium in Martian water than there is in earthly water, which is really a great product to use in nuclear reactors. The current market value of that is $10,000 a pound. And if you can collect that at five times the rate you can collect it here and launch it back cheaply through Mars' gravity well, which is only three-eighths that of Earth's, that could be another viable long-term trade deal. We don't know what kind of resources we're going to find there that could in fact kick off the equivalent of a California gold rush. But I'm confident that they're there. And the point is, if you can get there firstest and bestest, you're going to be in a position to take economic advantage of those opportunities because you saw them first. If you have a couple of small missions that are shown to be profitable, now you have a track record that's going to attract those larger investors.

One of the innovative features of the Fund is its noncash provision. (For securities laws reasons, the Fund can't be involved in direct barter, but it can invest in companies that barter with others.) Companies with products and services to offer the missions can contribute those in lieu of cash and will be credited just as if they'd invested money. Contursi and Olson don't even exclude NASA from "playing." Olson explains, "In my model, if NASA wants to play, they can play, but they're an equal investment partner. Even if they want to do it in a barter kind of method where they provide us some launch pad support and a booster rocket in exchange for a few shares in the whole deal, we can work out that deal as long as they deliver on time." Ever heard of such a thing? Olson thinks his way is a lot better than the conventional one. "The problem with some other people's proposals is they take on NASA as a partner and they've got private investors over here and what happens to some of these public/private combinations is four years later there's another administration and they have a different agenda and all of a sudden they pull the plug and say, 'We don't want to be a part of this anymore.' That's happened many, many times in our history and the private investors end up holding the bag. I want to avoid that. In this case, you can still have NASA as a partner. They're getting shares in exchange for what they're giving back. If they don't want to be involved any more after the first couple of missions or they decide to have a different agenda, they don't have to play any more. We have other people who can play, we have cash flow going, they still have their shares. Or they can even cash out and sell their shares to somebody else who might have a longer-term return."

It doesn't have to be just NASA, either. It can be the Russians, who seem to be providing a lot of the wherewithal to get people into space these days (like leasing out *Mir* and taking millionaire Dennis Tito to *Space Station Alpha*).

> They've got a lovely booster called Energia which is sitting on a shelf and not being used because nobody's got the money to develop it. If it's in good working order and we can use it to launch a big payload, and they're willing to provide that to us, and maybe even a launch base for it in exchange for a piece of that action, it would be worth it to them to do that. And this is a way by which they can play and be a part of the thing. The nice part of all this is that the Colony Fund concept will work whether government programs go to Mars or not. If NASA finally gets its head out of its ass and goes there, everybody wins, and the Fund will be there waiting. If NASA falls on its face, the Fund will still be there, playing a key role in bypassing government short-sightedness to finance the opening up of a new world. Either way, we win.

Olson, who can talk just as fast as Robert Zubrin (see Chapter 8), is modest about his idea.

> I just thought different, to borrow from Apple's slogan. That's what we have to do to go to Mars. Zubrin thought different when he came up with Mars Direct because

before him, the conventional wisdom said it would cost a half trillion dollars in thirty years with huge *Battlestar Galactica* ships and taking everything with us to go to Mars and stay there for a whopping thirty days and come back with nothing to show for it. Zubrin changed all that. He even turned Carl Sagan's mind around. Carl Sagan himself signed off on Mars Direct as the way to go not long before his death. It was almost like undergoing a religious conversion for a guy like Sagan. The Planetary Society, which he founded, was always big on pushing for more government-style programming.

I don't know what clicked in my head. I'd moved to New York, I had been living in Denver for a long time. All of a sudden I realized, I've got this idea, I'm in the middle of the financial capital of the world. Every dollar of the free market that flows in the westernized free-market economy flows through New York City at one time or another. If there are people here who can't figure out a financial plan and make this thing fly then nobody can. Let's see what we can make happen. My brain started exploding. These things started coming fast and furious and the actual structure of the Colony Fund proposal took shape during December 1998. I went on this jag, I was just writing furiously and getting this all out. And at the same time I went to my first local Mars Society chapter meeting. That's where I met Paul [Contursi] and a couple other people, and then over the ensuing months we sat down and got together and started hammering out some actual details and making this proposal a viable one.

I'm a computer consultant. I spent three years as an MIS director at a brokerage in San Francisco, so I started with that. Paul's been in that a lot longer than I have, so he gave me some interesting perspectives, especially when I realized that in the financial services world, a lot of these things are artificial constructs, Paul was the guy that could actually confirm that for me. He'd seen everything pass by the desk. Every kind of bizarre investment vehicle you can imagine. Whoever heard of a zero-coupon bond twenty years ago, or a junk bond, or any of these other hash things that they put together? And this happens every day. So there's no reason we can't do it again. No one else has tried the concept of venture capital for the masses. Yeah, it's a little riskier, it's a longer-term commitment. Is it an IPO?—no. Is it a junk bond?—no. Is it strictly a venture capital fund as managed by investment bankers?—no. It's a little riskier than that, so it's not like a savings bond where you get a little greater return. We don't know what you're going to get out of this. My gut instinct on this is, you put $50 billion in a pot and you stir it for thirty years in any investment management kind of pool, what are you going to get? Well, traditionally in the last fifty years, people put a certain amount of money in the pot, they're going to get ten times their money back at the end of that time if they hang in there. Even with economic ups and downs in the last half century, anybody who's started with this amount here ended up with a huge amount over there. This is the miracle of compound interest.

Part of it is also that once you've bootstrapped an investment pool to about a $10 billion level, not all of it will be going strictly into Mars rockets and infrastructure. Only half of it will, even though it's a venture capital pool for that purpose. You keep it viable by taking the other half and putting it into more traditional investment vehicles as a short-term hedge and getting a nice rate of return back from them.

We're going to raise long-term seed capital for a long-term colonization plan. We figure it will take about $50 billion to get it started. By the time you get to the stage where you're actually colonizing, sending people out there to develop an economic base on Mars itself, there's a lot more than $50 billion in that pot. But in the beginning you've sold fifty million shares at $1000 each, whether you've sold it to individual investors or to corporations putting $100,000 in the pot. Or governments could get in, and not just NASA. Maybe the Sultan of

> Brunei will decide, "I want to leave a legacy for my son. Here, I'll buy 100,000 shares—name a town after him." I have no problems selling shares to those kinds of groups. This allows everybody to play!

* * *

"I wouldn't mind being part of the first missions," confesses Olson.

> Those mission plans involve a two and-a-half year trip time. Ideally if I was still in good shape—I'd be John Glenn's age at that point [Olson is forty-six]—I'd go, I'd be part of the first hundred. I'd want to go because of that pioneer spirit. Try something new, wipe the slate clean. Live by your wits. Start over. An environment like Mars doesn't suffer fools lightly. Even though there won't be many people there, it will be a very social and intellectual environment to hang out in. Colonists are going to be pretty independent people, a pretty and very diverse interesting group capable of creating new ways of doing things. That's why I think it's very ironic that there are people already advocating devising governmental-type structures and things before we even go to the Red Planet. [There are people planning and hotly debating such things as you read this.] It's almost like setting up a bureaucracy before the first colonist leaves. How many colonists will want to go to Mars if they know there's already somebody waiting at the other end to check your passport and vaccination card, if there's already a tax collector and a cop waiting for you when you get there. I think that's why people left for frontiers in the past.

Insurance

It's an incredibly boring subject, insurance. Ask any science fiction fan about space insurance and you'll get a blank look. Any science fiction author who wrote about such mundane drivel would sink into oblivion faster than you can say, "Beam me up."

And yet, without insurance, you aren't going to fly. In fact, there isn't going to be a space industry at all because no one will put up the money to develop it. Here's how it works. (Thanks to Kirby Ikin of GIO Space in Sydney, Australia, for the following explanation.)

Let's say you build a spacecraft or launch vehicle. When the day comes, you count down, three, two, one, zero, and off it goes. Except it doesn't. Kerplunk, it fails. You've lost your mission. Whatever it is you're carrying doesn't get wherever it is that it's going, and your vehicle becomes unusable either permanently or until it's repaired. Everyone who was counting on you is not only disappointed but has lost money. So have you, in fact. What does that do to your business? Your reputation? Your relationship with your customers?

Enter your friendly space insurance underwriter. You buy a package from this extremely helpful person, and guess what: If your launch fails, you're reimbursed for your losses! On you go to fix the problems. You're not out of business. Whoopee.

In the late 1990s, the space insurance industry could cover about $1.3 billion in losses per year. (One failed launch can cost $400 million.) Claims from the September 11, 2001 terrorist attacks have reduced worldwide capacity to about $300 million.

In practice, it's not quite that simple, of course. When insurers are faced with high risk, they do one of two things: charge more, or decline to insure you entirely. (Anyone remember how hard it was to get insurance in California after the 1994 Northridge earthquake?) And the situation can change from acceptable risk to high risk quickly, threatening not just one or two projects, but the entire industry. All it takes is an unacceptable failure rate. Or new technologies, which may be uninsurable altogether. Even if cheaper launch vehicles are developed, their insurance costs may be so high as to offset the cost reductions. Space is not for the faint of heart, or at least the faint of pocketbook.

Insurance isn't just a good idea. Financiers require it, just as mortgage companies do when you're buying a house with their funds. They don't want to lose their money, and they don't want to lose your expertise. So if you're planning on going into the launch business, you've got to factor insurance in from the beginning.

There's another reason you'll want insurance. Let's say you're just about done building this great launch vehicle, you're nearing launch time, and you run out of money. Bummer. But not if you've purchased a policy that covers you under just such circumstances. It isn't cheap, but it may save your bacon.

> Eighty-five percent of space insurers reside outside the U.S., while most launches are conducted by U.S. companies.

Here's another little trick you can do with insurance. When you get investment money in, you designate it as a convertible debenture. Okay, not a household term, but an easy-to-understand concept when you remove the gobbledygook. What happens is this: If your project succeeds, the debenture, which is nothing more than a loan, becomes part of your investors' equity in your company. That means that they own a stake in the company, which will be converted to cash when your company either is purchased or goes public. If your project fails, Heaven forfend, the loan from your investors remains a payable debt. What happens if you can't pay that debt? Well, if you've planned properly from the beginning, you've taken out an insurance policy that will pay the debt for you if you can't. You've also insured the cost of reconstructing your failed mission. And you've made your investors happy, because investors want their money back quickly, which isn't always possible, especially in such a high-risk business.

> In 1999, the maximum payout capacity was $1.2 billion. (A shuttle loss exceeds that.) In the mid-nineties, it was $650 million.

Insurance also can help reassure your investors in another way. Let's again say that you're developing a new launch system. All things new have to have a first time, and until that first time and afterwards, the system is unproven. What's a customer's incentive to sign up with you ahead of time? We know what your incentive is to sign them up—a guaranteed market, but why should they play, considering the system won't be available for several years and might be obsolete by the time it is available? You know your investors are asking the same questions, and they're asking what will happen if you build a vehicle and no one comes. Your best answer for allaying their fears, and yours, is a guarantee. No such thing in business? Think again. You can insure your ability to pay your investors if you don't get any customers. And if your investors are guaranteed not to lose their money, they'll be more inclined to give it to you in the first place.

Okay, so I'm oversimplifying a little. Because new technology is risky, your underwriter will charge more, just in case, especially if he or she can't understand the technology. On the other hand, if there are lots of insurers offering their services, then you're likely to get better prices. The trick is to introduce your new technology when the insurance market is soft. And if you can do that regularly, you probably bought Microsoft stock right after the company's IPO.

Right now, the main type of launch insurance is for satellites. What's interesting about the satellite insurance market is that it can be cheaper to rebuild a failed satellite than to insure it. Also, you need to insure only one event—one launch—because the vehicles are all expendable. The same could not be said for vehicles carrying space tourists. The humans-to-space insurance market will present new challenges, because of the potential for lost life and because the vehicles will be reusable. Now you'll have to insure vehicles over time. In order to do that, insurers will want to know how reliable they are, and that information will take time to amass. Until reliability is known, insurance will be expensive.

In the case of American technology, the U.S. State Department has thrown another monkey wrench into the insurance works. Because it doesn't want the Chinese and other foreign governments to get the technology, it won't let rocket makers divulge what went wrong when launches fail. Without knowing what went wrong, insurers can't assess the risk, and without being able to assess the risk, they won't write policies, and they certainly won't pay off on existing policies because they won't be able to tell if the failure resulted from insured perils.

A few years ago, the idea of corporate sponsorships or companies financing space exploration missions would have been as unthinkable as explicit sex on American broadcast television. Times have changed. In the days of shrinking government budgets and the ascendancy of the bean counters, sports events are now plastered with corporate logos; television, movies, and even crossword puzzles are replete with product placement; even U.S. public television is no longer commercial-free. Space advocates have seen the writing on the bulkhead and are adopting similar strategies for financing their projects.

It shouldn't have come as any surprise that in 1999 Pizza Hut announced plans to paint its logo thirty feet high on the 200-foot *Proton* rocket that launched the Russian *Zvezda* service module to *Space Station Alpha*. The cost is reputed to be on the order of $2.5 million for the rights to place the ad. The firm also will sponsor a pizza party for the first crew that will live on the station. The company has done its homework: It will work with a "space chef" to come up with a palatable pizza that can be prepared on the station (a good move, since one's sense of taste changes in weightlessness). In 1996, Pepsico Inc. paid $5 million to the Russian space agency to have cosmonauts float a soda can replica outside *Mir*. Why the Russians each time? Because U.S. law prevents advertisers from putting ads on anything the American government sends into space.

But a 2000 lawsuit relating to e-mail spam may have made banning ads in space unconstitutional. A county Superior Court judge in Washington State dismissed a case in which a defendant was charged with violating Washington's anti-spam law. The judge said that the law, the toughest in the U.S., violates the interstate commerce clause of the U.S. Constitution, imposing undue restrictions and burdens on businesses that outweigh the law's benefits to consumers.

The European Space Agency is planning to plaster company logos all over its *Beagle 2* Mars life-seeking lander in 2003. This interplanetary mission will be the first one to be commercially backed; if successful, it could be the vanguard of a movement in that direction. European Space Agency officials claim that without sponsorship, *Beagle 2* wouldn't have been realized.

When space activists get together to brainstorm about how space tourism can be financed, you hear a lot of the same talk: Look at all the money out there in this or that industry; we can get in on that, make money, and finance our space ventures. The kind of money we need is a pittance compared to these industries' takes and expenditures.

For example, they cite figures like the following:

- The American aerospace industry brings in about $400 billion a year.
- The U.S. freight transportation industry makes about $450 billion a year.
- The U.S. adventure travel industry generates about $5 billion a year.
- The U.S. travel and tourism industry has combined revenues of $515.5 billion a year.

- The American advertising industry pulls in $180 billion a year.

- The U.S. gambling industry runs at a $50 billion a year profit. There's $500 billion a year in betting activity and $10 billion in new casino investment.

The Moon Society says that the cost of its Moon base would approximate the amounts of money spent by large corporations on advertising related to major entertainment functions like Super Bowl games, theme parks, and TV shows. What would you rather have: a Moon base or Super Bowl advertising? The Society also says that Disney's 1993 budget was four times higher than that of *Space Station Alpha*, and in 1994, Disney's revenues exceeded NASA's budget.

Garnering these amounts of money sounds good on paper, but it hasn't happened. One reason is that space people don't have the expertise needed to tap into the advertising, gambling, and tourism industries and the like. Space people are isolated, marginalized when it comes to the world at large, at least those who don't work for Boeing or Airbus or Lockheed Martin, and most of those who do are caught in big machines. The space community has a few stars, big names like Buzz Aldrin and film director James Cameron, but so far no Bill Gates types—no one with business expertise, billions of dollars, and a desperate longing to make space happen. What the space community needs is a bridge to the rest of the world, involvement from the rest of the world. It needs alliances, partnerships, joint ventures, or maybe it needs to offer its services as a subcontractor and let the business experts build it.

You wouldn't think of these otherwise impressive rocket and planetary scientists as being ineffectual, but when it comes to building a humans-to-space industry, that is, a financially sound market-based economic sector, they are. They know how to do what they know how to do. They don't know how to make lots of money, to run huge businesses, to get from here to there. I've come to the conclusion that what it will probably take is either rich people or the government. A couple of guys tinkering around in their garage isn't going to get us to space the way it got us a computer industry.

Now in the person of Walt Anderson of MirCorp, we might have something. At least, I thought so until I heard that the space station *Mir* was to be deorbited. Anderson is a wealthy space enthusiast who makes things happen. He started the boisterous Space Frontier Foundation (See Chapter 16). After several years of fits and starts, he worked out a deal with the Russians to save their ailing space station *Mir* and lease it out for commercial purposes, including tourism. MirCorp is funded through investors and by RSC Energia, the Russian space company that owned it previously.

Anderson saw *Mir* as a *consumer* product, at least in the short term. "Consumer" as in media and entertainment. Start with the consumer and build up the scientific and industrial markets later, Anderson advises. Quite a different approach from the traditional NASA one. Unfortunately, we didn't get a chance to see this one through. The tie-in with media and the public is just what space activists have been touting and could have been the first significant step in making space happen. The precedent has been set. Next time people won't be so startled.

How Government-Financed Space Projects Work

Congress spends money in a two-step process. First, it passes an authorizations bill, which is an officially approved shopping list. Then it enacts an appropriations bill, which actually allows Congress to write the checks. During the appropriations process, the amount of money for each item is often reduced or deleted altogether.

Lobbyists work hard on members of Congress and their staffs to include items favorable to their employers. Sometimes staffers pressure each other to conform to their own agendas. These powerful aides are the ones who actually write the legislation.

The bills are often so complex that the legislators themselves can't know everything that's in them. Staffers advise the legislators about what's in the bill and how they should vote.

As you can see, this process is intensely political and capricious, which is why it's so hard to forecast NASA's budget or that of Air Force space projects. Congress giveth, but Congress also taketh away. Completing a ten-year project like going to the Moon is, therefore, a chancy endeavor, and a bit of a miracle if and when it happens.

Source: Based partially on an explanation in G. Harry Stine, *Halfway to Anywhere*, M. Evans & Co., New York, 1996, p. 132.

THE AUTHOR'S OPINION

Funding is the most difficult aspect of getting new space industries going. In today's market, I can't see venture capitalists getting involved. There's too much risk, the horizons are too long, and the profits won't flow back at rates they expect to see.

Here are the methods that I find the most practical and probable.

Bootstrapping

Get one small thing working and paying back, then reinvest the profits in the next small thing, and the next, until you have enough cash to do something bigger. Along the way, you build credibility and a clientele.

Angel funding

Wealthy enthusiasts will always want to do something interesting with their money. The problem is that most of them aren't quite wealthy enough. But their backing does help, and they may be the key to some winner companies getting a start.

Tax breaks

These provide incentives, but they don't deliver cold, hard cash. They supplement other forms of financing. In the space industry tax breaks shouldn't run indefinitely; once the industry is off the ground, the credits or deductions should be lowered and eventually phased out.

Governmental organizations earmark some seed money and become paying customers

These methods are good for openers, but eventually the business is going to have to have other customers and start financing itself.

Space investment bank or investment fund

I like Tom Olson's Colony Fund approach. It gives ordinary people around the world a chance to invest in an industry they care about and to contribute to the future. You can't give $1,000 to NASA toward a Mars mission, but you could put it into one of these funds. There are all kinds of funds for all kinds of purposes these days: socially responsible funds, economic development funds, etc. Why not space funds? Now, there's much to be done in order for a space fund to work. If people contribute at Olson's rate of $1,000 apiece, you'd need a million contributions to raise a billion dollars; fifty million to meet NASA's estimated cost of $50 billion. That's a lot of people! (Olson says that with a minimum of only $100, he'll attract lots of investors, some of whom will buy on impulse.) Of course, some people might invest more than $1,000. But considering the number of people he'd have to reach, marketing would be a big issue, and a large expense. (Tom Olson can be reached at info@colonyfund.com.)

The other alternatives look less attractive to me at this time. Debt financing requires collateral and manageable risk. I'm not sure either is possible yet. I can't see enough money being raised through sponsorships, but maybe I'm missing something. Tax-free bonds are okay, but I'm not sure that in today's political climate you'd get something like that through—maybe some small ones. Raising money through other endeavors and diverting the profits to space, well, that's sort of like bootstrapping. Could work. Certainly Bill Gates has diverted a lot of his software profits to medical and educational causes—billions of dollars. But this alternative depends on the first business being wildly profitable, and there's no guarantee of that.

Will the humans-to-space industry get the funding it needs? I think so, but it will be a rough go. Once experienced business people like Dennis Tito become involved, the industry has a much better chance than it's ever had. And that is exactly what it will take—business expertise.

But it's also going to take will. The motion picture industry isn't a particularly sound business proposition. Most movies lose money. The same situation applies to the publishing industry, where most books don't turn a profit. Why do people continue to baby these industries along? Will. They like participating in something they enjoy, and they like bringing entertainment and information to the world.

The space industry will be so expensive that it won't be able to take hold on will alone. The financial cases will have to be so strong that people are willing to invest. But they will have to understand that the risks are extremely high, and paybacks may be slow. And that's where the will comes in.

I am not so naïve as to suggest that desire (and vision, by the way) will overcome a weak business case. It will not. But where investors are teetering on the edge, it may be the deciding factor.

If the funding doesn't come, whether in the form of government support or private money, there's no chance for the industry to work. It's as simple as that.

— 8 —

Robert Zubrin: The Fast Track To Mars

Going to Mars is just about the most ambitious yet doable space project we can consider these days. Lots of people think Mars is a more attractive destination than the Moon, which is much easier to get to. What's so special about Mars, and why should we bother with it? Read on for a most intriguing take on the matter.

Robert Zubrin signs copies of his book, *The Case for Mars*, "See you on Mars." He isn't kidding.

The man who founded the immediately popular Mars Society in 1998 is no crank. Cranks don't have detractors, and Zubrin has many. He's also got fervent supporters—disciples who'd follow him, well, to Mars, via his credible but controversial plan to put humans on the Red Planet permanently within ten years, first as explorers, then as settlers.

Mars. It inspires thoughts of water tower-like monsters terrorizing earthlings, little green men, and barren, dusty landscapes of the kind no human would seek out. At the same time, Mars recalls thoughts of the 1997 Pathfinder mission, with its adorable little R2-D2-like rover, pink sky, and "rock" heroes like Buffalo Bill. Is this inhospitable place—the only possible home for humans that we know of other than Earth and the Moon—worthy of the passion it inspires in Zubrin and his followers? Why would anyone advocate setting up permanent residence there, let alone devote his life to such a goal? Some people think the man needs to get a life.

But it's Zubrin's contention that humans must go to Mars because we need to get a life. And soon. If we don't, we'll stagnate, shrivel up, and just about die, he argues. Without a frontier, a *tabula rasa* (blank slate) in which to experiment and expand, Earth society will ultimately cease to function. We're losing our willingness to take risks, we're becoming banal and homogenized, and more than that, eventually "one world will be just too small a domain to allow the preservation and continued generation of the diversity needed not just to keep life interesting, but to assure the survival of the human race." Nothing is more important than the creation of a new frontier. There's that word again.

Not only is Zubrin passionate about his cause, but he adopts a different take on how it should be realized from that of some of his fellow noisy space advocates. The fossil-hunting, rooted-in-history engineer believes that government should play a major role in transforming us into a spacefaring civilization. Mostly he believes that government should establish a permanent human presence on the Moon, Mars, or both—at least to start. Then private enterprise could take over. Without government help, he argues, there's no economic justification for a Moon or Mars base, and no one will invest money in them. Competition isn't the way to get to Mars. People go places so that they can live in a certain way, he says, not to conform to a business plan. Mars will be a colony of technologically adept people living on the frontier, a hothouse for invention. The technologies that are developed there could be licensed for use back on Earth, and then Mars could begin to make money. But you have to open up a new world first. To support his arguments, Zubrin cites past colonization efforts, which he says have never been conducted for profit, but by governments or religious organizations not primarily profit motivated. (He's wrong about this. The British, French, and Dutch empires were established in order to create new markets and obtain cheap resources. In addition, most of the colonization of India and Canada were privately financed by companies such as the Hudson Bay Company, the Dutch East India Company, and the British East India Company.)

Author Kim Stanley Robinson (see Introduction) agrees that governments should be the ones to finance and implement going to Mars. "I basically think that it's more interesting if it's government because then in effect it's all of us doing it together as a public project. I don't like the privatization of everything, from the genome to land to city spaces. Ever since the frontier was closed, I think privatization has been happening these last couple hundred years, and I don't think it's actually healthy for the body politic. The privatization of space just makes me shudder. It's like talking about privatizing Antarctica or something," he says.

There are also practical reasons for governments to get involved, according to Robinson. "Now obviously businesses are part of life, and there are businesses that are going to make their business by doing stuff in space, but I think they're going to be subcontracted, and I think the big organizing capability and people setting the rules, that it should be government that's doing it. And in the end they do have the biggest fund of capital. And also I think a lot of things in space are not immediately profitable, and so private industry is not charging up there for the good reason that they can't justify it to their stockholders over the short term. It's especially true of Mars, but it's even true of space right above us."

Every time I've seen Robert Zubrin talk, he's received a standing ovation. He's the biggest draw at every space conference he attends—sometimes, it seems, more so than moonwalker Buzz Aldrin. One Mars Society member, who is also a sometimes fierce critic of Zubrin, admits that "Zubrin is an incredible engineer and thinker. *The Case for Mars* literally gave me my dreams back and changed my life."

Zubrin is also the biggest media star in the space community. Wendell Mendell, a planetary scientist and lunar expert with NASA's Johnson Space Flight Center in Houston, relates the following story about Zubrin's mediahood:

> This is about an appearance I made at the Space Frontier Foundation Conference in 1996. I had come to be on a panel because I'm the Moon guy. Whenever you want to have some sort of argument about Mars guys

> and Moon guys, you look it up and I'm the Moon guy. So I was with Zubrin and of course, you're never exactly with Zubrin. It's more "in the presence of." I showed up at this panel and there's going to be the asteroid guy, the Moon guy, and the Mars guy. And I come down and I walk into the room, and there are television cameras. You know, big setup, pros, guys with earphones, great big cameras and tripods and businesslike appearance. And I turned to Rick Tumlinson [head of the Space Frontier Foundation; see Chapter 16] and said, "Maybe I should change out of this T-shirt. There's going to be television presence here." And he says, "Don't worry about it. It's not for you."

There are several reasons for Zubrin's success with the media. One, he's written books (*The Case for Mars* and *Entering Space*), which automatically confers magical status and credibility. (John Lewis has written books as well—stylish and fascinating ones—but he's failed to follow some of the publicity and branding steps; hence, his image is more that of an academic than a leader. See Chapter 12.) Two, he's an impassioned speaker who knows how to warm up the audience with anecdotes, whip up a frenzy as he makes his case in step-by-step fashion, and swoop in for the kill at just the right moment to wild applause and cheering. Three, he's great with sound bites, and though not telegenic in a Hollywood sort of way (he's dark and slightly exotic looking, but in no way dashing or romantic), he gives good camera. Four, he's chosen a sexy subject: Mars ... soon. Urgently. And so the pent-up demand for space explodes like a rocket lifting off when he speaks.

Zubrin's Mars Direct plan is now so famous that his name will forever be associated with the planet. It was born out of the disillusion following the 1989 publication of "Report of the 90-Day Study on Human Exploration of the Moon and Mars," nicknamed "The 90-Day Report." This discouraging-to-activists NASA report concluded that the U.S. would require thirty years to build up its space infrastructure and $450 billion before humans could go to Mars. The study was requested by President George Bush in his July 1989 speech committing the U.S. to a new era of space exploration,

in case they don't land where they're supposed to. If the crew lands outside the range of the rover, they still have the second return vehicle. (Zubrin thinks it's unlikely they'd land so far off target.) The ship includes a solar flare storm shutter in the center where the crew can go when an alarm tells them a solar flare is on the way (these deadly events last for several hours). The shutter would consist of a space surrounded by five inches of water, food, or human waste, which Zubrin says is sufficient to shield people from the several thousand REMs (a REM is a unit of radiation delivered to a biological organism) a flare can generate.

As far as the danger of cosmic radiation is concerned (found everywhere in space as well as on the surface of Mars, it's delivered in a constant flow rather than an acute event), Zubrin says that it would take meters of water—an untenable amount—to stop it. However, he concludes, the risks aren't that great. The crew would receive only 50 REMs per year while traveling, and they could protect themselves adequately on the planet by putting sandbags on top of the habitat. With an increased fatal cancer risk of 1 percent attending the travelers, that's a risk worth taking, Zubrin thinks. (He compares the health risk with that of smoking: in the U.S., you have a 20 percent chance of developing a fatal cancer if you don't smoke and a 40 percent chance if you do. So what difference does 1 percent make? If we recruited the crew from smokers, we'd reduce their chances of dying of cancer, he jokes.) As far as gravity is concerned, the craft will be designed to make some to alleviate the ill effects on the crew's health and keep them from having to exercise two hours a day.

The crew sets up shop and explores the planet until the next launch window opens a year and a half later. The ISRU unit will have made enough fuel for the rovers to cover over 14,000 miles during that time.

On the next launch from Earth, another crew return vehicle and ISRU unit are delivered. These are landed a few hundred miles away (driving distance from the first landing site) with the mandate of opening up a new site and readying it for a new crew.

The cycle is repeated every twenty-six months.

The dangers are many, and unlike Moon missions, abort possibilities are slim. (You could go around the planet upon reaching it rather than landing, but that's it. A trip to Mars means no turning back.) A "rescue" mission would take three years to reach the crew, and even communications are delayed to the point that any telerobotic operation of systems, tools, or gear from Earth is impractical. This can be disastrous, as the recent experience of a female Antarctic explorer has proven. The woman, who found a lump in her breast, was not able to get medical care for months, and this was on Earth! In many ways the crew will be on its own, faced with even more isolation than on Earth's exploration missions. Put the wrong person in that crew and things could get very rough. Zubrin's answer: "Compared with the stresses dealt with by previous generations of explorers, mariners, prisoners, soldiers in combat, and refugees in hiding, the adversities that will be faced by the hand-picked crew of a Mars mission seem extremely modest." Besides, he argues, nothing great has ever been accomplished without courage; it will be risky to go to Mars no matter when, no matter how.

As far as legendary Mars dust storms are concerned, Zubrin admits that it's not a good idea to land in one. However, he believes that such storms pose little danger on the ground because the atmosphere is so thin that the force from high-speed winds is about ten times weaker than it would be on Earth. (This is debatable. Physicist Alan Jones, who works at AT&T Research Labs in Cambridge, England, thinks that the dust particles themselves may add to the force of the currents.)

To make all this happen, Zubrin founded the Mars Society in 1998. Boulder, which is close to where Zubrin lives and the site of the first Mars Society meeting, could not be less like Mars. Green and eminently hospitable, it's an unlikely home for one of the Red Planet's fiercest advocates. That didn't faze the 700 people who showed up there to attend the Society's first conference and sign on to Zubrin's vision. The founding declaration, ratified at that conference, lays out the reasons we *must* go ("must" being a word Zubrin invokes at every chance):

- We must go for the knowledge of Mars. To find out if life ever existed on the planet, and to show whether the origin of life is unique to Earth (a proposition that can be disproved by going to Mars, but not proved). We cannot find out what we need to know robotically. Only people, capable of traversing rough terrain, lifting heavy objects, sorting through material delicately, and making decisions, can explore thoroughly.

- We must go for the knowledge of Earth. Because it is believed that Mars was once wet and warm, we can learn much about our own planet and the changes it's undergoing by studying Mars.

- We must go for the challenge. Otherwise, our civilization will decay. And besides, going to Mars will give nations a chance to cooperate in an endeavor of huge importance.

- We must go for the youth. Going to Mars will challenge young people and inspire them to get scientific educations, which will benefit humanity in numerous ways.

- We must go for the opportunity. It's a chance to shed our old baggage and start over.

- We must go for our humanity. We can bring life to Mars and Mars to life. We are the only creatures on Earth who can do this and in the process make a profound statement as to the precious worth of humanity and every individual within it.

- We must go for the future. By opening up a new world, we make the potential of Mars a reality. A new branch of human civilization is waiting to be born.

The purpose of the Mars Society is to further the goal of the exploration and settlement of the Red Planet. This will be done by:

1. Broad public outreach to instill the vision of pioneering Mars
2. Support of ever-more-aggressive, government-funded Mars exploration programs around the world
3. Conducting Mars exploration on a private basis

The Society adds, "Starting small, with hitchhiker payloads on government-funded missions, we intend to use the credibility that such activity will engender to mobilize larger resources that will enable stand-alone private robotic missions and ultimately human exploration."

The Society is making progress on all fronts, though number 2 remains the most elusive. The Society in general, and Zubrin in particular, makes much more noise than many other associations of far larger size. Since 1998, the press has been filled with news, features, and interviews spotlighting Zubrin and his ideas. (The Society itself remains a supporting player, as does Zubrin's wife Maggie, administrator of the organization, and an intelligent, articulate doer in her own right.) Part of the reason is the Society's hard work. Part is Zubrin's book, which he promotes internationally—in person. And part of it is that Mars captures the imagination as healthcare and foreign aid do not.

Progress toward number 3 is especially interesting. To simulate the Martian environment and conduct research relevant to a Mars mission, the Society constructed a research station on Devon Island in Canada's Nunavut Territory. The project is modeled on a Jacques Cousteau approach: real exploration supported by funding generated through publicity and media. Mars-like Devon Island is located circa 75° North. It's mostly uninhabited polar desert, appropriately punctuated by a 15-mile diameter meteorite impact crater (Haughton Crater). The hope is that researchers will learn enough to be able to develop appropriate field tactics for exploring Mars itself. The $200,000 facility, funded by Flashline.com, The Discovery Channel, FINDS (Foundation for the International Nongovernmental

Development of Space), and others, was raised in the summer of 2000. (Interestingly, FINDS is headed by Rick Tumlinson, who has since fallen out with Zubrin rather grandly. See Chapter 16.) It will cost $1.3 million to operate over five years. The U.S. Marine Corps has contributed cargo flight support, with funding shared between the Society and the NASA-led Haughton Mars Project (led by Dr. Pascal Lee, who is exploring Devon Island's geology and biology as a precursor to Mars exploration). One of the initial parachute drops to the building site was a disaster. The parachute and its cargo separated a thousand feet off the ground; the cargo, some domes and equipment, completely destroyed a trailer, a habitat, and a crane that had been delivered earlier. (Critics of Mars Direct might seize on this event to point out that if Zubrin can't land cargo properly on Earth then ....) In fact, one of the journalists covering the event asked Zubrin if he saw a parallel between the "failure of your mission" and that of the *Mars Polar Lander*. Zubrin replied, "There's a parallel in that we both hit a rock. But the difference is that we have a human crew here, and we are going to find a way out of this."

They did. Using makeshift materials from nearby Resolute Bay, Mars Society members, Inuit kids, and the journalists who were on site, salvaged and raised the facility in nine days, some filled with freezing rain. The first crew, soon to be joined by others, consisted of Zubrin, British Antarctic Survey scientist Charles Cockell, Mars Society Webmaster Marc Boucher, NASA scientist Dr. Pascal Lee, general construction contractor Frank Schubert (who oversaw the raising of the facility), and Bob Nesson of the Discovery Channel. In reporting on their activities, they engaged in Mars-to-Earth simulated time-delayed communication with their Mission Control in Denver.

The 2,000 Mars Society members (30,000 on the mailing list) are as passionate as Zubrin himself. Many are engineers who like to debate small points of Martian infrastructure and scientists deep into geological and biological research. Monitor some of the Society's mailing lists and you'll find proposals for everything from

Martian timekeeping devices to how to use lava tubes as living quarters to how criminals should be dealt with to who should govern and how. This is either forward thinking or escapism, depending on one's point of view, as author Kim Stanley Robinson observes. "In science fiction I see this a lot. There are people who are using science fiction to get more out of life and make life mean more, and there are people who are using it to escape into a tiny little bubble world where they don't have to confront life as much. And I suspect that space itself functions somewhat that way."

In contrast with another space advocacy organization, the Space Frontier Foundation, which tends to attract hard-headed, practical business people, Mars Society members appear dreamy. Many are science fiction fans who love to imagine what life might be like outside the confines of the familiar. You won't find many rocket entrepreneurs in the Society (though Zubrin himself is one; he co-founded Pioneer Rocketplane, an X PRIZE contestant; Zubrin maintains his research company, Pioneer Astronautics, but is no longer associated with Pioneer Rocketplane). Foundation conference-goers dress in suits and business attire; Mars Society convention attendees show up in T-shirts and shorts. The Foundation holds its events at hotels; the Mars Society on college campuses. Is it any wonder that the latter attracts so many students? It's got student chapters all over the world, and they are active.

Zubrin feels that the Mars Society is doing much more than the space organizations that have been around since the seventies, like the National Space Society and the Planetary Society. This is evidenced by the Society's Mars Arctic Research Station, being built on Devon Island in the Arctic. No other space organization has itself rolled up sleeves and dug in a simulated space environment, though individual members of the various groups do research via their NASA or private jobs. Like other advocacy organizations, the Society lobbies Congress, but on a grass-roots level rather than through professional lobbyists.

The Society grew out of the Mars Underground, a movement founded in the early 1980s at the University of Colorado at Boulder and elsewhere. Zubrin was one of its founders. Members included scientists and engineers who desired to promote the idea of going to Mars and to keep technically up to date. Several "Case for Mars" conferences were held, at which Mars research was presented. Soon, NASA employees were joining the movement. With the publication of Zubrin's book *The Case for Mars* in 1996 and the ensuing excitement it generated (he received over 4,000 messages from people who wanted to help him realize the Plan), the impetus grew.

The Mars Society itself originally was funded privately to the tune of $400,000. Now it gets money from its conferences, dues ($50 per year), donations (tax deductible), and agreements with vendors who sell wares in its online Mars Society Mall. It's a nonprofit, which means that when it comes to lobbying, certain restrictions apply. (No substantial part of the organization's activities may involve attempting to influence legislation through propaganda or otherwise. The definition of "substantial" is the rub. Some courts have held that it may mean as little as 5 percent of the organization's activities, so Congress enacted a provision that allows organizations to measure percentage on an expenditure basis, ranging from 5 to 20 percent up to a cap of $1 million.) In fact, the Society does lobby, but not so much that it violates the IRS code. If it did, it would be subject to fines, and its nonprofit status would be revoked.

Zubrin's plan has been partially accepted by NASA; they've used it as the basis of their own varying plans. Even so, Zubrin climbs the walls because of NASA's to-ing and fro-ing. To say that he is impatient is a grand understatement. If you want to see what a sense of urgency looks like, go hear him talk, which he does in a manner reminiscent of expelling angry bullets from a gun. He talks so fast and with so much authority that you almost forget he's not in charge of an existing mission. He's so worried about the difficulty of holding together a political consensus for any length of time—like keeping the Red Sea open long enough for the Children of Israel to cross—that he seems almost frantic at times.

Many people disagree that there's such a hurry, even people on the Mars Society board. Author Kim Stanley Robinson, who holds a seat there, feels that the only reason to rush might be—and he's not even sure of that—to learn enough about Mars to be able to save Earth's biosphere. The issues associated with Mars are many, and going would be extremely complicated. As compelling as Zubrin can be, in order to convince the powers that be, he's got to address some knotty problems.

Mars is the only planet in our solar system, with the possible exception of Pluto, on which we could stand and survive for any length of time. (Mercury is too close to the sun, Venus' atmosphere would crush us, and the others have no surface. Some other bodies in the solar system like moons and asteroids, which possess hard surfaces and atmospheric pressures and temperatures suitable for humans, might make appropriate targets, though. So far we're only sending probes to them, but eventually, humans could follow to places like Jupiter's moon Europa or the asteroid Eros.) Mars is so far away from us—thirty-eight million miles at its closest—that with current technology it takes six months each way to travel there. Once you leave Earth for Mars, you soon reach the point of no return, which means whatever emergencies arise, you must deal with them yourself. The distance also means delayed communications; on Mars itself, the time delay for radio communications is a full four to thirty minutes. Long-term exposure to radiation during the trips is a critical health factor we have yet to learn how to deal with. Living in close quarters during the round trip and presumably on the surface—at least to begin with—presents physical and psychological challenges.

Of course, Mars has no human-friendly atmosphere, which means that even if the temperature, which rarely rises above 65° F and can fall to -225° F, is balmy by Martian standards, you can't walk around in shirtsleeves. You're going to need a spacesuit, helmet, and air supply. Radiation continues to threaten human health even after the crew has left deep space. Anyone living on the surface would be killed by one big solar outburst. And the gravity is one-third that of Earth's. It's great to bounce around and carry heavy loads with ease, but we don't know what the long-term effects of Martian gravity on the human body will amount to.

There are health issues associated with merely landing on the planet. All those g transitions throughout descent cause intense stress. What kind of shape will the voyagers be in when it's time to open the airlock and exit onto the ground? If everyone is prostrate, what then? There will be no one to carry the stretcher and nurse them back to health as there is here on Earth when astronauts return from near space.

But there's a subject more likely to inspire debate than even these: the discovery of indigenous life. It's one thing to find evidence of ancient life now gone; it's quite another to find living organisms, whether plant, animal, viral, or whatever. Some scientists (and Zubrin) argue that Martian microbes couldn't infect us because we and our planet are not designed to host them. Others are not so certain. If we land on Mars and return to Earth, will we inadvertently introduce previously unknown contaminants into our environment as was done by the Spaniards when they arrived in America? Could that already have happened when Martian meteorites hit Earth, as they occasionally do?

Even the most ardent space advocates worry about the opposite: What will happen to alien life if we invade its ecosystem? (This process is known as "forward contamination," while bringing alien life to Earth is called "backward contamination.") What if even our robotic probes aren't sterilized carefully enough and some Earthly microbes end up on Mars or other worlds we explore? What if our sample return missions leak out of their confining quarters back here on Earth?

There are ways to minimize the risk. A sample mission could be carried out entirely on Mars rather than bring Martian material back to Earth. A Mars-bound mission is easier: Rocks and soil could stay on the surface of the planet and be analyzed robotically there. Results would be transmitted back to Earth for further study. The safety advantages of such an approach are great: no backward contamination possible, and no humans in space. Another alternative is to return the samples to earth orbit, possibly *Space Station Alpha*, where, as Barry DiGregorio, founder of the International Committee Against Mars Sample Return and author of *Mars: The Living Planet* points out, "qualified scientists ... could determine whether Martian soil contains pathogenic organisms and then certify them as biosphere-safe before transfer to the Earth's surface" (*Space News*, February 22, 1999, p. 21).

A mission that involved returning samples to Earth would be complex and costly and present many openings for catastrophic failure, never mind backward contamination. A rover-carrying craft would be launched to Mars, where it would go into orbit around the planet. Rovers would descend and spend weeks collecting samples. The samples would be launched back to the orbiter, requiring a tricky docking maneuver and cargo transfer. (The necessity to orbit would require that the craft carry a lot of fuel. No possibility of making its own fuel from Martian materials would exist.) The samples could orbit the planet until another ship could be sent to retrieve them, possibly after another round of samples is gathered in a later mission. Again, the mechanics would be sensitive: rendezvous and retrieval and insertion of the samples into an Earth-entry capsule. The craft would return to Earth orbit and release the samples in a protected capsule for re-entry to the surface and probably a crash landing. The capsule would be taken to a high-security facility, where, theoretically, any bacteria or other micro-organisms contained within the samples couldn't contaminate Earth. Many scientists feel that this is the only right way to do it, that only instruments on Earth can yield highly meaningful results—they're more sensitive and flexible than robotic manipulation will allow. And much of the material can be socked away and analyzed later, when the instruments have advanced.

But other scientists around the world are questioning the need for a sample return mission at all. In the European Space Agency, some feel that the instruments being developed for the 2003 Mars Express mission are so sophisticated that they can take geological and chemical measurements that obviate the need for a sample return project.

Zubrin believes that fears of backward contamination are groundless for the following reasons:

- The surface of Mars is too cold for water to liquefy; Mars is sterilized by ultraviolet and cosmic rays; the planet contains an antiseptic mixture of peroxides, all of which mean that indigenous life can't exist on the surface.

- Even if underground life exists, it's impossible for it to threaten us because "pathogens are specifically adapted to their hosts, and there are no highly developed animals or plants to support a pathogenic life cycle in the Martian subsurface groundwater."

- Earth already gets about 1,100 pounds of Martian meteorites each year, which would have already infected us if they could. (Zubrin says that the trip between Mars and Earth is insufficiently traumatic to sterilize meteorites. DiGregorio conjectures that spaceborne epidemics indeed might have taken place on Earth already.)

Concern was institutionalized back in the 1960s when the Outer Space Treaty, a document signed by more than sixty nations including the United States, was ratified. The treaty says that spacefaring nations will avoid "harmful contamination and also adverse changes in the environment of the earth resulting from the introduction of extraterrestrial matter." When astronauts brought back Moon rocks, scientists carefully contained the minerals. Viking spacecraft that landed on Mars were sterilized ahead of time even though scientists thought that the planet was inhospitable to Earth microbes because of the low pressure, radiation, and low temperature, which would keep any water on the planet from assuming liquid form. (In fact, sterilization added a full 10 percent to the cost of the mission.) When we discovered through the Viking mission that the surface was even more inhospitable than we thought, the International Committee on Space Research relaxed future sterilization requirements except in cases where we were attempting to detect life on the planet. (The thinking was that Mars was too hostile for Earth microbes to grow and spread, but that they could contaminate Martian soil and rock return samples.) The 1997 Pathfinder probe was loosely treated, having been assembled in a clean room and swabbed with alcohol.

Because scientists are uncovering new evidence in support of present and/or past life on Mars, they're having to rethink their approaches. When the strategy was decided upon, we didn't know that life could exist even on Earth in extremely savage environments, such as under intense pressure at the bottom of the ocean, in airless caves, and buried in ice and permafrost. We also had no evidence of the presence of liquid water on Mars; thanks to images from the Mars Global Surveyor, we now do. If it could happen here, it could happen there.

Is there any way to take sufficient precautions if we land humans on Mars? That's a toughie. You can't sterilize the people or the air they breathe, and there's no point in our going if we can't open the door of the craft and go out of it. You could keep the crew in orbit while they operate sample-gathering equipment, but that isn't a permanent presence.

The real prospects for near-term human exploration of Mars do not look especially bright as of this writing. In the late 1990s and early aughts, NASA changed its plans for visiting the Red Planet—robotically only—more times than pop-star Madonna has changed her hair color. In early 1999, NASA and its European partners intended to establish an almost continuous robotic presence on Mars until about 2010, including sample returns and other missions that would look for evidence of past and present life on the planet. (The impetus for the increased pace of exploration was the 1996 announcement by scientists at NASA Johnson Space Center and Stanford University that a meteorite found in Antarctica in 1984, which came from Mars, showed evidence of vestiges of ancient Martian bacterial activity.) Craft were to be launched at twenty-six-month intervals to take advantage of every window available for reaching the planet. (Twenty-six months is the shortest amount of time possible between launches; that interval marks the point at which the planets are positioned for the shortest amount of travel time: ten or eleven months, at least using current propulsion methods.) The missions were to be called the Surveyor Program. (Mars Global Surveyor, which returns high-resolution images and atmospheric and magnetic field data and maps Mars' surface features, began orbiting the planet in September of 1997.) As of November, 2000, NASA's plans had devolved into four missions between 2001 and 2007: mapping and imaging in 2001, water-seeking rovers in 2003, more imaging in 2005, and a long-duration science rover/competitively selected Scout mission in 2007. The first sample return mission would not launch until 2011 at the earliest, with 2014 more likely. When that will happen depends not only on funding but also on technological advancement and the success of the earlier missions.

The Europeans and the Japanese are involved with Mars too, though the freezing of the European Space Agency's science budget in 1995 limits its options. Nevertheless, the agency plans to launch a combination orbiter/robot lander called Mars Express in 2003. The object of the $144 million (150 million euros) mission is to search for water. France, which is the largest financial contributor to the ESA, plans to collaborate with NASA on a 2007 or 2009 Mars sample return mission that will cost it between $260 and $390 million (between 2 and 3 billion francs), about a fifth of the French space agency's budget. Italy has been negotiating its possible involvement in the mission with NASA as well, and NASA has approached other countries like Canada, Germany, Japan, and Finland, hoping to gain their participation. Japan has already launched an orbiter called *Planet-B*—in 1999. The craft will study Mars's atmosphere and ionosphere.

Part of the reason for the scaled-back approach in the U.S. is the spectacular failure of two of its high-profile projects, the *Mars Climate Orbiter* and the *Mars Polar Lander*. In the case of the former, scientists made embarrassingly stupid mistakes, causing the mission to fail and the public to heap ridicule on Jet Propulsion Laboratory, the organization responsible for conducting the mission. The *Lander* and its two probes, which were supposed to set down on the South Pole and investigate the composition of the surface and subsurface, disappeared without a trace. The *Orbiter* cost $125 million; the *Lander* $165 million. Neither the public nor the U.S. Congress is amused, especially when you add in cost overruns and delays on *Space Station Alpha* and the drastically lowered flight rate of the shuttle in recent years. Some people wonder why they're giving tax money to NASA at all.

Many people feel that the demise of the *Orbiter* could have been avoided if the two groups working on the project had deigned to talk to each other. (One project team was calculating in meters; the other in feet.) The fate of the *Lander* remains a mystery, making it hard to ascertain whether JPL made mistakes or whether unforeseen pitfalls were to blame. However, in my opinion, the cause doesn't matter nearly as much as that we keep on trying. What the public and the politicians don't understand is that failure is a necessary part of success. Engineers know this. That's why aerospace engineers live by the motto, "Build a little, fly a little." Or they would if they weren't so afraid for their jobs. The best ones will tell you that failure teaches valuable lessons—much more than does success. The aircraft industry has progressed spectacularly as a result of this practice. The space industry will only succeed if our risk-averse cultures learn to tolerate risk.

Even if we could do so, garnering the political will and the funding for any human exploration of Mars would be a gargantuan task—in any country. Among spacefaring nations, only the U.S. has sufficient money, which could run into the hundreds of billions of dollars. Experience with *Space Station Alpha* shows that a mission mounted by a consortium of countries would be a logistical nightmare, though a joint mission might be far more politically palatable to the world than a U.S.-only one. Private missions are almost out of the question because of the scope, expertise and infrastructure required, cost, and lack of profitability. Mars is no promised land, though to listen to some activists you'd think it was Canaan itself. (Of course, it wasn't easy to reach Canaan, either—it required tromping through the desert—but at least it had air.)

Author Kim Stanley Robinson thinks that because the U.S. space program doesn't have much of a constituency, nothing will happen until or unless that changes. But we also need a compelling reason to go.

> I think what will get us into space and off to Mars is making a public relations push that makes it clear that going into space is not just a hobby or a specialized interest of people who happen to be interested in space, but that it's a useful tool for Earth management. If that

> was made clear, if we could make that case, then all of the people who are interested in their children's welfare might become more inclined to support going into space because it'll become obvious that in the future we need to be up there in order to keep the Earth in balance over the long haul. There are a lot more people interested in their children and grandchildren's welfare than there are people interested in the stars or whatnot. It makes it an environmentalist cause, part of the Green program, and even that, I think there need to be a lot more people interested in that than there are.

There are at least two other things that could spark that interest: discovery of life on the planet, and another political space race à la Apollo. Perhaps—perhaps—the discovery of a very dangerous near-Earth asteroid would also impel such a project. None of those possibilities has anything to do with profit—only a little thing called survival. We'll just have to wait and see.

THE AUTHOR'S OPINION

You've got to either love or hate Zubrin. He's so right some of the time and so wrong the rest of the time that people tend to fall into one camp or the other. I'm in the middle. I think he's terrific because he's technically clever and gets the juices going and has raised Mars' profile tremendously, and I think he's reckless and wrong as can be about many things. I think he's so driven by something that most people don't understand that he doesn't know how to moderate his statements and make them saleable to a wide audience. I also admire him tremendously for getting his Arctic research station up and running and doing real things that pave the way for our going to Mars. And for popularizing the idea of living off the land. It's a brilliant way to go.

I can't see a way to make Mars profitable, at least not until there's some significant infrastructure in near space. The problem is that there's nothing between here and there but the Moon, and even then, only sometimes. Going there now would be like sailing across the Atlantic was for Columbus, except that there would be no air or food when you got there. And Columbus thought he was looking for a route to India—a place he knew to hold exactly what he and his government wanted. Great riches probably do not await anyone who finds the way to Mars.

For a long time I thought Zubrin was wrong about advocating Mars as a government project. You'd never get taxpayers to go for that, I reasoned. And I might be right. But now that I know something about how the aviation industry got going, I'm doubting my former conclusion. No transportation industry except the automobile industry got going without significant government help, either in the role of operator or as customer. It will be a tough road, though. Right now taxpayers don't want to spend money on anything, and legislators aren't in the mood to buck them. Such a push would take leadership on the scale of Kennedy or Khrushchev; we just don't have that today.

I like Kim Stanley Robinson's idea of Mars being a public project, but I see problems with both that and its opposite. Whoever doesn't get there is going to be resentful. If the U.S. or Russia sets up a base there, people all over will see that as a threat—we've dominated the Earth, now we're going to dominate space as well. If the effort is a multinational one, it will become bogged down in the same sorts of disputes that the U.N. does. Even the sixteen-participant *Space Station Alpha* has all sorts of management problems. How could we possibly set up a multinational settlement on Mars and make it work? I guess what I'm saying is that I see the question of *who* as a knotty problem to which I have no elegant solution.

The issue of contamination could be a show stopper. I believe Zubrin is too cavalier in minimizing fears of backward contamination. He's got to be more methodical in dealing with these important issues. Otherwise, his credibility will be damaged.

It's obvious that Zubrin would like to be one of the first hundred on Mars. He probably deserves to be. But he's not the ideal leader. He's too polarizing. You need peacemakers, the Kofi Annans and Jimmy Carters of the world. Now there's an interesting thought.

— 9 —

Wendell Mendell: Mooning Over the Past

Why did NASA and the Russians abandon the Moon in the seventies? Read on for an insider's look at the obstacles involved in drawing NASA's attention back to Luna after the dust settled.

I caught up with Wendell Mendell at one of the National Space Society's conferences.

In 1969, humans landed on the Moon for the first time. In our 35,000-year history, we'd never walked on another world. It was a thrilling time—the beginning of moving out into the solar system, we thought. But after that, we went back only a few times to root around and explore. Our last foray to our closest neighbor was in 1972, thirty years ago. The fire still burned within us for a few years, but when it became apparent that NASA and the Russians had turned their attention to other things, we, too, began to forget, and went on with our lives. The dream fell dormant. Can you recall the last time you heard someone talk about going back?

Actually, there is such talk, but it's pretty muted these days. There's much more noise over sending humans to Mars, thanks to Mars proponent Robert Zubrin and his associates in the Mars Society (see Chapter 8). Most people seem to think "been there, done that" about the Moon, an airless, dead world, which appears to most to have nothing to offer.

According to some activists, they couldn't be wronger, as the expression goes. The Moon holds a variety of resources, and it might act as a useful test site for pre-Mars expedition training and system testing. Its one-sixth-Earth gravity and atmosphere-free sky makes it an appealing a tourist destination: You can fly, do sports, and observe the universe through powerful telescopes. The Moon is only three days away; Mars is six to eight months from us. Get the picture?

There's no better way to set the stage for today's situation regarding the Moon than to let Wendell Mendell, a lunar expert and planetary scientist from NASA Johnson Space Center tell the story. Mendell explains that after Apollo, NASA's emphasis changed, and the Moon was gradually forgotten—until recently.

> Let me talk about where I was in 1981. I was at the Johnson Space Center. I was a planetary scientist with a Ph.D. I was eighteen years into my career. I had an Apollo patch on my belt. I was very proud. I was doing what I considered to be decent work in an area which at the time was considered to be sort of peculiar—let's just say I was twenty years before my time. But in June of 1981, several external events hit. The space shuttle had just made its maiden flight. The space shuttle program was behind schedule and over budget. There was a lot of anxiety in the space agency over promises that had been made about the shuttle and its role in space. *Skylab* had been allowed to fall to Earth and incinerate. President Reagan had been elected on a platform of very much less government. I was a government employee. Very much less government meant that I might not have a job. There was also a space policy which, I'm not sure if it was in place in the middle of 1981, but it was soon to be in place, and it was that the only launch vehicle allowed in the U.S. inventory was the space shuttle. We would basically get rid of all expendable vehicles. If you had satellites to launch, you

would have to reserve a place on the space shuttle. At that time that's the way the future was shaping up.

President Reagan had not yet named the new NASA Administrator, although he had named the NASA Deputy Administrator, who had written a memo to his head lieutenant, and this memo speculated that perhaps NASA really needed to focus on getting the space shuttle up to performance standards. As a result, things like planetary science would just be stopped and put on hold for a decade or so until the space shuttle was up and running smoothly. I was a planetary scientist, I was a government employee, I would be hit from both sides. And so my organization went into immediate strategic planning mode as to how we were going to react.

The organization to which I belong was created at the Johnson Space Center during the Apollo program and consists of world-class scientists who study rocks and other geologic samples of things. The scientists that I work with need something in their laboratory. Of course in the '70s they were working on lunar samples. As funding for lunar sample science declined, they focused more on meteorites and began to get more involved in other kinds of planetary science. But still there was a very strong interest in the Moon as a research topic. Yet the field of planetary science itself had moved away from the Moon. There was a strong feeling among the scientific community that we had already spent $25 billion in "then" dollars—$100 billion in "now" dollars—studying the Moon and now it's time to study the rest of the solar system, and those of you interested in the Moon could wait a few years.

But we were very interested in getting a small satellite called the *Lunar Polar Orbiter* up because it would be very inexpensive, yet it would complete the mapping of an entire planet that the Apollo program had begun. It would be the opportunity for planetary scientists to learn about an entire body. Of course, the only way to get a lunar orbiter up at that time was to place it on the space shuttle. So a colleague and I did something which we had never really done before, and that was to walk across the Center and talk to engineers. [Mendell laughs.] You would be astounded at the wall within NASA between the scientific part of NASA and the human space flight part of NASA. There's just literally no communication. Engineers at the Johnson Space Center would have no idea that there were scientists there, and we're still something of a mystery to them after forty years in business.

We had a very simple question. How big a satellite could we put on the space shuttle? It turned out we could get quite a large satellite, and we were excited about that, so we thought, well, maybe using our influence as insiders we could get the launch of a lunar polar orbiter to get to the Moon. But you first had to talk with people about why you would ever want to do that, and science wasn't always the best reason. One of the things I did learn from these engineers was that they were thinking about something called a space station, and that beyond the space station they thought maybe there ought to be rockets at the space station called orbital transfer vehicles. And the whole idea was that there was to be a space transportation system piloted by human beings. In fact, if you looked at the JSC phone book, you didn't find "space shuttle." You found "space transportation system office." I'm sure it was confusing to people on the outside, but it was a sign that the thought was there would be a space transportation system that would carry from the Earth starting with the shuttle, ultimately onto orbital transfer vehicles to other destinations, particularly geostationary orbit, where there would be communications satellites.

And then I discovered from one of the engineers a fact of orbital dynamics that I should know but didn't. That was that by the time you got up to geostationary orbit [24,000 miles above the Earth, where satellites move at the same speed as the Earth, and so hover always above the same point on the planet], you were basically out of the Earth's gravitational field, and that if you just did the transfer a little differently, you could go to the Moon with the same amount of energy as you could go to geostationary orbit if you departed from the space station. A light bulb went off over my head and I said, "If you're building this space transportation system to have routine launches to take satellites to geostationary orbit, why can't we just one day kind of point the other way a bit and take something to the Moon?" They hadn't really thought very much about that, and I said, "After all, when you do finish this—in about 1999 they were going to finish this, it would all be ready and operating—in 1999 you're going to be thinking about the next project, aren't you?" And I said, "It will probably be people on the Moon. If you could go to the Moon regularly, then that would be a nice thing to do. And by the way, you're building these orbital transfer vehicle things to carry these communication satellites. They might be too small to carry people to the Moon. Have you ever thought about that?" Well, no, they hadn't. That didn't seem really logical to us.

At that point, we started a little campaign inside the Center in order to convince people they ought to think about the Moon in the future. They ought to be designing their system so that it would have that capability, or at least an option for it, in 1999 when it was all up and running. That ought to be on the agenda. Well, we ran into some difficulties. The first difficulty was that we were told to sit down and shut up. Here was the reason: NASA is a corporate entity, and any corporate entity is concerned about steady-state activity. You don't want to finish something and have to fire everybody, then start something else and hire everybody back. You really can't do that in civil service anyway. So if you're finishing something, you need to think about the next thing that's coming up. I think they decided on the space station because the space shuttle needed a place to go, and a space station was the logical next step for the space shuttle. They were talking in Congress about the space station, and if some wiseacre stood up and began talking about the Moon, Congress might get confused and wondering is it the space station or the Moon? "So the best thing for you to do is sit down and keep quiet until we can kind of get all this figured out. And then later on we'll talk about it," they said to us.

Well, that didn't sit very well with us, and we persisted. What I really want to emphasize is the type of thinking that was going on in NASA at the time, and particularly a problem of advance planning. So first of all we went to NASA headquarters and talked to the three major parts and showed them our idea, and the Office of Manned Space Flight said, "That's really interesting. I think you really have something there. The problem is that the Moon is a planet, and we don't do planets. The Office of Space Science does planets." So we went down there and they said, "That's a really interesting idea, and we think you're thinking along the right lines, but we don't do people. Sending astronauts to the Moon is out of our charter." So then we went to the Office of Advanced Technology and we said, "We've got this advanced idea." And they said, "It's really interesting but we do technology, we don't do missions." So there was no place in NASA that you could go to and knock on the door and say, "I'd like to tell you about this idea. Maybe we can find something in the budget to work on it."

We had no official standing in the agency as future planners. Ours was a problem that I refer to as the scarecrow problem. That is, when the scarecrow went to see the wizard of Oz, he asked for a brain, and the wizard said, "You don't need a brain. You really need credentials." He hung a medal around his neck and gave him a diploma, and suddenly the scarecrow was qualified to run the Emerald City. Our problem was that nobody knew why we were talking about the Moon. In fact,

"Who are you guys? And who authorized you to talk about the Moon?"

And in fact, if you went to try to find something that said "Advance Planning" on the door, that person was thinking about the space station. The last time an office studied lunar bases was about 1972. The last documents I can find all stop about there.

So what we needed were credentials. In NASA or the space program, the way you get credentials is to convene a blue ribbon committee to consider whatever it is you're advocating. The idea was legitimization. So NASA had this problem of why should we be talking about the Moon, because in NASA, if you are not part of a funded program you have a real problem convincing people that they should pay attention to you. It's a real serious issue.

I have to tell a story. I went to a lunar and planetary conference in 1983. We had a special session that we put together about returning to the Moon. I wrote a couple of papers, and one of them addressed all the reasons people gave me why we can never go to the Moon again. It was too expensive, the economy of the world's changing to a service economy and there would never be net growth again in the economy—it would go to zero. In the early '80s I remember the economy was near zero growth, and there was a theory that it would be that way from now on, so you could never do these sorts of things within the budget. Or they would say, "The Japanese are beating us in technology," or they'd say "The Russians are leaders in space now." There was a whole list of reasons why we would never be able to go to the Moon again. So I wrote an abstract with a [long] title that said "Why are you talking about a manned lunar base now when the shuttle fleet needs a fifth orbiter, when the space station project is sputtering, when planetary exploration has had no new start since 1978, when the national economy is in the worst shape since the Great Depression, when the Japanese are pushing for world leadership in technology, when the Soviets and the Europeans are challenging American pre-eminence in space, and when video games are corrupting our youth?" And I had a NASA engineer come to my office and tell me that people at the Center had received my abstract, and there was concern that I was commenting on video games, which had nothing to do with my job.

Fortunately the next day, President Reagan appeared at the Air Force Academy. And he gave a speech. In his speech he said, "You know, our children are playing all these video games, and that's allowing them to hone the skills to fly jet planes." Two days later that same NASA engineer came to me and said, "You were right." The point was legitimization. So what we did was to come out and form a blue ribbon committee.

By this time, through a series of circumstances, we had begun a collaboration with Los Alamos in terms of planning to go to the Moon—we held a workshop in the spring of 1984 at Los Alamos. Los Alamos was a good place to collaborate with in those days because President Reagan had picked a lot of ex-Los Alamos people to be his national security apparatus and Jay Keyworth, the president's science advisor, was from Los Alamos, and people high in the National Security Council were from Los Alamos. So if you were working with people in Los Alamos at high levels, you had entrées into places that were unusual for NASA scientists.

We conducted a two-step program to legitimize the subject of lunar bases. The problem was we got a little bit of money, and we were in this dilemma. If you really want to get something done, you need a small group. On the other hand, if you want to form an advocacy, you need a large group. In one workshop, you can't do both. It's contradictory. So we decided to hold a closed, invitation-only workshop with a small group of people and flesh out the ideas, and then have a public symposium based on those ideas and open it to everybody. And that's what we did. The public symposium was the 1984 symposium held at the National Academy of Sciences. The key thing was this theme, that with scientific research, lunar resource utilization, and settlement as the three things that you would do on the Moon, we ought to be thinking about the Moon.

The word used in the workshop was colonization, and I found out very quickly that it was politically incorrect. Because colonization has to do with imperialism and exploitation and settlement has to do with families and warm feelings and so on. These were pretty advanced concepts for NASA. They could kind of understand scientific research, but this idea of resource utilization still sticks in the craw of people in various places, and settlement in those days was something you didn't want to talk about in a meeting. You would really gloss over it because that was just something that was like science fiction.

Much later President Bush [the elder] made a very important speech in 1989 about going back to the Moon and exploration of Mars, finishing the space station, and so on. And that began something in NASA called the Space Exploration Initiative. The Space Exploration Initiative died a fiery political death. It's fairly important that we understand why. NASA is profoundly influenced by what I call the Apollo meta-event. People can write about the Apollo program in a number of different contexts. In terms of its politics as a Cold War exercise, in terms of its effect on our culture and changing our thinking into a spacefaring nation, in terms of its effect on human history, that human beings for the first time left the planet and stepped on another one. In terms of technology, that NASA had the reputation of being a technology engine because of the Apollo program. In terms of the science. There was a profound effect in thinking about the Earth and its environment and science, and from the fact that we could manage these large complex projects. After the Apollo project there arose this idea that only in the United States was there the skill and technique to be able to manage large, complex projects. Its mark on the structure of the U.S. space program is almost indelible. So, it's implicitly assumed that anything else we do in the future will be the same.

In the Apollo paradigm you have success which was defined by an event. Now in relativistic space-time physics, an event has four dimensions—three spatial coordinates and time. In this case it was to land a man on the Moon—a place—and return him safely to Earth before 1970—a time. So, that was the measure of success. And the important thing is that other attributes of the measure of success had secondary importance, such as cost and safety. Now there's going to be a lot of people who argue with me that we paid a lot of attention to safety, we tried to make things as safe as possible, and yes that's true, but I still think the thinking in Apollo was to do the project on time and to beat the Russians and that sometimes we chose to take calculated risks.

This goal was chosen specifically by President Kennedy to have an unknown solution. That's the reason he picked landing people on the Moon, because nobody had done it, and we thought it was far enough down the line that we could get ahead of what we thought was a Russian superiority in certain kinds of launch technology. Because it had an unknown solution, the activity was technically complex and risky. Since it's risky and complex, it's extremely expensive to do. As a consequence the only place you can find the amount of money to do this event is public funding. So, it has to be managed by a government agency and within that government agency over time institutional

> complexity grows. The reason for that is that there is public oversight, public eye, federal rules and regulations, and so on. So the space program over time—and this is a maturing factor in any large organization—becomes more complex and beset by rules.
>
> Another part of the Apollo paradigm has to do with the engineering design. The space environment is treated as a threat and not a resource. Habitats are self-contained cocoons in which the human being should stay, and in fact they should stay inside them. They should not go outside them. Mission designers hate for astronauts to go EVA [extra-vehicular activity, i.e., going outside], particularly in the old days. For astronauts in Apollo to actually go out on the surface and do things was frowned upon. Now the only reason that has changed is because it's been driven by the space station problems in design recently; there are predicted to be about 600 EVAs in the construction of the space station. But believe me, in Apollo that would never have been thought of. The mission duration has to be minimized in order to contain the risk because only bad things can happen to you in space. The longer you stay, the more bad things can happen, and in fact, Earth is the only safe haven. These ideas dominated the way that people thought about the human space program, and particularly ideas about the Moon.

It's only recently that these ideas have begun to change, though in many ways, NASA is still bedeviled by the same structural problems, and there's still no plan inside the organization to go to the Moon. However, now there's a difference. Now private people and organizations are looking at ways to get to the Moon without NASA. Some want to send people, some plan to visit robotically only. The only rub? How to make it an economically viable proposition.

THE AUTHOR'S OPINION

Mendell's story about the compartmentalization of NASA goes a long way toward explaining how the idea of returning to the Moon got lost in the shuffle. If you look at NASA's Web sites today, you see the same tendency to organize by function rather than by subject. This is bureaucracy at its most bureaucratic. And that, combined with a lack of political leadership, is why the U.S. hasn't been back.

Read on for an explanation of the Moon's assets and a story of one group's effort to establish a settlement there.

—10—

Greg Bennett: To the Moon

Could private enterprise actually build and sustain a settlement on the Moon? Sounds far-fetched. But the Moon has resources most people aren't aware of, and maybe, just maybe, they could be used to economic benefit.

The bulk of my interview with Greg Bennett took place at a McDonald's in Las Vegas. We had wanted to go someplace "nicer," but every restaurant was too crowded to fit into Bennett's short lunchtime window. No matter. You can talk just as well over french fries as you can over a Caesar salad.

Greg Bennett thinks living in Las Vegas is peachy. That the terrain there resembles the Moon and the culture that of some space tourist resort escapes him. He just plain old finds it a nice place to live.

Bennett thinks the Moon would be a nice place to live, too—some day. In fact, he's so keen on getting us there that in 1994 this rocket engineer started the Artemis Project, a for-profit enterprise whose charter is to establish a permanent, self-supporting Moon settlement—get this—based on raising money through the entertainment industry. Well, why not? Nothing else has worked. The cost of the project? About $1.4 billion in 1994 dollars. (NASA's total budget is about $13 billion.)

The idea has grown gradually. In 1994, Bennett—also a science fiction writer—posted a message in his forum on the now-defunct Genie computer community, saying, "Hey, anybody want to talk about a private Moon base?" They did. And how they did. And the international Artemis Project was born.

In just a few years, Bennett and his colleagues have come up with a detailed plan and scrupulously well organized, impressively well written Web site (www.asi.org). Like others who were coming to the same conclusion at about the same time, they acknowledged that no government or organization would finance such a project and looked for a new source of money. It hit them: The entertainment industry had money oozing out of its ears. Here was an established business with action they could not only tap into but merge with theirs. The building of the Moon base would comprise entertainment, before, during, and after. Funds from entertainment projects would be used to bootstrap it, then the base itself would generate revenues from *being* entertainment. Then, lunar resources themselves would keep the project going.

These ideas are not revolutionary any more. What is new and unique is Bennett and colleagues' enshrinement of fun as a first principle. If it ain't fun, it ain't worth doing. Bennett writes on the Artemis Web site, "The Apollo program was run with engineering precision, its drama hidden by the need for a government agency to present an unflagging image of confidence to the world. In stark contrast, The Artemis Project is designed to be entertaining from the start." The *project*. Not just the result, but the *process*. And not just for the participants, but for the public. "Show business is what space travel means to Joe Sixpack; and Joe is our main audience." In person, Bennett explains further. "You have to be entertaining to sell entertainment. And the question is how big an audience do you want to have. No way you can please everybody. But the basic business plan pretty much calls for appealing to a wide general audience."

It's taken decades for the space community to realize the necessity of appealing to Joe Sixpack. Public support at the voting booth is all well and good, but where you really need it is at the wallet. And government space programs provide no way for that to happen. If you don't believe me, call NASA and ask them if you can contribute some money.

Ironically, private enterprise stepping into the space arena probably will make it possible for Joe Sixpack to participate—

much more than when governments do space. Government money, which represents funds from all of us, hasn't gotten us—Joe—there, and may never do so.

On the Web site, Bennett explains, "The Artemis Project is designed to be a profit-making enterprise from the first launch. For two decades we've tried wheedling the government to do it, and we've tried the begging approach. That didn't work. But we've never really tried to use the economic power of the entertainment industry and do it as show business." Bennett is unaware of the irony of his living where he does when he says, "We plan to pay for the initial stages of the project through shameless commercialism."

Since so many people these days are talking about the same kind of thing, one might wonder if they'll all be competing for the same assets. Bennett admits that yes, that might be the case. "To some extent. There are several groups, if not companies, who are talking about tapping into the entertainment side, the fact that space flight really is just another form of entertainment at this point. Of course, then, most things are. First of all, you've got six billion people on the planet, all of whom are seeking entertainment, and you only need a small chunk of that audience to accomplish what we want to do. And to some extent, if one group does it, it reinforces the medium. Somebody had to make the first space opera TV show, the first detective movie. And now if you make a detective movie, there's a huge audience for that kind of thing ...."

Bennett may be right. The 1999 hit *Double Jeopardy* with Tommy Lee Jones grossed over $116 million in the U.S. alone. *The Pelican Brief* in 1993 grossed just over $100 million in the U.S. Also in 1993, *The Fugitive,* with Harrison Ford, brought in $184 million! Of course, gross and net are two different things, and over 70 percent of major films fail to make a profit.

Entertainment. Las Vegas. Science fiction. You'd think a guy who got behind these things would be one of those glitzy types. He isn't. Bennett doesn't want to be an icon. He doesn't even want to be a minor celebrity. All he wants is to be a manager.

If you try to build a charismatic organization organized around a given personality, then the success of the organization varies with the individual. It also helps, if you can avoid it, one of the big problems on the volunteer side. People tend to become envious of all the ego boost that flows to the charismatic individual. The only good thing that comes out of it is that if you create a celebrity, people will follow the movement because they want to know about the celebrity. I try to avoid being an icon without a lot of success. People keep making me one despite my desire not to be one. Both inside Artemis and outside.

One of the results of this starmaking is that people want to join the project and the Artemis Society International, an ancillary think-tank also started by Bennett. Such problems. In fact, right now he'd rather keep a low profile. That's not the ultimate goal, but for now, he wants to avoid the limelight.

If you have a lot of publicity, that creates a burden of support. And everybody would call me. Not right at the moment, people have gotten used to it [his being unavailable], but for a while there, more than half the e-mail I got was people writing to gripe to me about the fact that I hadn't responded to their e-mail. When it comes down to it, there isn't time to open the number of messages that I've got. Much less read them, much less even provide a quick, polite answer. So I had to give up on that. The obvious answer is I need to delegate that responsibility to someone else, and there's no one to delegate it to. So, we've backed off till we create the infrastructure to handle large membership in the Artemis Society, then do a recruiting drive, which is where we are right now, or about to be, and then look for people who can handle maintaining the Society.

Since the interview, Bennett has, in fact, accomplished this goal by founding the Moon Society to take care of membership services.

The deliberate low profile is so different from that of the other space advocacy organizations that you can't help either admiring Bennett or thinking he doesn't know what he's doing. Since few of them have succeeded in getting real projects off the ground, it's possible that Bennett is a management genius after all. Time will tell.

His approach differs in other respects as well. For example, the project adopts no social agenda. "The whole thing about social causes is that so many organizations have been spoiled by distracting social causes. And I can't think of any benefit to it other than attracting a lot of people to what you're trying to do, but if you do that then you're not attracting them because they're interested in what you're doing." So, Bennett doesn't advocate anything, or at least he doesn't think he does. "I don't care what people do when they get there as long as it's a society that works," he says.

This statement isn't quite true. "Works" is a subjective term. In Bennett's world view, despotism doesn't work. "If the only way to do it were to create an Islamic state, then I'd lose interest," he says. Does the new society have to be an American-style democracy? "That's how I grew up," he says. "We're not doing it to push free-market economics. We're doing it because empirically it works. Empirically, a despotism model for a government is not enduring. This democracy game is a relatively new invention. It's been tried to some extent a couple thousand years ago and did not endure. I don't know. It seems the best thing we've got going right now."

But the Project and the Society *are* international, and Bennett is keen to expand on that aspect of it. At the moment, the only thing holding him back is the resources. "We have someone in each of the members of the European Union now, Mexico, and a few other non-English-speaking countries, but of course, all of our communication is in English. We don't yet have a program that reaches out to people who don't speak English. We've tried to get folks to volunteer to translate our stuff into other languages, and they start the project and realize how overwhelming it is, and if nothing else it tells you that we don't have and probably won't have the dedication that people put into translating the Bible."

Include everyone. That's the Artemis approach, at least if everyone isn't a totalitarian. "Joe Sixpack is just a representative of the common man. The average person in the street who doesn't really exist. And the point of the average person is if you want a society that works, you can't just take a little chunk of the human society and expect that society to endure. You need everybody from the street sweepers to the president. And also because if you want to do something as a popular movement and you want popular support for it ...."

So in sum, don't advocate divisive philosophies, and enlist the participation and support of as many people around the world as possible.

To do what exactly? Artemis' goals are to:

- Demonstrate that manned space flight is within the reach of private enterprise
- Bootstrap private industry into manned space flight
- Build a permanent manned base on the Moon
- Exploit lunar resources for profit

More specifically, this is what is supposed to happen:

- Start with a few very small robotic explorers
- Construct a small lunar exploration base
- Explore the Moon
- Construct lunar industrial process pilot plants, expand the industries
- Expand the base; build a subsurface village
- Expand the lunar fuels industry
- Begin tourism (manned launchers will now be lower in cost)
- Build subselene city (below the Moon's surface)

- Erect far-side scientific outposts
- Build surface transportation systems

Now the economy will be self-sufficient.

Once the base is built, it will be owned by those who invest in the space-flight company that hauls people and things up there. A company founded by Artemis, The Lunar Resources Company, will be one of those investors.

The plan is detailed and voluminous. At its core is, of course, money: where it will come from and how it will be spent. "We face three major issues: technical, political, and financial," Bennett and colleagues say in the plan. "All these issues must be resolved before we can send our first crew to the Moon. Of these three issues, the money question will take the longest time to resolve."

Really? Money more than politics? More than the complexity of engineering?

Yes. Artemis believes that the technology is mature enough that the company can get the lunar base going without investing in a lot of new development. It plans to use proven off-the-shelf components for almost all parts of the spacecraft. To augment this asset, it has people with experience in both the commercial aerospace industry and government aerospace.

In a grand sweeping statement, the Project proclaims, "The Artemis Project pays for the initial lunar base primarily by exploiting the entertainment value of the grand adventure of space flight." Five billion dollars, if you want to know, just from the first flight. Artemis arrived at this number by looking at similar mass-marketing ventures that link movies and TV shows with associated merchandise and services. "Once we're on the Moon," the company says, "we continue developing the lunar community with entertainment and tourism." And energy, on which people spend more money than anything else. Tourism takes the number two spot.

Bennett points out that in 1993, Disney's budget was four times higher than that of *Space Station Alpha*. In 1994, the company's revenues exceeded NASA's entire annual budget ($13 billion), while their profit exceeded the *Space Station Alpha* annual budget. And the entertainment industry's market share is growing, "as we apply increasing automation to the industries that provide our basic needs. Whole new industries have been added to the tally that didn't even exist when we started studying this problem—the entire range of video games, virtual reality simulations, and the Internet were barely in their infancy the first time we ran a spreadsheet for the Artemis Project," Bennett proclaims on the Web.

Here's how they figure their revenues, in millions of 1995 U.S. dollars:

Motion Pictures	1,200.0
Videotapes	1,000.0
Retail Sales	400.0
Toys	400.0
Video Games	375.0
General Merchandise	200.0
Scientific Data	200.0
TV Coverage	198.0
Magazines	126.0
Product Endorsements	125.0
Retail Franchise	50.0
Educational Data	40.0
Clothing	30.0
VR Experience	25.0
Other Publications	20.0
Models	12.0
Books	5.0
Software	5.0
Games	4.4
Total	4,415.4

Motion Pictures includes documentaries, some distributed in IMAX theaters. The $1.2 billion for motion pictures represents the "net from a series of movies released over a period of ten years, each movie with half the attendance of the previous one."

Some of the numbers (especially publications, merchandising, simulations, and motion pictures) are fairly solid, says Bennett, based on research into their respective industries. Others (educational and scientific data in particular) are just educated guesses.

Bennett and colleagues explain,

> We might have significantly underestimated the potential revenues available from some areas such as product endorsements. That's where we'd list businesses which elect to sponsor the project in return for promotional consideration, without actually producing a product directly related to the project. Nothing in this table reflects that sort of sponsorship because we don't have any basis for estimating it. [According to the *New York Times* on July 1, 2001, recent lunar meteorite sales have brought about $18 million per pound. Denise Norris of Applied Space Resources (see Chapter 14) is estimating that lunar material is worth about $4.5 million per pound, or $10 per milligram.] Also, merchandising of lunar samples returned from the Moon will probably be a large source of income. We didn't include selling moon rocks in this table, however, because we have no way to determine their market value. In past few years, a few lunar samples have exchanged hands for enormous prices, but we can expect that once a steady supply is available, the price will drop.

For comparison's sake, consider these numbers researched by the author:

- Amusement parks, U.S. annual receipts, 1997: $7.3 billion (Source: U.S. Statistical Abstract)

- Videotape rentals, U.S. annual revenues, 1997: $7.2 billion (Source: U.S. Statistical Abstract)

- Toy sales, 1999: $24 billion in North America, $31 billion in the rest of the world (Source: Toy Manufacturers Association)

- Video games, 1999: $7 billion in North America, $8 billion in the rest of the world (Source: Toy Manufacturers Association)

For Bennett to make the amount of money he's talking about, he would have to capture:

- Toys: 1.6 percent of the market
- Video games: 5 percent of the market

These are quite ambitious and not particularly realistic numbers, especially in such fiercely competitive markets.

Even if Bennett could adjust the numbers to make them more achievable, how would he get involved in producing these products? The plan says vaguely that we (Artemis) "intend to bring on board existing entertainment companies as participants." In conversation, Bennett is less clear. When asked how he plans to get documentaries made, he says, "Find a movie producer who'll make them. Do we have one today? No time to find one. 'Course this is also assuming that there's a movie producer who would be interested in doing this, and interested enough in the larger goal that he wouldn't run off and do his own thing. People keep talking about Tom Hanks. Tom Hanks seems to be interested in doing his own thing. It could well be that the only way to do it would be to put together a business plan, go out and get the capital, hire a director, and produce it."

Not an easy thing to do without a sympathetic angel investor or several. Movies are not the kind of business that appeals to venture capitalists. Bennett's got his work cut out for him.

COST

The Artemis Project estimates that the "reference mission" will cost U.S. $1.4 billion. As they put it, "The cost is well within the range of the money that large corporations spend on advertising related to major entertainment functions such as Super Bowl football games, theme parks, and popular television series ....

This level of investment is quite common in the business world. For example, one new deep-water oil rig typically costs about $1 billion."

The cost breakdown, in millions of 1995 U.S. dollars, goes like this:

Launch & Recovery*	$800.0
Spacecraft Hardware	350.9
Project Management	86.8
Ground Test & Verification	65.7
Spacecraft Design	38.1
Software Development	30.8
Operations Planning & Integration	24.9
Sales & Marketing	20.6
Launch Vehicle Integration	3.7
Total	$1,421.5

*Launch and Recovery is for two space shuttle launches, at $400 million each, including various related expenses. This is an overestimate, because Artemis probably will not use the shuttle for most of the mass due to expense and a prohibition on launching cryogenic fuels. The group is looking at a cheaper launcher, perhaps an existing Ariane.

In fact, each shuttle launch costs a billion dollars. Marketing should typically represent a third of a company's budget, which would place Bennett's figure more in the $400 million than the $20 million category. We've already discussed costs for developing new airliners (see Chapter 5), which way exceed Bennett's figure of $38 million. In fact, it's likely that Bennett's costs will reach several billions of dollars rather than his $1.4 billion estimate.

The Moon is the only body besides Earth that humans have visited in person. At 240,000 miles away, it is also our closest celestial neighbor. Still, debate has raged over whether it's worth going back, let alone establishing permanent settlements. Compared with Mars, the next-closest permanent body (excluding near-Earth asteroids and comets from this definition), it is poor in resources. Atmosphereless, without rich soils from which vital life-supporting elements can be extracted, and with a two-week day/night cycle, the Moon ostensibly offers little of interest to humans. But for those willing to look more closely, the Moon proffers gifts of high value.

Solar Power

Because it's undiluted by an atmosphere, sunlight on the Moon contains five times more energy than that which we receive on Earth. Correctly located, solar power collectors on the Moon could produce heat and electricity for domestic or in-space use.

Important Elements

Samples gathered by Apollo astronauts show rich supplies of oxygen, silicon, and calcium, plus the four major "engineering metals": iron, aluminum, magnesium, and titanium. With the exception of iron, which can be harvested with a magnet, these elements are tightly locked together in the Moon's crust (regolith); we need to find out how to isolate and extract them—outside the laboratory, where conditions may not be the same as they are in a controlled environment. The same goes for the many elements present in lesser quantities, like alloy ingredients and color pigments for use in glass, glass composites, ceramics, cement and manufacturing stuffs, and building products.

Water

We still don't know for sure, but there's a good chance that water ice is present at the poles, where the sun can't reach. The Pentagon's *Clementine* moon-orbiting spacecraft found some kind of ice at the South Pole in 1996—from afar. NASA's 1998–1999 *Lunar Prospector* went on to find evidence of water ice at both poles, again from orbit. However, scientists have been unable to confirm that the large amounts of hydrogen detected in permanently shaded craters does, in fact, come from water ice. The substance could be hydrogen in another compound. *Clementine* didn't touch the surface of the Moon at all; the *Lunar Prospector* crash-landed in what scientists think was a crater. The theory was that such a crash landing would throw up plumes of water ice, but that didn't happen. For that reason, and because neither craft was able to dig down into the deep craters to take conclusive data, we still don't know what is there. The presence of water on the Moon would make a huge difference to the viability of a colony there. We'd need it for life support, agriculture, and industry. Without it, we'd have to import it—a much more difficult and expensive way to go.

Energy from Thorium and Helium 3

The *Lunar Prospector* probe sent in 1999 found significant amounts of radioactive thorium on the Moon. Because thorium becomes fissionable uranium 233 in a fast-breeder reactor, the Moon could support a nuclear fuels industry. It also could be used as a propulsion for travelers to Mars, and it wouldn't have to be transported through Earth's atmosphere, putting the planet at risk. Rare on Earth, helium 3 is the fuel of choice in fusion reactors (of which we have no production models).

Lava Tubes for Habitation

As on Mars, the Moon contains subsurface lava tubes. Some of these may be intact and hollow, capable of providing shelter from radiation, micrometeorites, and solar flares, as well as protection from the extremes of daytime heat, nighttime cold, and choking dust.

Mars Training Facility

Some people want to skip the Moon and go straight to Mars, but Moon advocates point out the value of first testing technologies and low-gravity effects on the human body much closer by. While conditions aren't exactly the same on the two bodies, the Moon could provide critical testing of the complex systems needed for journeys to Mars, something that ought to be done before sending humans off millions of miles with little opportunity to turn back if things go wrong. Wendell Mendell (see Chapter 9) says that you could work out the bugs using almost-current technology, and you could recover from failures more easily, giving the crew a better chance to survive. He feels that a long-term lunar stay could provide the medical data to help assess risk for going to Mars, addressing issues such as low gravity, cosmic rays, solar particle events, and possible psychological problems. It also could accumulate data on integrated system reliability, maintenance rates, and degradation of performance in deep space. It's just plain easier to go there: There are more launch windows, and it's closer.

John Lewis (see Chapter 12) isn't quite sure that even with these resources, a lunar base is a good idea. He argues that there exists no political or philosophical argument that lunar colonies are desirable. Therefore, why we should create one is a knotty question. He advocates starting with an examination of lunar resources and known technologies to see if a colony might make logistical and economic sense as a self-sufficient entity. In other words, let's get a foot in the door and see what happens from that. Feasibility will determine whether we proceed.

The Moon vs. Mars debate is one that divides space advocates who see the pie as finite and the Moon as a diversion. Bennett doesn't believe in the idea of a circumscribed pie. He's a member of the Mars Society and supports going to Mars as well as the Moon. He sees the Moon as the front door to the universe.

NASA scientist and Moon base expert Wendell Mendell agrees. He thinks that exploration of the Moon will accelerate human exploration of Mars and that if it helps us achieve cheap access to space, one project doesn't have to preclude the other. It's logical to conclude that if both tracks are pursued privately rather than through governments, it doesn't matter. Each group can do its own missions to wherever it wants. So the debate is really about allocation of government funds.

In Bennett's view, that's irrelevant. NASA doesn't have the money to do either one, and other national space agencies harbor even fewer resources. So why argue? Let's just get on with it. Let NASA be political. It should be; that's its job. Politics is all about keeping things in check.

Ever the optimist, Bennett doesn't see all these issues as major problems. His biggest bugaboo is that rocket scientists aren't commercial business people. But his new boss, Robert Bigelow, a Las Vegas hotel magnate who founded Bigelow Aerospace in 1999, is.

And that, in Bennett's view, changes everything. (See Chapter 3 for more detail on Bigelow's plans to build space hotels.)

. *. . . .* .☾*. * . *. . *.

Artemis isn't the only group trying to get us to the Moon to stay—not by a long chalk. The list of aspirers includes the Space Frontier Foundation, the Space Studies Institute, the National Space Society, and commercial companies like the Lunar Research Institute (the now-famous company run by Alan Binder that sent *Lunar Prospector* to look for water), Lunar Enterprise Corp., run by a fellow named Steve Durst, David Gump's Lunacorp, and Transorbital. Gump has mortgaged and remortgaged his house to support his enterprise, which will offer a ridelike experience. The company will put a telerobotic vehicle on the Moon that will send signals back to Earth and simulate the experience of being there. Radio Shack is a sponsor. The attraction will simulate the feeling of every pebble the rover travels over. Transorbital has its first customer—and maybe the second—and is accumulating money to put microsatellites around and on the Moon, as well as lunar sample returns. In addition, Europe, Russia, and India are talking about lunar missions. But Artemis is the only one talking about putting humans there.

There is actually a savings and loan—or there used to be; it's gone bankrupt—that has a license in Texas to operate on the Moon. Authority was based on an extension of a Texas law that says a savings and loan can reserve a location if it can demonstrate that it is within a certain number of miles of people who need its service. The attorney for the business, one Art Dula, successfully argued that when you go into space, the same principles ought to hold, but the distances are really much larger. Therefore, for that kind of location, it would be reasonable to extend the law to astronomical units rather than miles. The Texas legislature, notorious for all kinds of bizarre behavior, passed the extension.

. *. . . .* .☾*. * . *. . *.

If Bennett succeeds in establishing a base on the Moon, would he live there? "I will if I can, if we can get it done fast enough. I don't know if my family would go with me. It depends on them. I suspect my wife would think it would be fun. It would be something different. Of course it would have to be a nice place to live. Las Vegas in the sky."

Whether he goes or not, Bennett feels that humans as a species should.

> It's the frontier. The history of our species is based on needing a frontier. Although there is one society that has existed continuously throughout history, and that's the Egyptians, who only became part of the world culture when they had no choice. Well, China too. Very few human societies without frontiers have endured for long. Besides I think it would be really cool to pressurize a big dome on the Moon and put on wings and fly.
>
> The moment we establish a permanent exploration base on the Moon, people will want to go there. As the lunar community grows, so will the market for commercial space transportation. Besides transporting holiday adventurers to and from the Moon, we will need regular flights for the people who are working there. We will need cargo flights to resupply and expand the lunar facilities. Many people are pushing for commercial passenger and cargo spacecraft to happen. We complement that effort by working the other side of the equation—by establishing facilities on Luna that need low-cost access to space, we are pulling on the same goal from the other end. We are creating the market for commercial spacecraft.

THE AUTHOR'S OPINION

Greg Bennett is a careful thinker. However, he has never started or run a business and lacks experience with the entertainment industry he wants to tap into. He's also too busy making a living to get his project off the ground.

I think the idea of the Moon as a test bed for going to Mars makes a lot of sense. Once we're there, the rest—the engineering metals and the helium 3 and all of it—will follow. The Moon is far less expensive to go to than Mars. Let's set our eyes there first.

—11—

What's Up with NASA?

NASA's doings have affected humanity's relationship with space for forty years. How is it that NASA became so powerful? What is it like inside the agency? NASA scientist Wendell Mendell and I discussed his employer at a National Space Society conference.

The National Aeronautics and Space Administration (NASA) was established in 1958 by President Dwight D. Eisenhower for a single purpose: to oversee the development and testing of rocket propulsion systems by the private sector so that the U.S. could beat the USSR in space. When that mission was accomplished, the agency needed new objectives, so NASA planned colonies on the Moon and Mars and orbiting space stations. Its mission became space exploration, and it started looking for cheaper and less expendable ways to launch payloads into orbit and return them to Earth. Existing vehicles like the *Saturn V*, designed by Wernher von Braun for the Apollo missions, cost $3,800 per pound and could be used only once. They were obviously too expensive. The space shuttle was developed as a low-cost, partially reusable alternative. Unfortunately, it turned out to be almost three times as expensive as the *Saturn V*.

To accommodate large military payloads, NASA had widened and lengthened the payload bay, nearly doubling the size of the shuttle. But President Richard Nixon deemed the cost of doing that too high and cut the shuttle's budget almost in half, forcing engineers to scrap many improvements that would have reduced launch costs. In the late 1970s, the agency attempted to bring down per-flight costs by requiring all payloads to fly on the shuttle, whether commercial or not. The idea was to create a steady market for space shuttle transportation services, but it didn't work. Instead, by dampening competition, the agency caused costs to remain high and effectively kept new technologies from being developed.

Eventually, the Air Force spoke up. Relying on one vehicle wasn't the best thing for national security. As a result, NASA changed its mind. Commercial cargo could be launched by a nascent commercial launch industry but only by using government-developed expendable rockets. Even with the change, many commercial U.S. firms found that it was cheaper to launch from China or Russia, which they did. Now you can launch through Arianespace—a European consortium—as well, and in fact, the Europeans have captured the global launch market.

NASA is now working on both the shuttle and its replacement. The latter requires technology that doesn't yet exist. In the meantime, the agency is charging ahead as fast as it can, which isn't very, with *Space Station Alpha*. Originally envisioned as a laboratory, the Space Station is much more expensive than originally planned. NASA spends almost $2.5 billion per year on it. It has spent $60 million to support Russia's participation, and it planned to spend $448 million in 1999/2000 for the same. Cost estimates keep changing. As of 2001, the station is supposed to set American taxpayers back a cool $30 billion. Wendell Mendell (see page 227) explains why the cost and schedule have ballooned. "ISS [*Space Station Alpha*] is political. Not in the same way as Apollo was, but political just the same. For one thing, it's seen as pork. Anybody who has a NASA center in their constituency supports it. And recently," says Mendell, "a request came down from the top of the center to everybody to please list all the counties in the nation where you have contractors working for the space station. One of the reasons space station's stayed alive is because they've had something going on with the space station in every county in the United States. That's what's called political engineering. And the political engineering is one of the reasons the space station costs so much and has taken so long."

NASA's involvement in and support of commercial space endeavors has been controversial and checkered. Free-market space organizations like the Space Frontier Foundation are forever calling for NASA to stop operating the shuttle and the ISS and do two things: become a good customer, and explore the farther regions of the solar system. Its doing either is slow in coming.

At the Space Frontier Foundation conference in October of 1999 (see Chapter 16), Dan Goldin spoke out in favor of commercializing the ISS, or so it seemed. Some critics, like Jeff Foust, editor of an electronic newsletter called *Spaceviews* (now part of Space.com), allege that Goldin couched his language in hedgy terms. In his speech, Goldin said that NASA's plans for ISS commercialization would soon be backed by legislation that would allow NASA five years to establish market prices for commercial use of the station. NASA could initially charge below cost to stimulate demand and later increase prices and keep revenue above cost. The profit would be reinvested in other commercialization efforts. This strategy wouldn't be as bold as some advocates would like, but it was seen as a step in the right direction. Goldin said that eventually he'd like to turn the station over to an entrepreneur if the private sector sees an opportunity. However, he said, if there's a lack of commercial interest after ten years of operation and ISS no longer meets NASA's research needs or those of its international partners, it should be deorbited. How many billions of dollars for ten years and only meager interest?

Goldin asserted that NASA has become a strong proponent of space station commercialization because, no surprise, it doesn't have enough money to run routine operations and explore the solar system at the same time. Foust claims that he's seen other presentations in which rhetoric exceeds real actions. Others have pointed out that even though NASA has become more receptive than ever about commercial use of the station, it has introduced a number of constraints for such use, ranging from government control of physical access to a limited power and communications capability. Although NASA has made encouraging statements about space station commercialization, it's not necessarily the leader in this effort. The Canadian Space Agency has signed an agreement providing for the first commercial use of the station by allowing private company SpaceHab to use one of Canada's experiment lockers on the station. The agency is exploring other commercial possibilities. Russia also has commercialization plans in the works.

THE ROLE OF THE U.S. GOVERNMENT IN COMMERCIALIZING SPACE

Some people believe that the U.S. government has been increasingly supportive of private sector space development in the last few years.

Legislators like Dana Rohrabacher (Republican from California), James Sensenbrenner (Republican from Wisconsin), and John Breaux (Democrat from Louisiana) all have introduced bills that provide incentives to and support for private firms developing reusable launch vehicles.

Congress passed the Commercial Space Act in 1999, which encourages commercial use of Space Station Alpha. It also allows vehicles to re-enter the atmosphere. (Until the Act was passed, private firms could leave, but they couldn't come back.)

NASA is moving some shuttle operations to the private sector, specifically United Space Alliance, a consortium comprising Boeing and Lockheed Martin. Houston Mission Control also has been contracted out. Portions of Space Station Alpha may be privatized as well. Why the change? Shrinking budgets and redefinition of NASA's role.

Dan Goldin, former NASA Administrator, made a concerted effort to redefine NASA's mission.

The government may be backing out of the indemnification business.

Government is providing regulatory and investment incentives that allow private industry to compete on a level playing field for government payloads.

Cities and states foster space business as well. Houston City Councilman Rob Todd, whose district includes NASA Johnson Space Center, wants to take advantage of his city's $2.5 billion budget. He thinks cities can offer:

- Business mentoring
- Incubators
- Investing the city pension fund in aerospace startups
- Regulations that don't impede business development

Todd wants to see a federal tariff schedule for liability like that in the airline industry. Such a document, which limits liabilities for certain losses, can help the industry calculate what their liability exposure is ahead of time.

The reason for these seeming contradictions within NASA is that the agency is entrenched and filled with people who have never had to make a profit. Even Goldin, who worked at aerospace giant TRW, was never responsible for seeing that the company made money. Change comes slowly if at all under such circumstances. Workers feel threatened, believe that they'll lose control, fear for their jobs. And so, like all security-conscious people (which is most of us), they dig their heels in and fight to preserve the status quo. What will or won't change once the new Administrator, Sean O'Keefe, gets his feet wet remains to be seen.

Not Wendell Mendell, planetary scientist and lunar base expert at NASA Johnson Space Center. Mendell is a space enthusiast who manages to straddle the thin line separating government and private activity. He supports both and wants to see his two worlds unite to achieve common goals. Sometimes he pushes his employer a little too hard, but so far, he's always managed to emerge unscathed. A likeable Teflon gadfly is what he is. He has a formidable grasp on how NASA works, which he uses to advantage: While others get mired in bureaucracy, he scoots in and out of the system doing work that interests him. On the side, he supports commercial space development.

Looking out through Mendell's eyes affords a useful picture of NASA and its view of commercial space. If you have an unrealistic picture of the agency, which most people do, he'll set you straight. For example, he has long advocated that we use the Moon as a technology test bed for going to Mars. At the same time, he's quick to point out that NASA has no plans for doing so. In fact, NASA has no plans for going to the Moon at all. He explains, "Everybody thinks that NASA is kind of a cut-and-dried thing, and that if you just call the right number, you would get a piece of paper that says what NASA plans to do. For example, in the early days when I was talking about lunar bases, students or professors would call up and say, 'My class wants to do a lunar base design project. We'd like to come out and interview you, talk to your people, get some ideas. Because what we'd really like to do is to work on some kind of problem that is relevant. We don't just want to do something we make up.' And so the class would come out, and we'd talk, and the very first thing we would do is spend an hour convincing them that there was not a NASA plan for a lunar base. That, no, there is not a drawer that you pull out and there's the blueprint with the budget and the schedule. We haven't thought through it that much. The information's not that detailed." Surprise! Once the shock wore off, then Mendell and colleagues could inspire them by telling them, "You're coming in at the bottom of the problem. You can contribute ideas that are fundamental as opposed to just kind of dotting the i's and crossing the t's." They liked that.

Normal people are not in the habit of examining government agency budgets. Space activists might look at NASA's, but only the most rabid of them. Mendell advises them to take a peek. "If you want to know what NASA's doing, you look at the budget. Because whatever NASA's going to do, there's money being spent on it. There's no money being spent on going to the Moon. There are no people working on lunar science or lunar strategies. If you called up NASA and wanted to talk to people working on lunar strategies, you might get me. And there just wouldn't be anybody else."

He explains why you'd get him, if you got anyone at all. It's the nature of what scientists like him do.

> It's because I'm kind of a free agent. The people I work with are like professors at a university. Guy sitting in the

office next door to me measures very obscure isotopes of elements in rocks that are radioactive. By measuring the abundances of certain of them he can tell you when that rock was formed billions of years ago. And so in order to do his work, he has to have a grant, basically, to pay for the equipment and the people in the laboratory, and he applies to NASA under a research program. He writes a proposal, and that proposal is placed in a big stack of proposals from universities and other places. And those proposals are submitted to a scientific review committee of outside scientists who read them all and decide which ones are good science and which ones are mediocre science and which ones are bad science. And they rank them. And then the program manager who has X amount of dollars goes through that ranking and tries to fund the top ones based on how much money he has. And they never have enough money. The cutoff's always about 40 percent of the way down, or 50 percent of the way down, and there are good proposals that don't get funded. So the guy in the laboratory next to me has to be his own boss. He's not told what to do by his supervisor. He is the world's expert in that. And so our organization is based around people working on their own projects, but it's also based on people bringing in money. Although in the government, that's not as crucial as it would be in a university. In a university, you bring in money or your salary's not paid. A tenured professor does get his salary paid even if he doesn't bring in money. So a civil servant is a little like a tenured professor. Salary's paid, but if I don't bring in money or don't do something constructive with the vision, it's a bit of a pain because in the civil service it's very difficult to make people do things. But I'm given a certain amount of freedom to choose what I do, and the things I do are considered by my colleagues to be constructive and contributive to the general intellectual welfare of the organization, so I do them. If they were considered to be against the welfare or not constructive, then I wouldn't get resources like computer support or

> travel money or trips I want to take wouldn't be funded. In fact, sometimes that's the case, and I fund my own trips.

This state of affairs is not generally known among the public. Movies like *Apollo 13* and JPL press conferences paint a picture of teams of scientists all working together on some very important project. In fact, that way of working belongs far more to the engineers than to the scientists. Mendell says, "Most of NASA's engineers are a different culture than scientists. They work in hierarchical, militaristic organizations; orders come down, money appears at the top and is distributed down through the hierarchy based on function. And they don't do anything until the money comes down and they have a requirement. The scientist *never* does not do anything. The scientist is *always* creating something or working on something or making something. But an engineer waits for the assignment."

Which makes engineers sound awfully passive. But that's part of the NASA culture. "The problem is that if you say anything that's kind of off the mainstream, people don't know how to deal with it. They don't want to go to their boss and say, 'We ought to be studying this.' 'Who told you that? What's the upside potential?' Even though it might be a good idea philosophically."

He offers an example that shows how ideas percolate at the agency.

> NASA just recently got an instruction from OMB [the federal Office of Management and Budget] to do some planning for the next decade, and he's [the guy who's in charge of it] been through all the ropes since the Space Exploration Initiative [Bush-the-elder's proposed program back in 1989], and he thinks he's seen every idea that every human being and a few aliens have had about how to go into space. And he wants to know what is there that we can bring to the Office of Management and Budget that looks like it might help create a robust civil

space program and do something very significant by 2010. What is the answer, what do we bring to them? Well, there's something that's been floating around that Bruce Murray began promoting about a year ago called robotic colonies. Maybe because JPL loves robots. Anything that has lots of robots JPL's in favor of. So, the idea is that the settlement of a planet occurs in certain stages. First of all, you kind of go by and look at it. Then you put something in orbit around it and kind of study it. Then you start landing things that are robots, and then normally the next stage is you land human beings to do something. Now they're talking more about an enhanced robotic stage. In other words, you have not only just a robot do a few things but you actually have some kind of a robotic complex that you develop that does more elaborate things than just picking up something or pointing a beam at something. That kind of grows. And then that precedes the human presence. It's a little vague, but it's getting to be popular.

I'm associated with something called International Space University, and their main activity is a summer program for young professionals from around the world. We have about 100 who come. They spend ten weeks, and one of the activities that threads through those ten weeks is something called a design project where they're presented with a problem. They generate about 500 pages worth of report on the design project. Then they attend lectures. I've led about four of the design projects. In 1993 in Huntsville I was asked to lead a design project to install some kind of advanced observatory on the far side of the Moon, and I wrote the problem statement in such a way to explore the possibility as to whether or not a really high visibility scientific project would be used as a justification or rationale or lead-in to a human mission. The idea was that if the answer was really super-sophisticated you would really want human beings there to set it up and make sure it ran right before you went there and let it operate.

So the students did a sort of a business plan along those lines. They said, "Okay, first of all you need to make sure you find the right site. Reconnaissance, orbital satellites. And then it would be good to land some simpler versions of the experiment on a spacecraft, take data, make sure you understand what the environment is like, what kind of quality of data you can get, and so on. Then design the really humongous thing and put it down there." In that secondary phase I made the point that these robotic lander missions would not be horribly expensive and might be well within the reach of many countries outside the U.S., even places like India. I also told them that you could make it an international program. As we've been getting this idea of how these robotic colonies work, it suddenly hit me that we have this whole document that discusses that and has an international aspect to it, and it might be really interesting to shape into an answer to this question. I was just thinking over the weekend that I need to get out my old charts and go through the presentation with those guys and tell them about it and let them think about it. That would be on the Moon. So when you say "Is NASA doing anything about the Moon?" well, no, it's not doing anything about the Moon, but we're having discussions about all kinds of possibilities. We really don't know the answer yet, although officially the answer is that we're going to take human beings to Mars as soon as we can.

That's a little misleading because you can go to the Web and you can find NASA's human Mars mission scenario. But what it is is a Bob Zubrin kind of story (see Chapter 8). It basically shows you how big the rockets are and what the habitat looks like and what the fuel production unit is. There's very little discussion of how you get to that point. The ordinary reader assumes the Apollo paradigm. That is, the president has said we're going to Mars, and there's been this big effort and we've had these people design the rockets and these people design the habitats, these people train the astronauts, these people build processing plants. Then you bring it all together under NASA and you fly to Mars.

The problem, and people who are inside understand this, is that it's a real leap from operations that are a few hours or days away from you and operations that are years away from you. There's reliability issues, there's psychology issues, there's a lot of stuff. And you really ought to build your experience base from where we are now into the time frames that are appropriate for that mission. In my mind, the Moon is the ideal place to do that. You can scale your activity to cover the whole gamut. Goldin's concern is that if you go and do that then there will be a whole aerospace lobbying group built around investments in keeping that activity going and that will soak up all the money in the NASA budget, and you never get off the Moon. So part of the issue is we can allay that fear if somehow NASA's not responsible for everything. For example, if you somehow get private industry involved and NASA does certain elements of it that have to do with the risky parts of the technology development, and the industry is investing in the infrastructure out of some other some source of revenue, then NASA can be nimble enough within its budget to move on. That's why the space shuttle has been commercialized—the fleet is owned and operated by a joint venture between Boeing and Lockheed Martin called United Space Alliance. That's why Goldin wants to have the space station commercialized, so NASA won't get locked into a long-term commitment to operating it. And the same could be true of the Moon. Now I don't know exactly how you're going to commercialize the Moon or what the answer is, but the principle is important. And you need to look for solutions where that principle is. What you fight against in the NASA culture is the sense, particularly in the old guard, that NASA is the owner of space and that everybody who does anything in space is either doing it under our flag or is basically a

> motley crew. NASA thinks it owns space. And that's important. It really doesn't matter whether NASA owns space or not. What is important in our particular problem here is that NASA thinks it owns space. Now Dan Goldin knows better, but the people who are two levels down in the management structure don't buy that. They think they own space.

Mendell believes that the triggering event would be the creation of an economy in space where services are performed and are paid for by entities with a physical presence. That's different from the current satellite industry. "Right now we have a telecommunications satellite in space, but the services are paid for by people on the ground. But if there was something in space that was launched in order to provide a service to customers also in space, then to me that's the beginning of an economy."

How did Mendell come to this unorthodox (at least for NASA) conclusion? "The thing that opened my eyes was going to the Space Frontier Foundation's conference in Los Angeles in '96. I got invited there to be on a panel, and I happened to be in Japan and was flying back that weekend. I stopped in Los Angeles and went to the conference. I had no sense of all the people who were building these privately financed launch vehicles. I didn't really know about it. I knew about it intellectually, but I didn't really put it all together. And that really opened my eyes. And a lot of things I've written since then have been heavily influenced by that experience."

Mendell is unorthodox in other ways as well. He writes articles and gives talks that are rife with criticisms of his employer, and yet nothing ever happens to him, despite the fact that the people who watch him know exactly what's going on. This is a subject that Mendell finds extremely funny.

> Two or three years ago I had a friend who was on the faculty of the Air War College at Maxwell Air Force Base. Everybody who's on his way to general goes through it

twice. The second time you go through it they're all colonels. My friend was a civilian, and she taught space policy there to these colonels. She invited me there for a panel, or a lecture. So I gave my lecture and then, of course, went to lunch with my host and some other faculty and the base commander. And the base commander sat down next to us. He said, "I want to thank you very much for coming to visit us here at Maxwell Air Force Base. We really enjoyed your talk. You were very frank in your comments, and we appreciate that. I do want to know how you keep your job."

The other funny incident was that one of the guys I work with was getting married. He was going to the Space Center in Houston, he was having his wedding reception there. He and his wife are big space buffs. So I went to the wedding; the wedding was at their house. I went to the house and he said, "Were you the one that Goldin was talking about in the paper, the *Wall Street Journal,* the other day?" And I said, "What? What do you mean?" "Were you the planetary scientist at Johnson that Dan Goldin mentioned in his speech?" I said, "What?" And my friend said, "Yeah, I remember now, he read a quote from the *Wall Street Journal* and said, 'This really makes me mad.'" I finally put the pieces together. Somebody else came in my office a few days later and said, "I was at this Colorado Springs conference, and Dan Goldin got up at lunch and held up the *Wall Street Journal* and read the article and said it was a quote from a planetary scientist at Johnson Space Center, he didn't read the name, and then he read the quote, and said, 'And this makes me mad.'"

Yeah, it was me. And it turned out, I sent a note up to Lori Garver [Associate Administrator for Policy and Plans], who was working in his office at the time, and I said, "Excuse me. Is Mr. Goldin mad at *me*?"

What had happened was that a reporter from the *Wall Street Journal* who was based in Tokyo called me one day because he had run across the Japanese construction company Shimizu, and heard about their lunar base plans, and all of that, and it really intrigued him. He wanted to know is that for real, and they may have given him my name as a person who knew about the thing. And I said, "Yeah." So I talked to him quite a while on the phone giving him the background, the Shimizu involvement, my interactions with them and so on. He said, "Great. That's a really interesting story." Well, I never heard from him. Then about nine months later I was in Vienna with the Space University, it's '96, and he showed up. He said that he's still interested in this article. He hadn't written anything about it, but he still was working on it and blah blah. And then he called one more time to ask me questions.

One year after that initial phone conversation, the article appears on the front page of the *Wall Street Journal*. And there's a quote from me from a year ago. It basically said that in Japan, there are all these civil engineers, and this Japanese construction company is working on studying lunar bases, and he asked me how many engineers in NASA are working on lunar bases. Now he said civil engineers. We're talking about constructing lunar bases. And my answer to him was "Somewhere between zero and one." That was the quote. Well, apparently immediately after Goldin read that article, he called the director of Johnson Space Center and asked him how many people were working on lunar bases and so on, and the director said, "Oh, about a hundred." And then Goldin used that hundred in congressional testimony about a week or two later. We were having meetings trying to figure out who the hundred were. We finally thought that if you added up everybody who had been to a meeting about it, you might come up with a hundred people. But it turned out that he was not mad at me. What he was mad at was the viewpoint in the article that NASA should be doing this, where in Japan the private sector was doing it. And Goldin's theme was, "Well, why aren't the private sector in the U.S. involved in this if the Japanese are?" He resented being told that NASA was supposed to do this. So it turned out not to be me. It turned out to be the whole tone of the article that he was upset about. So occasionally you get famous.

Even though the problem in this incident didn't turn out to be something Mendell said, in other cases he has served up NASA-critical comments and has managed to keep his job. He explains.

I'm anonymous. It's not like I'm the head of the space station program. I honestly don't know how I get away with it sometimes because for a long time I just figured that nobody paid any attention. That nobody read the paper. I write in fairly obscure corners of the world. You won't find anybody at Johnson Space Center who reads *Ad Astra* [the National Space Society's magazine], at least at higher levels. But then every so often, something happens that kind of gives me the clue that they're aware of my existence. But I have never been hassled, except kind of by lower-level people.

The International Space Year was 1992, and it was supposed to be kind of a hype event, where people could go to their governments and get them to do special projects for the Space Year and, therefore, get some money spent on space things in various countries. In preparation for it in 1987, I think, there was a conference in Hawaii, a Japanese-American conference to plan for the Space Year. It was put together by a Hawaiian Congressman who's since died. At that time, there was talk about putting a spaceport in Hawaii, so there was a Hawaiian interest in having space things happen. A couple of Japanese from the Japanese Scientific Space Agency, ISASS, came to Johnson Space Center and visited me. I didn't know them. They wanted to put some lunar base ideas into the International Space Year pot of things to do, but they needed somebody to come to the meeting and help support it, and they wanted me to do it from the American side. They were going to be doing it from the Japanese side.

Well, I was nowhere on the screen to be invited to this. They informed me that the guy who was inviting the American side was a planetary scientist actually in Hawaii, whom I know, and who's a fairly arrogant, secretive guy, and who regards me as a name on a list. I'm insignificant in the scheme of things. So I called him up and told him that these Japanese wanted me to come. And he said he was really sorry but there was a limited invitation list and it was closed and there wouldn't be a possibility for me to come to the meeting. I called the Japanese guy and told them his answer, so they got me invited by the Japanese side. In that process I turned in travel orders to go to Hawaii with the invitation letter, and I was called by International Affairs, who wanted to make sure that I understood that this conference wasn't really what it was advertised to be. That it really had to do with promotional activities in Hawaii, that it wasn't exactly a scientific technical conference, sort of intimating that I really shouldn't go. But I told him that I knew all that, in fact I knew more than he did about it, and after I kind of told him more about it, he said, "Oh okay. You understand then." I said, "Yes, I do understand, but I've got to go." And I went. But that's sort of the limit of it.

I admit I don't think a lot about it. I just do my thing. I'm not rabble rousing. Some of the things I say are opinions that can be challenged, and maybe should be, but I don't do it to be malicious. I'm trying to be constructive. And in fact in a lot of my speeches, I praise Dan Goldin to the sky. In my view of things, he has done a masterful job of changing the agency along lines that I thought desperately needed to be done. Now his change wasn't pleasant, and maybe he's not the easiest

> person to interact with or work for, but the net result, I think, is much needed.
>
> See I'm almost invisible in NASA. John Logsdon, he's sort of the dean of space policy, and whenever there's anything that happens in NASA, all the press in Washington goes to him and gets his comment on camera and Dan Goldin talks to him, makes sure he's fully informed and so on. He's a friend of mine through the Space University and through the lunar base effort actions when I first met him. I have been to conferences where he's introduced me as a person who's internationally known and recognized in a certain field, but totally unknown in NASA. I used to worry about it. There was a time when I was kind of upset that things would be going on that I had talked about or was considered to be an authority on, and nobody called me. Nobody asked me. Nobody in NASA called me or asked me or named me to the committee or thought I ought to be part of the effort, and after a while I kind of got over that. And then I thought about it a little more deeply. I realized that there are advantages. If you get too visible, then what you say has import to the outside world.

Mendell knows that there are things he'd better not do. "What I don't do is stand up on a stage and claim to speak for NASA. One year, I actually got on the list for permission to attend a major international technical conference at the last minute. Years later, I met a guy at a workshop in Italy who was the NASA representative to Europe. When I introduced myself he says, 'So, you're Wendell Mendell. Your name popped up on the list one year to attend the International Astronautical Congress, and I said, 'Who the hell is that?'"

The Author's Opinion

Poor Dan Goldin. He said one thing, then he said another. He was jumped on by everyone for not being commercial enough, or being too commercial, or saying one thing and doing another, or changing his mind. Despite Goldin having been the voice of NASA, he didn't unite the agency. It's probably impossible for anyone to do so. There's too much rivalry among NASA branches and too much insecurity among workers.

Poor NASA. How would you like to work for an employer whose budget was completely uncertain from one year to the next, whose existence and funding was totally at the mercy of politicians? It would make me schizoid.

There's no tradition of government agencies in the U.S. having to make money—except the Internal Revenue Service, if you want to look at it that way. The Commerce Department tries to help others to make money, and many agencies have to charge for publications and other products because their budgets don't cover producing them, but no agency is charged with having to make a profit. Is it any wonder NASA doesn't know how to go about fostering such a state of affairs, let alone doing that itself?

So a maverick like Mendell is just that. He's iconoclastic, open-minded, honest, and manages to maintain an infectious sense of humor about the agency in general and his place in it in particular.

—12—

John Lewis: Raw Materials From the Sky

As with space tourism, the idea of getting important metals and energy from space recalls scenes from science fiction. Doing so is farther off than space tourism, but it could happen in our lifetimes if the "cheap access to space" nut is cracked. John Lewis wants to see us move out into the solar system permanently by exploiting in-space resources. But first, he wants to help us get the vehicles to take us there.

John Lewis and I met at his office on the University of Arizona campus in Tucson. He surrounds himself with the accoutrements of Southwestern life, astronomy, and academe. But he also likes seashells....

In the 1950s when John Lewis was nine years old, his parents bought him a book that changed his life forever. *The Little Golden Book of Stars* contained a big table of data on the solar system, but it was littered with question marks. Lewis was astonished at the omissions. "We didn't know! We didn't know how big Pluto was. I looked at this and said, 'Good grief! Here is a field where people don't know everything. They need me.'"

And so a career was born. Lewis is the kind of guy who, when presented with a lemon, makes lemonade. Or, more to the point, who makes gold out of lumps of rock—specifically lumps of asteroid. A chemist and planetary scientist from the University of

Arizona at Tucson whose Ph.D. advisor was famed Nobelist Harold Urey, Lewis enthusiastically advocates "mining the sky"—he wrote a book of the same name—for metals and other useful elements as well as energy, resources which he believes are, for all practical purposes, infinite.

"But we have so many resources on Earth," argue Lewis's critics. "Why go through a lot of rigamarole and expense when we know how to mine these metals on Earth and already have operations in place?" Lewis counters with several arguments:

- If we wish to use these kinds of materials for space commerce and operations, we still face the problem of lifting them through Earth's gravity, which makes "exporting" them so expensive that cheap access to space will remain a brilliant dream.

- And the corollary: Using space-based resources will help defray the cost of exploration.

- Asteroid resources are worth a fortune! By Lewis' calculations, the total market value of Amun, a two-kilometer diameter (the size of an open-pit mine on Earth) asteroid of the M class, is worth $20 trillion in assorted metals. If we launched the same amount and type of material from Earth, it would cost $300 million billion. In the asteroid belt, there is enough iron and steel that it would be worth $3.5 times ten to the nineteenth at present Earth prices (that's ten to the tenth times more than $3.5 billion, or about $6 billion per capita). Add in nickel, cobalt, platinum group metals, gold, silver, copper, manganese, titanium, etc., and you get $100 billion per capita. Not only are these resources pregnant with economic potential, Lewis sees them as the way to universal prosperity on Earth and a way of financing major space programs and operations.

- Last, one of Lewis' most attractive ideas: mitigate the threat from Earth-crossing asteroids and extinct comets by using them up. In fact, he says, many of these nasty little bodies are easier to reach and land on than the Moon, and they are

> rich in water and useful materials such as natural stainless steel. Even better, we can extract these substances automatically and make propellant, air, water, structural metals, and nutrients for use in in-space operations.

But glutting the terrestrial market will cause prices to plummet. Lewis has thought about that. He cites a study by the U.S. Geological Survey analyzing the effects of importing large amounts of extraterrestrial precious and strategic materials on the market prices of gold and platinum. No problem, says the agency, as long as you don't import too much.

Lewis has another idea: Use asteroids as space transports. Why not? You don't have to launch them from Earth, they provide good shielding against cosmic rays (and solar flares?), and you only need small amounts of propellant to maintain orbit.

Lewis is also interested in the Moon. He thinks that the way to get and keep us there is to start by using Moon resources to defray the expense of exploring the natural satellite. You could make propellant and life support materials (air and water), power fuel cells that would make energy for use on the Moon, and mine helium 3 for use in fusion reactors (which don't exist commercially yet and have a long way to go before they do, but someday ...). Lewis also envisions making solar cells from lunar silicon and generating electricity, a scenario he feels might finally provide a credible rationale for a lunar base—the first scheme that seems capable of paying its own way.

Lewis' ideas are beginning to catch on. The Colorado School of Mines has set up an ongoing Space Resources Utilization Round Table to brainstorm about how to get useful products and services from space. Sixty experts from industry and academia showed up to the first round table, in 1999, including representatives from NASA, the Russian Academy of Science, and Merrill Lynch, the American brokerage firm.

Lewis likes people and figures the more of them, the better. In fact, he thinks people and their intelligence are humanity's greatest resource. Therefore, he concludes, populating the solar system with lots of them is a good idea. And we can do that by using the staggering wealth of materials in the asteroid belt between Mars and Jupiter. Lewis contends that such wealth could support billions of people living there at a high standard of living. The only "slight" problem that far out is energy. Solar energy is quite weak there, but the atmospheres of Uranus and Neptune contain copious amounts and high concentrations of helium 3 and deuterium, which can be used in fusion and fission reactors. (So do the atmospheres of Jupiter and Saturn, but extracting the gases would be difficult because of their immensely strong gravities.)

Lewis has spent a long time analyzing what's out there in the solar system and has come to the following conclusions:

- The Moon. This airless world provides great opportunities for resource use. At first, the resources could be used primarily to defray the costs of exploration. The primary market would be on the Moon itself. Oxygen extracted from rocks and polar ice can be used for life support and propellant. The ice could be exported for near-Earth activities, but this option might not be economically attractive because of the Moon's gravity, which, though one-sixth of Earth's, is still significant for spacecraft trying to escape it. Hydrogen extracted from the ice can be used as fuel and in industrial processes. Dirt can be used as radiation shielding against the Moon's high concentrations of cosmic radiation unmitigated by any atmosphere, and as protection against micrometeorites. Metals can be made as byproducts of oxygen production. Helium 3, which can be used in fusion reactors, is rich on the lunar surface and rare on Earth. All schemes for using lunar materials depend on our knowing not only what's present, but also how to capture and process them. Some of this we know from our experience on Earth, but some we'll have to learn. We'll also have to learn about the physiological effects of low gravity—as opposed to zero gravity—with which we have little experience (only a few humans having stood on the Moon and none on any other celestial body). Lewis gives both the American and Soviet space programs low marks for accomplishment in the human realm: We still haven't learned how to grow food in space, deal with weightlessness, or counter the effects of cosmic radiation.

- Earth-crossing asteroids. Many of the most dangerous of these little bodies are easier to reach and land on than the Moon. Lots of them are rich in water, as ice and in water-bearing minerals. These asteroids also contain natural stainless steel, nickel, cobalt, and the platinum metals. These resources could be extracted both from asteroids and extinct comets, used for propellants, life support, and structural materials for in-space infrastructure.

- Energy. Once we establish some infrastructure on the Moon and/or asteroids, we can build portions of solar power satellites, which could beam energy to Earth, in space. This activity would bring the cost of building such satellites down because the transportation would be lower in cost than if the components were built on Earth and launched from there. Or, we might build solar arrays on the Moon and export the power.

Which came first: those question marks in *The Little Golden Book of Stars*, or Lewis' insatiable curiosity? When you discover all the things that he's involved in, you might surmise that Lewis already has found an energy source and is tapping into it continuously. He writes science fiction, is fluent in at least five languages (German, French, Latin, Sanskrit, and Italian), and reads at least one more (Russian). He translates books, some with his wife Ruth, on subjects like antique dolls.

Lewis became interested in space resources in 1977 after reading an article in *Technology Review*. At the time, he was a professor at MIT. "The article by Tom McCord and someone else talked about retrieving metals from asteroids and bringing them back to Earth," he recalls. "You could use water vapor and carbon dioxide to foam metals—make a big poof of iron and nickel and send it to Earth. Because the density would be so low, the poof would just slow down in the atmosphere and fall relatively gently to Earth." Gently? Lewis concedes that the term is relative. "I'd hate to be hit by this particular kind of dandelion seed, but it would land at subsonic speeds and it would stay intact and you could target it fairly well," he says. But he tabled his interest when he realized that, although technically feasible, such an endeavor would not pay for itself. "I did some back-of-the envelope calculations on the economics of returning iron and nickel to Earth. It struck me that this was not the economic wave of the future, that the cost of the space operation would be so high that there was no way you could ever make a buck on returning cheap metals, common metals."

It didn't take him long to change his mind, however. "Shortly thereafter, about '79, two people fairly simultaneously had an idea that maybe the best target was expensive metals. If you had some way of separating the metals, you could bring back the expensive stuff and use the cheap stuff in space, where suddenly it assumes an enormous value because you didn't have to launch from Earth." Immediately recognizing the brilliance of this approach, Lewis wrote a proposal for the Space Studies Institute—Gerard O'Neill's operation (O'Neill wrote a seminal book in the 1970s called *The High Frontier* that details how humans could live in floating space colonies)—to support some work on the subject. Unfortunately, he couldn't get his ideas across. "My timing was bad," he laments.

No matter. Lewis' development continued apace. In 1980, he read *The Making of an Ex-Astronaut* by former astronaut Brian O'Leary. "He told it like it was about the astronaut corps at NASA and the NASA bureaucracy. If I thought I'd read the handwriting on the wall, that book confirmed all my worst suspicions about NASA. We had just [run] through a lengthy ability to launch heavy payloads because the shuttle came in late and over budget, and instead of managing sixty launches a year they were fiddling around with about three or four." Note: This is still the case.

> I was a very unhappy camper. So I basically decided that we needed to get rid of the dinosaurs we were using for launching space vehicles and replace them with high-tech, newly designed vehicles that were intended to be either completely reusable or cheap but expendable. We needed something that would bring the cost down by at least a factor of ten immediately and the promise to bring down costs by a factor of a hundred. They wouldn't necessarily have to be manned vehicles, though in fact there are fairly strong reasons that manned vehicles could be economically competitive with unmanned vehicles, but I was wishy-washy on that subject. It was dollars I was interested in. How many dollars to put a pound in orbit. There were other people who were barking up that tree too, but in 1980 there were precious few of them and they had zero to work with.

Lewis explains what a formidable force he was up against. "That was a time in history when NASA was perfect and anyone who criticized it was clearly a maniac. I wasn't able to get any significant critique of NASA into print anywhere until the *Challenger* blew up, and then suddenly NASA went from automatically being perfect to being the laughingstock of the universe."

Shortly after the *Challenger* accident in 1986, Lewis published a book called *Space Resources: Breaking the Bonds of Earth*. The book included an analysis of the problems with NASA. He figures the book sold about 2,000 copies. "But at least I got those words into print. And I'm still startled by people who come up to me and quote to my face things I said in that book." One person who read the book was Carol Meinell, who called Lewis immediately upon finishing it. "Now Carol was a person who flirted with the military side of space for many years. She's produced industry newsletters on military space activities. We met in person at a party and ended up writing a couple of articles that we published in an unlikely journal called *Defense Science 2000 Event Plus*." Meinell introduced Lewis to her friend George Koopman, an independently wealthy department store heir who had gone through life trying to decide what would be fun to do. Koopman had done special effects for the *Blues*

Brothers movie, but he wanted to do something "real." Lewis recalls that around 1985—before *Challenger*—he decided he was going to go into the private launch business.

> So Carol talked to him, told him about the space resources book. He went out and bought a case of them and would literally carry around a briefcase filled with copies of the book when he was meeting with potential investors. After using this ploy successfully several times, he called her up and invited my wife and me to attend the test firing of one of their rockets at Edwards Air Force Base. What they had done for themselves was a rocket design that was unlike everything that had ever been launched before. It was the birth of the private launch industry.
>
> George invited us out there to see a test-firing of his rocket—a hybrid that was designed to do away with the major causes of unreliability and cost in conventional rocketry—liquid fuel, which is highly explosive, lots of moving parts, all of which have to work or—phew.

Lewis whistles. Because of their complexity, liquid rockets carry little hope of achieving a success rate greater than 98 percent. Solid rockets are no better because they give off ozone layer-destroying aluminum oxide particles and emit toxic hydrochloric acid and chlorine exhaust. They are also liable to explosions and inadvertent ignition, and you can't turn them off once you've started them. If you jettison a problem booster, "you've now got the dead weight of the shuttle riding on one SRB [solid rocket booster] and the other SRB is now removed of the weight of lifting the rest of the stack, so of course it's accelerating much faster than you are. So it goes flying on ahead and incinerates you in its exhaust plume."

Lewis interrupts his Koopman story to explain why such treacherous devices as solid rocket boosters are used on the shuttle. "At the time they were designed they were the best solution to the problem. Remember that the fundamental design of the shuttle system goes back to the late 1960s, long before hybrids existed." Did the designers know how dangerous solid boosters were then?

"Yes. They knew about the dangers, but they figured they could control them. And they've done a pretty good job of doing it. Ninety-nine times out of a hundred they're successful."

He resumes.

> George Koopman said we combine the best features of both, solid and liquid, and leave out the worst features of both. And he did. We watched this test firing, which was stunning. Here's this big piece of rubber—tire rubber—burning tire rubber, and there's no smoke. Why? They burn it oxygen-rich so there's no smoke. The exhaust is carbon dioxide and water. There's no way we could build enough rockets to affect the carbon dioxide and water content of the atmosphere. So here's this environmentally benign rocket that is safe, reliable. It's obviously the right thing to do.

Koopman invited Lewis to join the board of directors at American Rocket Company (Amroc), his company. (For more on hybrid rockets and the ultimate fate of Amroc, see Chapter 13 on Jim Benson and SpaceDev.) "So here's this college professor, all the business acumen of a slug," he says.

> I turned to my wife, and she looked at me. I said, "George, why, why do you want me to join the board of directors?" "Well," he says, "we expect to be launching commercial payloads fairly soon, and we expect that we can launch them for somewhere between 5 and 10 percent of the cost of a normal launch. Now we think many of our customers are going to be scientists. Instead of asking NASA for $20 million to build a flying instrument we'll be able to ask them for $2 million or $1 million to build and fly the same instrument. And you know the scientific community; we don't." I thought about it a little bit and said, "I can just picture NASA writing a check for $1 million to a launch service. Where's he getting the money? How did the money get into his checking account? I'm thinking that the other customer has to be

> the government. What if the government doesn't like this? They'd regard it as competition." So I turn to my wife and she gives me this look—I don't get it. I said, "George, what's the real reason you want me on your board of directors?" And he leans forward and says, "I want to go out there and mine asteroids."
>
> We rode that marvelous machine for several years before we ran out of venture capital back in '95. It doesn't exist any more. The principal investor sequestered the technology that had been developed by Amroc, which had not been put in the public domain, hadn't been sent to the U.S. Patent Office to be published. It was all secret stuff. Rather recently one of the startup companies has bought the technology developed by Amroc. They have a team of engineers right now going over it and seeing what it would take to put it in the air.

(Lewis is referring to Jim Benson and SpaceDev, a fledgling California public space development company. See Chapter 13.)

For a special effects guy, Koopman was pretty good at running a rocket company. How did that happen? Lewis laughs.

> He told me that he was really interested in doing something real. He said "I've always been a space cadet. I was a member of the L5 Society." [The L5 Society was the original American space advocacy organization. In the mid-eighties, it merged with the National Space Institute to become the National Space Society.] He thought space was exciting. It was just a hobby for a long time, but eventually he felt secure enough financially to do what he really wanted to do. So he went around, talked it up, found people of the same mind and hired them and put the company together. And most of those people are now in very responsible positions doing similar things.

Interestingly, Jim Benson of SpaceDev, the company that purchased Koopman's intellectual property, has followed a similar route, from computer industry tycoon to space developer, demonstrating that you don't have to be a rocket engineer to develop space. Even more interestingly, Benson followed in Koopman's footsteps in another way. "In all of '96, I was on sabbatical at the University of Massachusetts," says Lewis.

> *Mining the Sky* had just come out. It's a much updated and differently written version of *Space Resources*. I was sitting there reading a science fiction novel and outlining this book *Worlds Without End*, which is about the plurality of worlds, and telephone rang in my little retreat, which was in a cute little cottage near Amherst, Massachusetts in the foothills of the Berkshires. I felt I was quite safe from any serious professional engagements there, and someone says "Hello, you don't know me but my name is Jim Benson. I just read your book and I thought you'd like to know that I'm going to start a business to go out and mine asteroids."
>
> It turned out that over the first few conversations with him I got some idea of his background and I realized it was another George Koopman story. He'd gone through two rounds of the software business, he'd retired in wealth both times, he had taken a few months of retirement. It had driven him right up the wall. His reason for going into space access was "I've always been interested in it." It happens that in his case he had some arguably relevant background. He actually has a bachelor's degree in geology from the University of Missouri. Hasn't used it in years, he hadn't thought of it as a way to make money, but it had opened his eyes to reading popular science literature and he had become enthusiastic about space and he had, I suppose—although I can't actually remember him articulating this—the same sour taste in his mouth about what the government was doing in space and trying to prevent private groups from having access to space up until the time of the Reagan Administration.

Have they really tried to prevent people? “Oh yes,” he said. “They’ve actively tried to prevent people. In the later years, I think it was ’94, the government realized that hybrids were a much more intelligent way of doing things than what they were doing. Now the cognizant NASA center for propulsion is Marshall Space Flight Center (MSFC), which used to be Redstone Arsenal—the army’s ballistic missile agency where von Braun and his people were located. And that’s how they got inherited by NACA/NASA. [NACA, National Advisory Committee for Aeronautics, is NASA’s predecessor agency.] The director at MSFC at that time was J.R. Thompson, who later became deputy administrator at NASA. He heard that there was this bunch of gutsy innovators out there in Camarillo, California who had a thing called a hybrid rocket, and it looked really good. In fact, it looked so good that it was possible we could build a replacement for the shuttle solid rocket boosters and *Titan IV* solid rocket boosters. It was safe and environmentally benign. Thompson’s response to this was, ‘We got to find out if this works.’ In fact, his response may have been, ‘If this technology works, we need to control it,’” Lewis speculates. “But I don’t know that,” he admits.

> Very shortly thereafter Marshall issued a request for proposals [RFP] for development of hybrid rocket boosters up to shuttle class. The request went out and it was to be three phases: A, B, and C. Phase A was pure research aimed at producing a preliminary design for an engine that worked. Basically it was not hardware-oriented; it was a paper study. Phase B got into hardware development. Phase C got into construction and production. For Phase A, the rules were that anybody could compete. When all the proposals were received, three would be selected for funding. The three would each produce a competing design and the best design would be declared the winner. Phase B would be competed openly, but there would be one winner, and that one winner would take the winning design and develop flight test hardware to demonstrate that it really worked. One final feature of this RFP that had never

> been seen before was the condition that the winner of Phase A would be ineligible to participate in phases B and C. The winner's design becomes government property because the government paid for it. How many companies on Earth have the ability to design a hybrid rocket booster right off the top of their heads: one. So who's going to win phase A? No question—it was going to be Amroc. So what does that mean? The only company on Earth that would be ineligible to build it would be Amroc. George called me up and says, "What do you think of this?" I said, "George, call up J.R. Thompson directly and say thanks but no thanks, we're not competing." He did, and within twenty-four hours the request for proposals was withdrawn.

How could NASA think Amroc was that stupid? "They were hoping Amroc was that desperate," he said.

George Koopman was killed in an auto accident in 1989. "We only attempted to launch once, at Vandenburg," says Lewis.

> The fog rolled in, and we waited for it to clear. One moving part in the rocket that wasn't built by Amroc failed because sitting there in the fog for long periods of time with the cold liquid oxygen up there above the valve—the valve got so cold that it iced up and the orifice under the valve leading into the rocket engine became plugged with ice. They hadn't anticipated that because they had been testing at Edwards Air Force Base in the desert. You know how you can solve this problem? The place where the liquid oxygen tube enters the top of the rocket, take a baggie and tape it over the orifice. Then when you turn on the oxygen it blows away. It was a one-cent fix. The damage to the launch

pad cost $5. Compare that to *Challenger* or *Titan IV* ... it's a very different way of doing things.

After the Vandenburg failure, Lewis decided that his proper niche in life wasn't in the private launch industry. "It was doing something a little more far out. Needless to say I'm still a strong believer in the private launch approach," he assures. "The economic arguments for it are just overwhelming. Here's the motivation right now that should get anyone reaching for their checkbook." Lewis explains that on a strictly technical basis, launch costs could come down astronomically today—if it weren't for government interference. "If you launch a payload using the shuttle right now, it costs about $10,000 a pound. If you use the most cost-effective expendable boosters you can bring that down perhaps as low as $3,000 a pound. That's modified dinosaurs—1960s-era boosters that have been redesigned. Not by the shuttle—the shuttle would never do it because of the overhead that carried that winged orbiter and the people and the life support and the safety issues, etc." Lewis posits a scenario.

> Let's suppose we design from scratch, as many of these startup companies are doing, an expendable launch vehicle which is designed to be functioning well within its performance envelope—it's not pushing the engineering limits or materials or whatever and, therefore, will be rugged, robust, and reliable and to a great extent, if not completely, reusable. Then how can we expect the launch costs to run? Well, actually this is fairly close to the Russian philosophy that they used in their first two generations of boosters, the so-called A-class booster, that's the *Vostok-Soyuz* launch vehicle—eight different upper stages, same first stage—and the *Proton*. In those cases, the actual cost of putting a payload in space is around $600 a pound. Here are the Russians out there saying, "Okay. We'll sell you a ride and we'll charge you $600 a pound." What happens? State Department says that's dumping. You must quote prices that are similar to what American purveyors of

launch services charge or we're not going to let you do it. You're wrecking the American launch industry. So the Russians now quote prices of $3,000 a pound or something like that and they compete. Historically, if you reduce costs by a factor of ten, demand goes up by a factor of thirty to one hundred. It hasn't happened."

Well, then, what if someone—an American—cracks the cost barrier? Says Lewis,

> First, every possible political obstacle will be placed in their way by the lobbyists of the major purveyors of launch services. All kinds of red herrings will be raised—safety issues. They'll try very hard to prevent the space insurance industry from insuring such launches on the grounds that these are new boosters and untried. Never mind that the existing ones have been tried and demonstrated highly unreliable. The insurance companies will come in there and charge enormous premiums for launch on these new cheap vehicles, which will diminish their competitive margin relative to the existing boosters. Fortunately most launch insurance in the United States is from overseas companies. Europe and Australia both.

Well, then, what would happen if the Europeans cracked it? "Then space would shrivel up and die in North America. We wouldn't have any way of launching anything cheaply," says Lewis. "They could. They could offer launch services for sale to us cheap, but we wouldn't be allowed to take them because the State Department would call that dumping and prohibit the signing of any contract."

Many people don't understand why the U.S. State Department, a diplomatic agency, retains jurisdiction over commercial space launches rather than the Commerce Department. "The Commerce Department is our friend in this matter," he says.

> The Commerce Department has been given centralized authority to deal with authorizing launches, there

> are some intelligent and well-meaning people there, and the regulatory climate was changed by an executive order by President Reagan, who wanted to see this logjam broken. He wanted to see the creative potential of American private industry liberated. So the process of getting approvals and authorizations for launch, which was originally distributed over heaven-only-knows, there were dozens of federal agencies, was centralized. The responsibility was given to one office and you could push the papers through that office very quickly instead of having to wait a year or two for an authorization. It's an office at the Department of Commerce. Commercial Space Operations or something like that.

However, the Commerce Department is in charge of domestic operations only. "A private company in the United States cannot negotiate an agreement with a foreign nation or agency without the approval of the State Department," Lewis explains.

> The State Department is in charge of U.S. relations with foreign governments. The State Department also holds the strings on technology transfer. Let's say you're sending a payload over there with modern electronics aboard that's going to be launched into space. The paperwork shows that it's going to France or it's going to Algeria or Australia or someplace to be launched, or to China. So you need an export license for it. "Export license denied: technology transfer!" says the State Department. Even if you say you're going to send armed guards along with the payload who'll guard it till it's launched, they'll say, "Denied. Technology's still being exported. How do you know that these guards won't all fall asleep at once and someone will steal a chip? What about a so-called launch failure? What if they steal the payload and then crash the rocket in the desert and say, 'Ooops, sorry, we've wrecked your payload.'" You know, they're paid to be paranoid, and based on what's happened with the nuclear weapons issue, I think probably

> paranoid plays very well before Congress and before the public right now.
>
> The way around this is for an American company to get into the private launch business.

If you question this statement saying that they were shut down by NASA when they tried, Lewis explains that it wasn't a matter of shutting them down; rather, NASA tried to steal their proprietary technology and failed. At the same time, Amroc ran out of money. "They would probably have lived longer if they had the money from NASA. NASA didn't shut them down, NASA couldn't shut them down. But what did happen was that they ran out of venture capital and fell into a bacterial spore state where the basal metabolism is reduced by a factor of 100,000 and nothing moves."

What about the dozen or so private launch companies working on new vehicles right now? If someone succeeds, that's it. Lewis agrees.

> Let's suppose they get enough funding to launch one payload. We played out lots of these scenarios when I was with Amroc, so I have a feeling about how it might work. You go to a government-supported university research group, and you say, "Hey—got any spare instruments sitting around?" They say, "Yeah, we have a flight spare here from *Galileo*." "How would you like to launch that and look at the Moon with it?" Or "How would you like to launch it and look at the Earth with it?" They say, "Well, I don't know—we can't afford a space mission." You say, "Well look, we're trying to get the first launch of our booster. We would like a little income stream to help pay for that, but we're going to eat a lot of the cost of the first launch so that we can demonstrate it works." Once we've had one successful flight, insurance premiums come down by a factor of two or three, and that makes it much easier for people to fly. You don't need to insure your payload—it's a flight spare you had no hopes of flying. So, you come in with a little bit of money—you ask NASA for half a million to refurbish this thing and to prepare you to collect and

> use the data from it. And then you pay us $300,000 for the launch, and that defrays our cost for the launch. But the whole thing is to get that first launch. Once you get the first launch anything can happen. If it succeeds, insurance premiums come down. You go to the government and say, "Look." Better yet, you go to universities and say, "Look, folks, there's the Discovery program at NASA where you can propose to do a mission—a university consortium proposes to do a mission—I want you to take a look at our booster and see if there are any missions that you're interested in that could be done with our booster. If so, you could propose to NASA and the total cost coming out of the gate is going to be about a third of what NASA's accustomed to paying for that sort of mission. You'll enter the competition, you will of course win because you'll have by far the best bargain." And that's what's trying to break out of the egg right now. NASA's contracting office will say, "We've never done this before—oh no. How can we do this, we haven't been doing this for thirty years. It's impossible." And Dan Goldin will say, "Blankety blank—do it!" And for a year or so they will ignore Goldin until finally it will become a scandal and Congress will investigate it and they will be compelled to do it.
>
> I had lunch with Goldin a couple of years ago here in Tucson. I said, "Mr. Goldin, tell me frankly, what is it like to deal with the entrenched NASA bureaucracy?" He says, "Oh, nothing to it. I give an order—they ignore it."

Why does all this launch stuff matter to Lewis? Well, no launch industry, no mining, and no solar power from space.

He's a man who navigates a centered course between liberals, who worry about pollution and resource depletion, and conservatives, who want to slash spending for research and education and ease environmental standards. He thinks we should balance long-term research, short-term applied research, engineering development, and the commercialization of new products. In other words, maintain a spectrum of activities focused on doing what's best for humanity's far future and for enriching all of us in the short run. The reason we're not doing this now? The rise of bean-counters into top management, a generation of professionals who know how to count beans but who don't know anything about where future crops of beans come from. And this shouldn't be because, for all practical purposes, the resources that are available to us are infinite.

> I don't see a future in which everybody gets poorer and madder. I see a population in which everyone gets better fed, happier, and what happens when a country achieves self-sufficiency? When they're self-sufficient in food and have a growing per capita income, the birth rate goes down. And this whole business about the population bomb. If you go back and start reading the books from the 1960s and 1970s about population projections, we're way below all those projections. People were saying ten billion people by the year 2000. There's nowhere near that. So why have the projections failed? Because the projections assumed if things continued to go on like this. Things can't go on "like this," and they never have. The grand rules are always changing. If you don't make allowance for new technologies and new capabilities and for policy changes, then the projections are utterly meaningless, and all of those scare projections were utterly meaningless. The population is well on its way to stabilizing at six and-a-half or seven billion, and but that's not going to happen unless things continue to change in Asia. And the most important things we can do to solve the population problem and the carbon dioxide problem is to increase the technological level of our societies. And, by the way, the second largest environmental threat, now that the chlorine problem has been solved, is probably methane from rice paddies.

For Lewis, seeking social and economic justice for all isn't merely compassionate. It's good business. And it's possible.

He believes that more than a sea change in human space activity is coming. "This is a tsunami," he says.

> It's like the early days of the personal computer industry. Remember the popular wisdom was there was a market for maybe five computers in the world? And the idea of personal computers was just sheer nonsense, they'd always be too expensive. But then along came a little company called Altos, and the Altos computer appeared, it started selling, and what happened? They ended up owning the world, right? No—they went belly up. What happened was that Steve Jobs and Steve Wozniak and whats-his-name in Seattle said, "You know, we can do the same thing cheaper and better." And then IBM watched them for a year and said, "I think we can do the same thing that Apple's doing but cheaper and better and with an open system architecture so that we could involve a huge range of other writers of software and producers of hardware." So, Apple dominates the industry today? Not. IBM dominates the PC industry today? Not. The IBM design does, but IBM doesn't. So along comes Compaq, and Compaq says, "Hey, you know, we can do the same sort of thing that IBM's doing, but we have smaller overhead, so we can compete with them successfully." So Compaq took over. And along comes Dell, and Dell says, "We can make computers to order without retail outlets, do it all by e-mail and by telephone and keep costs way down below anything Compaq can offer going through retail dealerships." So what happens? Dell took over. You see what's happening here? This is exactly what's going to happen in the launch industry.

There are those who would take issue with Lewis' computer industry analogy. For one thing, the capital investment required for space is exponentially higher—hundreds of millions of dollars to produce a commercial product. He disagrees.

No. That's the really, really strange thing. The cost required for Packard Bell to get into this business was billions of dollars. The cost required to develop one of these new boosters, probably less than point one billion in every case. This is small potatoes, which opens up several things. One is that splinter groups or work teams from a major contractor can say, "The hell with this. I'm tired of this blankety blank blank. We're quitting, and we're going to start our own company, and we're going to build what we really want to build, and enough of this business of pandering to these generals who want 1960s hardware with an extra bell and whistle on them. We're going to build something new and profound and high tech and cheap." This has happened over and over again. And you see what's happening. The big aerospace companies are in an exponential death phase right now; they're merging with each other hand over fist. It looks like the genealogies of 1970s rock groups. At the end what are we going to end up with? One or two giant aerospace companies will be all that's left, and they'll be so diversified that they won't have any expertise in anything. And they're making mostly civil transports because that's the continuing market. So what's going to happen to the people with the bright ideas? They won't be able to stand the corporate environment. In fact, most of them are out of the big corporate environment already, and they're operating in basements in Sausalito designing the next-generation launch vehicles, and they're capitalized at the level of a few million dollars and they're always wondering where the next launch is going to come from. If you're thinking about it from an investor's point of view, this is going to be the personal computer industry all over again, and you want to diversify your bets. You don't want to put all your money in one because you might be investing in Altos.

THE AUTHOR'S OPINION

Lewis is facing the same chicken-and-egg problem as many other space activists. It would be great to use asteroidal resources to support space-based activities, but a certain infrastructure in place must be in place to be able to exploit them. Lewis is already on page 2; the world is on page 0.

But that's okay. Many technologies have come about in the real world because science fiction writers popularized them. Arthur C. Clarke promoted communications satellites. Noninvasive medical devices like those used on *Star Trek* are coming into use, as are artificial body parts. Wireless phones resemble Trek communicators, or Dick Tracy's two-way wrist radio. Lewis' ideas could become self-fulfilling prophecies as well.

I think his idea to use up Earth-threatening asteroids is extremely attractive. It's making the initial investments and/or convincing governments that's going to be the hard part. With more and more scientific evidence showing that such bodies pose real threats, though, the argument becomes easier and easier to make.

At first I thought the idea of solar-power satellites was brilliant. Such devices would solve Earth's energy problems. Then I began to talk to people who surmised that not only would the expense of building such an infrastructure be prohibitive, but that people wouldn't stand for having microwave beams dot the landscape. I'm beginning to agree with them. I simply do not see governments allocating that kind of money, and it would take years of testing before the feasibility of such devices would be proven. You could repeatedly prove the safety of the beams, and there would still be doubt. So now I think it's better to shore up the solar and wind energy industries on Earth and see if we can't find a better way to power our world.

As far as the billions of people living in the solar system using indigenous resources scenario, well, Lewis is just a lot better chess player than I am. It could happen, but I can't see those twelve moves ahead. I'm just not sure how we're going to get there.

—13—

Jim Benson: Space Entrepreneur Extraordinaire

What's it like being a space entrepreneur? Do space entrepreneurs really make money? Are they really doing anything important, helping us get out there? If so, why haven't we heard more about them? See for yourself.

The bulk of this interview was conducted at Benson's office in Poway, California, near San Diego. SpaceDev is located in a little industrial park-like area off the beaten track. It's a functional place featuring little decoration. It seems that Benson has more pressing things to do than to shop for artwork.

In 1996 and 1997, Jim Benson attended sixty conferences. After a successful career as a software entrepreneur, Benson, then fifty, was looking forward to retirement. A few months later, he was climbing the walls. Two years later, he founded SpaceDev, a U.S. public company dedicated to providing commercial customers with small deep-space and Earth-orbiting missions from a fixed price list. No one had ever attempted such a thing. To this day, no one other than governments conducts deep-space missions, and each mission that governments do is on a custom basis. SpaceDev also makes low-cost spacecraft and vehicles for maneuvering in Earth orbit, and offers a different kind of rocket motor—a "hybrid," which is nontoxic, can't explode, and doesn't cost a lot. Jim Benson isn't talking about retiring any more.

.*. . .* .☾ . * . * . * . .

In 1997, SpaceDev (www.spacedev.com) made headlines when it announced plans to conduct an asteroid rendezvous mission in 2002. The project, called NEAP (Near Earth Asteroid Prospector), would involve landing a microspacecraft on the asteroid Nereus, characterizing the one-kilometer body, and ultimately claiming ownership of the big rock. Another Benson first. No one had ever tried to claim a celestial body before, and a property rights controversy was off and running. Benson had decided that he wanted to claim an asteroid to help set a precedent for property rights in space, provide a focal point for public debate, and create an incentive for commercializing space.

He got plenty of debate, all right. The 1967 Outer Space Treaty, written during the Cold War, forbids governments to claim property in space but says nothing about individuals or businesses. (The Moon Treaty does forbid such claims, but only a few nations signed it, none of them spacefaring.) Nevertheless, the United Nations, which oversaw the creation of both treaties, is not likely to be amused if and when Benson stakes his claim. (See Chapter 18 for more about property rights.)

In July of 1999, the company sold its first payload ride on NEAP to Dojin Limited, a Texas firm that wants to send "digital passengers" to the asteroid on a CD-ROM. Dojin's Cosmic Voyage 2000 program puts people's images, identities, and personal messages on the digital medium for delivery to, and preservation in, space. SpaceDev will receive $200,000 for delivering the package to Nereus. A researcher funded by the University of Arizona has also signed on, sending a camera akin to the ones he and his team designed for Mars Pathfinder. In return for providing a ride for the camera at no charge to the University of Arizona, SpaceDev will receive exclusive rights to sell the camera's photos and scientific data, and the University will receive a royalty. The company is now looking to launch by 2007 and considering alternate targets for the mission while gathering corporate sponsors and funding for the mission. Benson puts the cost of the mission at under $50 million.

JPL's Near Earth Asteroid Rendezvous mission, which served as the model for NEAP, is costing between $250 and $300 million.

NEAP Price List

Item Price	Price ($M)	Type	Description
N.1.1	$2.5	Data Set	Radio Science—Available
N.1.2	$12.0	Data Set	Multi-band CCD (from Peter Smith at the U. of AZ)
N.1.3	N.A.	Ejectable Payload	N.A. (Reserved by SpaceDev)
N.1.4	$9.0	Ejectable Payload	Available
N.1.5	N.A.	Custom/ Nonejectable	Payload Purchased by Dojin, Limited
N.1.6	$10.0	Custom/ Nonejectable	Payload Available
N.1.7	$10.0	Custom/ Nonejectable	Payload Available
N.1.8	$10.0	Custom/ Nonejectable	Payload Available
N.1.9	NEG.	Sponsorships and Logos	Mission sponsors and co-sponsors, spacecraft, launch-tower, Web site logos
N.1.10	NEG.	Rights	Collateral rights—media, documentaries, lift-off, streaming video, special access, etc.

NEG. = The price is negotiable

N.A. = Not available

While the company planned for NEAP, it simultaneously began to concentrate on other projects. Now it offers the delivery of science instruments and technology demonstrations to Earth orbit, deep space, and planetary bodies. It also designs and sells satellites for commercial or research purposes. It's designing and testing hybrid rocket motors that will power small vehicles for use in Earth orbit to do things like inspect satellites, rendezvous, dock, and refuel. The technology for these hybrids was acquired from the former Amroc. SpaceDev is also working to integrate hybrids into manned suborbital vehicles (read "possible space tourism vehicles").

SpaceDev is developing mission and spacecraft designs for three separate commercial deep space missions: a robotic lunar mission to provide live, interactive streaming video, with a Web server and e-mail forwarding from a commercial lunar orbiter; NEAP; and a 2003 Mars data relay orbiter possibly combined with a Mars probe carrier. All three proposed missions use the same spacecraft architecture and flight software—an approach not used elsewhere. SpaceDev, sometimes called a "mini-Boeing," has also partnered with Boeing to analyze the economic potential of a series of such commercial deep-space missions. No other entrepreneurial space company is so tight with, yet so different from, mainstream aerospace.

Nor are other space entrepreneurship companies public companies. By going public, SpaceDev has incurred special responsibilities, advantages, and disadvantages. Public companies can raise capital more easily and cheaply than private ones, and in so doing they face fewer operational restrictions. For one thing, they have more alternatives for raising money. They aren't limited to banks, VCs, and other private investors: They go to the public market place, where investors place fewer terms on them and are more willing to put money in because there's an ongoing exit option for them. They're also more willing to invest because there's less chance that their position will be diluted (see Chapter 7), and because they are publicly traded, their investment is more liquid—there is an existing exit strategy. As a result, public companies tend to be more highly valued than private ones.

Dilution for public investors will occur if the company issues more stock, which happens rarely, and usually only when it's about to acquire an enterprise that will greatly increase its value. Dilution for private investors occurs much more often because new stock is being issued all the time to make the company grow exponentially in its early stages. Having public capital available provides leverage when companies go to institutional investors. On the other hand, because the stock is publicly traded and held by outsiders, the founders and executives can lose control of the company. The extreme outcome of this risk is the hostile takeover, although there are ways to mitigate it. Another risk is that of shareholder lawsuits precipitated by large drops in the stock price. Class action suits can be filed when there is suspicion or proof that the company knew something negative and didn't disclose it early enough. That's why public companies have to be so careful about "forward-looking statements" and overly optimistic projections or interpretations either in speech or in writing. Any chance that stockholders are being misled can result in legal action. (Directors' and officers' liability insurance can be obtained to cover this risk, but it costs $15,000 to more than $100,000 per year.)

Public companies can grow more or faster than private ones because of access to larger amounts of capital, and they can offer their investors greater liquidity—just sell your stock if you want to get out. Market fluctuations provide opportunities for profit: Founders can buy up stock when the market as a whole is depressed, and sell stock when it's riding high. (Assuming, of course, that the fluctuations have nothing to do with the company's performance.) Private companies enjoy no such option. Initial public offerings can afford company heads more generous possibilities to increase their personal wealth. The stock they hold can be sold on the public market, though with significant restrictions on how much and when, and if the value has risen, founders can do brilliantly. Even if they hold onto their stock, it's valuable as collateral for new loans.

Public companies carry prestige with customers and employees. Well-financed companies find it easier to attract quality staff, and customers have more confidence in them because of their stability. The money that flows in as a result of the IPO can be used to make acquisitions, which can help the company grow faster. Alternatively, companies can use stock to make acquisitions. Private companies are in a less advantageous position to acquire because their stock is less liquid; to acquire they need large amounts of cash, which most of them lack.

On the downside, CEOs of public companies are under pressure to show short-term earnings growth, which can lead to poor decisions. If the results are poor, investors may be spooked, even if long-term profitability is likely. Public companies must comply with federal and state securities laws, which adds extra layers of overhead and cost. It's expensive to go public—the underwriters' commission can run to 10 percent of the offering! Out-of-pocket expenses can span $150,000 to $500,000. Producing the required reports, holding stockholders' meetings, and managing investor relations can eat up $25,000 to more than $100,000 per year. And those figures don't include printing and distributing annual reports, proxy statements, and stock certificates or the extra personnel required to deal with these issues (though some smaller companies outsource these tasks and save themselves considerable expense).

One of the more interesting differences between public and private companies is that for the latter, operations, financials, personal information about the officers, directors, and major shareholders are open to public scrutiny. That means that competitors, customers, and employees can all see what's going on. Read a 10-K (a public company's annual filing that includes balance sheets, performance, transactions, executive compensations, risks, and other information), and you may be surprised at how much you can learn about not only the company, but its entire industry. However, since there are other ways that competitors can learn about the company's dealings, these disclosures may not pose as much of a threat as it seems. They can get information from business credit agencies, customers, former employees, suppliers, news reports, press releases, and Web sites. And the information that's included in such documents is only that required by investors. Much of the detail, such as financials for divisions and specific products, can be left private.

Leisure doesn't become Benson, a hard-driving, ever-questioning environmentalist-turned-businessman. Even though he talks about retiring (again) to space some day, it's difficult to imagine him completely at rest. You can't be near the self-proclaimed former "Mr. Alternative Energy" without feeling that Benson is his own energy source. A man whose mind races at night, he's positively blossomed in his mid-fifties.

What would compel anyone to attend a conference every other week for two years? Logic. If that's what you have to do to get where you want to be, you just do it. And where did Benson want to be? In space. But only if he could make a profit doing it. Otherwise, no thanks. And so Benson spent two years investigating just what space-related endeavor could achieve that goal for him.

> I was sitting there with a nice bank account not having to work again for the rest of my life but also realizing that I'm fifty years old and I could live another fifty years. What the hell am I going to do for half a century? I'm a worker. I've been working since I was eighteen. And so I immediately—almost immediately—came to space because I've loved science, technology, and astronomy all of my life. And combined together, that's space, and all my adult life since eighteen it's been business, so ... space business. Space commerce. That's how I arrived at it.
>
> And I wanted something that I really would like doing. E.F. Schumacher's book *Small Is Beautiful* discusses "right livelihood." People should seek out right livelihood. That basically means getting paid for what you most enjoy doing. I spent almost all of '96 deciding what in space is possible from a commercial point of

> view, from a business point of view. And frankly came up dry. I read books all the way back to the seventies, eighties, nineties. Went on the Internet. Every idea that I could find. I ran it through a business filter and it fell on the floor every time.

Unlike many space entrepreneurs, who start out as rocket scientists and never quite make the transition, Benson has a business background, and it shows. At the Mars Society meeting in Boulder, Colorado in 1999, he was the only one who pitched an existing commercial plan. He was the only one who gave a press conference. He originally wanted just to invest, not run his own company. But while he was thrilled to meet rocket entrepreneurs, including the late astronaut Pete Conrad, at his first space conference—the Space Frontier Foundation conference in 1996—after looking at their business plans, he declined to invest in any of their companies. He could see that their business cases were lacking.

"So I was an investor looking for something to invest in. And I'll tell you what, three strikes against all of them, uniformly," he says. "Not one of them has a business background. Strike one. Every single one of their technologies, their reusable launch vehicles, has at least one pet technology that's exotic, untried, unproven, needs a lot of R&D, and therefore is expensive and risky. Strike number two. And in the case of one of them there are six technologies that all have to come together. Strike three is that they're all going for the brass ring the first time they ride a merry-go-round. All of them are probably going to cost a quarter of a billion to a billion dollars. How the hell do they think they're going to raise that kind of money with no experience and no track record?"

But Benson had started formulating some ideas of his own. He realized that NASA was performing what amounted to commercial missions to acquire science data, and there was no competition for its services.

> NASA's Office of Space Science, then run by Dr. Wes Huntress, was spending $2 billion a year collecting scientific data in space. And they do that by sending out *Galileos* and *Cassinis* and *Pathfinders*, near-Earth

asteroid rendezvous missions and so forth. That was the first thing, that there's an existing $2 billion a year market and I said, "Who's supplying that market?" All commercial stuff, no commercial missions.

I thought, "There's a market, right there." And the next question I asked myself was, "Can't the private sector usually do things more cost effectively than governments?" And if you have a $2 billion market and there's no private sector doing it, then it's ripe for the private sector. So that was really the beginning of everything for me and SpaceDev.

Actually, the beginning was earlier. "I remembered clipping this 1991 article about that asteroid 1986DA, characterized by Steve Ostro at JPL. In 1996, that's what got me started in the space resources direction, because when I did some rough calculations I came up with a very rough street value of $80 trillion for that." Benson points to an iron-nickel meteorite (former asteroid) on his desk.

If you analyze that in a laboratory, you'll find that the gold concentration is ten times richer than anything being mined on Earth, and the entire platinum group of minerals in there—metals—is 100 times more concentrated than anything being mined on Earth. We're talking about parts per million, parts per billion.

I found that little clipping, got in touch with Ostro, and he put me in touch with Dr. John Lewis [see Chapter 12]. I read the two Lewis books on resources, and that focused me right in on near-Earth objects and resources in space. This book [he holds up *Resources of Near Earth Space*] is by Lewis. It's an important one because it's a thousand pages of scientific and technical descriptions of near-Earth objects and resources in space. So it was those two guys—the world's leading radar astronomer, Steve Ostro, and world's leading planetary scientist in terms of near-Earth objects, John Lewis, who got me started thinking of possibilities.

> Twenty-five percent of near-Earth objects are the easiest planetary bodies in the solar system to reach. Easier than the Moon. The Moon's only three days away; these things take longer, but they take a lot less energy. Energy is money. Keep this in mind. When we reach Earth orbit, we're halfway to anywhere in the solar system. You've probably have heard that before: Robert Heinlein, scientist and author, said it first. That means that the energy that it takes to get up through our gravity into Earth orbit is about half the energy that would take us anywhere in the solar system. When our rockets or satellites make it to Earth orbit, we are halfway to anywhere. But our tanks are on empty, out of gas. We need filling stations in orbit. To make space happen we need the same basic infrastructure there as on Earth: diverse transportation, diverse communications, and concentrated portable energy.

Benson knows about energy. In the late 1970s, he wrote the U.S. federal legislation that set up the small-grant program for appropriate technology and worked with the late California Congressman George Brown to push it through Congress. In the second year of its operation, the U.S. Department of Energy got 20,000 requests for these grants, which were limited to individuals, communities, and Indian tribes and were intended to demonstrate appropriate technology using local resources to save energy or to produce reusable energy at the local level.

In the mid-70s, he wrote the *County Energy Planning Guidebook*, which he typed and distributed by hand out of the trunk of his car. "I traveled around for two years giving speeches all over the country as an environmentalist and I sold 25,000 copies of my book, half of them out of my suitcase. I'd give a speech and I'd go down by the stage and open up my suitcase and sell them to fans in the audience. We typed that thing seven times on a typewriter. It was terrible, repetitious work." Until he hooked up with a friend who helped him typeset the book. "I drove out to a friend who was living in a commune in Pennsylvania and up in his attic he had a typesetting machine. God knows where he got the damn thing. And we stayed up for thirty-six hours while he sat there and typed the entire book on a typesetting machine. I proofread it in real time. I mean he'd be one page ahead of me. I'd be proofreading, take a break, he'd make the corrections and go on again. We did the entire book in one sitting. For a few hundred dollars." Typical Benson. Find something you believe in, live and breathe it, and make it profitable. Twenty-five thousand copies sold is nothing to sneeze at, even today.

It's difficult to reconcile the hardheaded business image with that of rootless environmentalist. But, like his ersatz analog, Richard Branson, who traveled a similar route from anti-establishment journalist/publisher to head of an enormous corporate empire, Benson has only changed the externals. Even as he fought the establishment, Benson calculated carefully.

> The idea was that I had killed off a couple of nuclear power plants with a study that I did in New York for the Council on Economic Priorities. And I thought the public needs to know how they can do this kind of analysis in their own community. It was basically a formula that you go through. You ask all these questions and find out how many houses there are and the average size of the house and the square footage and the insulation, and you multiply this out and you get the kilowatt hours and natural gas. Eventually you come up with how much energy the community's using, how much it's costing, where it comes from, who gets paid for it, where the money goes, if you put in insulation how much you'd save, how much money would stay in the community instead of going to the big power plant financiers. And you know it was kind of a radical thing, but I hoped eventually that we'd get all 2,000 counties done. And then we could just put them all together and we'd have a citizens' national energy plan based on the 2,000 counties but done by the people. What a daydreamer. A couple hundred were completed though.

Although Benson can be a daydreamer, he's so practical that he circumscribes his fantasies. Unlike most people who are wild about space, Benson will not readily admit that he wants to go there, and it's difficult even to get him to articulate what his dreams are. At first, he claims that his dreams consist of developing or acquiring thirty practical space products he's identified as necessary to get people living, working, and vacationing in space.

> Some of them are very little things, but they're really important, like stepping stones. If you have this one, it would be easier to do that one. Not all of them are on the same track, not all of them are in series, in sequence one after the other. There are several branches running in parallel that, if you want to do reusable vehicles you have to start small and go this direction with these step-by-step technologies. If you want to do space transportation … you know I mentioned the three things: energy, transportation, communication. Each one of those threads has its own set of required technologies to get us where we are today, to where we want to be tomorrow. We have a corporate slogan: "If we want to go to space to stay, space has to pay."

This is the stuff of which dreams are made? Benson backs up. "The technologies are just a means to an end. The end is exploring the universe." Does he want to go? "I wouldn't mind." He wouldn't mind. Ask most other space enthusiasts and they say they'd go in a New York minute. "When I first got into this I thought, 'Nah, Story Musgrave just retired at sixty-two, I'm fifty-two, ten years, yeah, maybe that could happen. Glenn went at seventy-eight. So I'm fifty-five now, I've got another thirteen years.' You know, I don't think, in thirteen years will they have retirement communities in space?" The author corrects his math. "Twenty-three years, yeah. A quarter of a century? I think that's possible. I hope this doesn't sound negative. I'm just thinking about it while I'm talking. You know twenty years ago was just the beginning of the microcomputer revolution, and look how it has changed the global economy and lifestyles. If there are retirement communities in space that have no gravity, and your bones are getting old and brittle and you've sort of done everything on Earth. Why not go up to orbit and spend the last five years—maybe it would stretch into twenty? If you didn't have gravity to be tugging at your heart and everything else. So maybe that's the way to go—maybe you retire to space." Always thinking glamorously.

Benson returns to 1996.

> I had this little budding idea. I had looked at NASA's near-Earth asteroid rendezvous mission (NEAR), which was launched in February '96. And what struck me about it was the $250 million cost. I thought, "There's a box out there in space that's half the size of that credenza, a quarter of a billion dollars. How?" I mean, my mind won't wrap around a quarter of a billion dollars going into a little box three feet on a side. And this is what triggered me to ask myself, "Can the private sector do this better?" So the first thought when I went to the October '96 Space Frontier Foundation conference, what I had in my mind was just this real crude idea, a private near-Earth asteroid rendezvous mission that would at a minimum see it, size it, characterize it, touch it. I was thinking in terms of something really inexpensive, about the size of a basketball. And by scientifically characterizing it and analyzing it and sending valuable science data back to Earth, we would pay for the mission. And then touching it, which could even mean crashing into it, if it was a purely privately financed mission—no government money or subsidies—then wouldn't I basically be in a position to claim ownership of the thing, since my corporate robotic representative had traveled to it, assessed it, and landed on it?

And with that question, environmentalist Benson had started down a controversial road.

> That's what started me off on the whole property rights issue. This is not a new issue, but I'm the one who popularized the idea of property rights in space through this concept of unsubsidized private risk, our private mission going out to another small lifeless planetary body, assessing it, landing on it, and claiming ownership of it. So that still remains a real high priority to me personally. Can you imagine life or an economy on Earth without property rights? Why not have the same rights and incentives out there in the infinite universe beyond Earth orbit? Without economic incentives, would the Old World explorers have discovered the New World? Space is a place, not a government program. Some people don't like my approach, but then I don't have much respect for people who do not put their money where their mouth is. Anyone can be an armchair philosopher, but it is doers who make things happen, change paradigms, get people thinking in new ways, and not clinging to the past or fears of the future.

Benson's high priority doesn't sit well with everyone in the entrepreneurial space community. Some other entrepreneurs are begging him to drop the claim thing. If he does claim the body, he could be tied up in court for a decade, possibly making it difficult for others to visit celestial bodies. Applied Space Resources CEO Denise Norris, who plans to put rovers and a time capsule on the Moon, provide vicarious experiences for the Earthbound, and bring back soil samples for sale (see Chapter 14), puts the argument this way.

> In the meantime, any rational space power will not allow its citizenry to get any probes near any celestial body where they can create the same type of circumstances. Because God forbid Jim Benson went, that meant that when ASR landed our lander in the midst of

that battle, boom, ASR can turn around and claim the Moon because governments cannot. And that, I believe, is the fundamental reason everything is going to get put on hiatus until they resolve the issue. My great fear is that none of us will get to space simply because of the shortsightedness regarding how the international community is going to react. The Outer Space Treaty is federal law, so it can be actually executed in federal court now. If I objected to Jim Benson's plans, I could sue him in federal court over them today. But I don't want the fight. Sure, we can have a de facto claim if we get there first, and later if we do win the battle and allow whoever gets there first to claim a celestial body or a certain territory or volume, all the people who have their stuff there still have de facto claims. At least we'll still have an ongoing space presence until we get to the ability to fight that battle.

There are people who think we shouldn't even be in space taking soil off the Moon. They're concerned about the common heritage of mankind. Current law is ambiguous about whether a private individual can claim a celestial body. Governments cannot extend sovereignty to those bodies, so U.S. laws are not applicable to the Moon except for U.S. entities on the Moon. So, if I got into a battle with a private Russian entity on the Moon, I couldn't take a case to court—there are no laws governing that situation. However, private individuals might be reasonably able to claim a celestial body by occupation, either robotically or manned. This is a loophole. There's no difference between robotically and manned—people have established grades to what occupation means. The law doesn't distinguish between an asteroid and a moon as a celestial body. Allowing a precedent to be established that would allow a landed robotic or crashed robotic craft to claim a celestial body would imply that any celestial body could be claimed the same way, which means the Moon, planets, asteroids, comets, everything but the Earth.

We know we're going to have to cross this Rubicon. We do not have to cross it today. Nobody in this industry has the financial backing to fight City Hall right now on it. Bureaucracies can always dig deeper into the taxpayers' pockets to fight these battles. A U.S. citizen leaning toward the sympathies of the internationalists or the Third-World economic order can fight this battle in U.S. courts receiving the funding of non-U.S. governments. This type of maneuvering has gone on before.

Do you think the political climate in the U.S. is going to be that they want me claiming the Moon? And here in the U.S. we're about as close to a true free market economy as you can get these days, shy of some very small but not very powerful economies. The case starts, the arguments are made. I can predict that the U.S. will not grant any launch licenses until it's resolved because even though I say I won't claim the Moon, I might change my mind once I get there. So, pending a better space property rights regime, I think getting into a battle is ridiculous at this point. It's only going to damage us getting into space. Jim doesn't get richer by claiming an asteroid; he's not going to bring back iron ore or platinum or gold from the asteroid. He might think he wants to mine it, but the point is he can go mine any asteroid, and the noninterference provision protects most of his claim area. So Jim has no real economic necessity from a business model to claim an asteroid. He doesn't have the money to fight a battle of this magnitude. All it takes is one smart politician who makes it his life to think of things like this to realize the implications of Benson being allowed a precedent to claim an asteroid. On top of it, there'll be a race to get probes to every planet in the solar system to claim those, if Benson's allowed. So they're just basically going to stop everything, and that's my fear—that it will just stop. We shouldn't fight this today. First get credibility, public opinion. Show the public how space puts benefits in their pockets, money in their pockets, platinum on their tables—whatever it is. Helium 3 into their medical diagnoses or possibly some day fusion plants—then go out and push the issue because if public opinion's on your side, the politicians will be on your side.

Once Benson had found his quest, everything else faded away.

In November of '96, I came here to San Diego for my second conference. Mike Simon produced it in association with the National Space Society. During one of the lunch breaks I was sitting at a table with a couple of guys, and I mentioned this asteroid mission concept—after reviewing those weak business plans I'd given up on investing in anyone else. I mentioned this near-Earth asteroid "see inside it, characterize it, touch it" mission, and it turned out that sitting opposite me was Dr. Jim Arnold, the founding director of the California Space Institute, which was formed by [California] Governor Jerry Brown. Jim was the founding director, and also then the current director, because [former astronaut] Sally Ride had just finished her two years of tenure as director. He's also a world-renowned lunar scientist and chemist. Sitting next to him was Dr. Mike Wiscurshin, the director of the California Space Grant Consortium. All the money that comes through NASA into California for student space projects goes through Mike Wiscurshin. So here's Mike; here's Jim Arnold. I told them about my project and they said, "Hey, we would love to do a feasibility study for you. We'll put some undergraduate students together because they're free, we'll get some aerospace industry mentors in the area because there are lots of 'em, General Dynamics used to be here until they sold out to Martin and moved to Colorado—but many experienced engineers stayed behind. And we'll put three professors on it." And that's what they did. From January through August of '97 we

> had about twelve students, ten aerospace industry mentors, and three professors. And they did a preliminary mission feasibility and a preliminary spacecraft feasibility and a cost study. Very quickly it was obvious that I was right. That the private sector can do this mission very inexpensively, for about $25 million.

NASA's Near Earth Asteroid Rendezvous project, which involved sending a probe to the asteroid Eros, was Benson's model. Looking carefully for ways to beat NEAR's $250 million price, Benson and his team noticed that with NEAR's five instruments, the mission averages $50 million cost per dataset returned. "So the going price is $50 million per dataset for a near Earth object science base. So we thought, 'Okay, we can beat that.'"

Benson called NASA Administrator Dan Goldin's office and said he'd like to meet with him and explained why. He was astonished when the door not only didn't slam in his face, but he was referred to the number three guy at NASA headquarters, Chief of Staff Mike Mott, a retired Marine colonel. "Colonels run the Pentagon. I thought to myself, 'Retired Marine colonel. Hatchet man. Chief of Staff. This guy's going to eat me alive.'" So Benson went to see Mike Mott.

> I was just to the part where I said we want to fly a nongovernment-subsidized privately financed mission to another planetary body, and we want to obtain and sell to NASA valuable space science data at a fraction of the cost of a NASA mission, and we want to land on the asteroid, we want to land on it, we want to claim ownership of it to set a precedent for property rights in space, and Mott said, "Stop right there." And I thought, "Here I go. That didn't take long." And he said, "Jim, NASA thinks it's smart. We think we know what's exciting. And we try to tell the public what to be excited about. And we're usually wrong. What you're talking about is potentially the most exciting thing since Apollo. And I'll help you in any way I can." And he did. I think he figured that if SpaceDev were to pull this off, the

> whole awareness and excitement about space would go up and NASA would get more money from Congress. But what was valuable, Mike helped me meet with Dr. Wes Huntress, who was then in charge of this multibillion-dollar per year space science program. Later Mott recommended to Dan Goldin's staff that I brief Dan about NEAP.
>
> I went in to see Wes personally, and we did a conference call to Dr. Jim Arnold, who knows him. Arnold is one of these guys who knows everybody in planetary science. He's highly respected, a highly competent professional, and a really nice guy. So we got him on the phone, and the late Dr. Gene Shoemaker [astronomer after whom Comet Shoemaker-Levy 9 is named] on the phone in Dr. Huntress' office. Gene was a fellow geologist and was scheduled to be the only scientist to walk on the Moon. When Gene developed a health problem, he got Harrison "Jack" Schmitt selected, and we finally got a geologist on the Moon. I'm a geologist, you know. That conference call was the last time I ever talked to Gene—he was killed in an automobile accident several weeks later in Australia. I didn't even get to thank him for the call. So, Arnold and Shoemaker told Wes why this kind of near-Earth object scientific data is so important, what it's going to mean to the scientific community, and that if I was right, it would be worth a try for NASA to use a commercial mission to get three times the science per dollar. When they were off the phone, Wes turned to me and said, "Jim, every day I get people walking through this doorway. And they're always promising something. But the bottom line is they always want something from me—a NASA subsidy or handout. You're the only person who's ever come in here and hasn't asked for a subsidy. And because of that, I'll do anything I can to help you."

Benson makes it all sound so easy in a way that belies all the work that got him and his team to that point.

It was by working with Mike Mott and Dr. Huntress, who is now on the SpaceDev Board of Directors, that I was able to pull together a briefing for Dan Goldin in July. In July '97, we had a videoconference in NASA headquarters with Dan Goldin sitting on my right, and associate administrator Wes Huntress on my left. Next to him was Alan Ladwig, associate administrator for policy and planning. Next to him associate administrator Peggy Wilhide. So I really had three of his top people in there. We had John Lewis on the phone and we had Jim Arnold and all the UCSD students and mentors on the big wide screen in front of us. It was a great briefing. It lasted for about an hour and Goldin was very excited about it and promised to help. It was Wes Huntress who helped most because he ran the Office of Space Science. He changed the Discovery Program around by adding in a concept called "missions of opportunity." If you were a scientist, you could say, "The French are flying a mission; I'd like to put a space science instrument on it. NASA, will you please pay me to take advantage of this mission of opportunity to build my instrument and I'll give it to the French and they'll fly it?" Huntress moved missions of opportunity into the Discovery Program, and redefined them to include private commercial missions for the first time ever. He had Carl Pilcher on his staff send a letter to us that said our proposed NEAP met the qualifications to be considered a commercial mission of opportunity. Wes, knowing we were under time pressures, also sped up the whole Discovery process by six months, partly for SpaceDev, I believe. All of that set the stage for us to go out and recruit universities and science teams to look at our NEAP commercial price list, to consider buying an instrument ride or to buy a dataset, and then write a simple mission of opportunity proposal. Seven universities sent in letters of intent, and three submitted full proposals.

Not bad for an outsider. Benson sees his lack of space industry experience as a plus.

> One of my other strengths, I think, is that I am new to this field and only know what I have been able to read and glean. I came to this field in late '96, but I feel perfectly comfortable in questioning everything, because I don't know, and I found that it's really worthwhile in life to always question your assumptions. If you get into an argument with somebody you respect, and you don't resolve it, you better go home and wonder why. Because maybe you're wrong. Maybe I'm wrong. And I do that. I ask myself if I am missing something. Am I wrong about that? Do I need to go learn something more or drop an assumption? Do I have a bias or prejudice that I didn't know I had?
>
> It's kind of fun for me and kind of aggravating for my engineers. But questioning everything sure has reduced our costs. I always want to know what's it going to cost, whether it is a clean room or a momentum wheel. When I get all the relevant information, reasonable decisions can be made. Tony Spear, the Mars Pathfinder project manager who I asked to do an independent review of the UCSD NEAP mission and spacecraft plans likes to, as he says, "Get all the liars around the same table." So, that's what we try to do—get everyone together and discuss issues face to face. We have another way of getting to the bottom of a controversy, introduced to us by our Chief Technology Officer, David Smith, a thirty-year JPL veteran. He gets the staff together and goes through an exercise where we identify "Fact and Opinion." It is absolutely amazing how much of what people think is fact is actually just their opinion. Once they see it is an opinion, they are then more open to accepting new ideas and possibilities.

As CEO of a small company, Benson constantly walks a tightrope between handling the minutiae of corporate life and coming up with the new products and services that will expand his company. Much of what he does is no different from what other company heads do.

> My typical day is from 8:00 in the morning until 6:00 to 8:00 in the evening five days a week, and then I use Saturday and Sunday to learn new things and to get caught up on e-mail, reading, and document reviews. You've got people that you have to interview and hire, positions to fill, you have payrolls to make, you've got money to raise and products to sell, you've got all the little things that are always coming up that need to be dealt with, while at the same time you're trying to develop new ideas, new products that are practical, simple, needed, and desired that will also be profitable and reliable, and working with the engineers. Sometimes the engineers come from a history of having worked on larger projects, so you have to make sure that they have the right corporate culture in terms of doing things smaller and simpler and less expensive.

Then there's the ever-present demand of looking for new business. Benson and his staff go trolling. They attend conferences, walk the halls of the Capitol, visit the Pentagon, meet Air Force people, and, now that they're known, get invited to present new ideas to the likes of the National Reconnaissance Office. It doesn't hurt that he has thirty-year industry veterans on his staff, people who know people, and know what programs are in place and what people need.

But because he's a space entrepreneur, the rest of what Benson's daily experience diverges significantly from that of other CEO's. After all, it *is* rocket science. "We've got almost thirty employees, twenty-five engineers. Half of them are rocket scientists and launch vehicle engineers, the other half are spacecraft, satellite engineers. In ninety days we put together a rocket motor test firing stand. We built it, we instrumented it, we tested it. And in parallel we designed, built, and tomorrow we're test firing, our first rocket motors. So, the people are motivated, and they're very excited about it now that we're actually getting our hands on things and building things."

Engineers love to build things. It makes them happy. They call it "bending metal." Benson gleefully relates how his engineers tested the rocket motors outside SpaceDev's back door. "The initial rocket motors that we're testing are very small. They're about five inches in diameter and about a foot long. We're literally firing them out of our garage into the parking lot." It's a lot more fulfilling than sitting at a desk running calculations.

Another thing that SpaceDev's engineers have built is a one-kilogram S-band transponder—a radio—for Earth-orbiting missions, but they also slipped in the capability to communicate out to the Moon with it. Benson's engineers built it because it would have been too expensive to purchase one. Now they're not only using it themselves, they're selling it to others. "This is a little radio that only weighs about two pounds, and it has more features than anything available. It's lighter, smaller, and half the price. If we had bought one from a traditional supplier, it would have been too expensive for the budget. We spotted the need for a product, not only for ourselves, but for anybody else who's doing Earth-orbiting missions. Part of managing a company is looking for new product ideas and then getting customers to pay for their development."

And then there's what Benson refers to as "the dark side," the fact that the U.S. State Department has taken over the administration of licensing satellite exports from the Commerce Department. Until March of 1999, the U.S. Commerce Department was in charge. However, following publication of the Cox Report, a congressionally mandated look at technology transfer issues, concerns arose that sensitive launch technology had been given to the Chinese during review of failed Chinese satellite missions, and responsibility was transferred to an extremely slow-moving State Department. Satellite companies complain that this change is hurting their business and U.S. competitiveness enormously. Benson finds himself in their company, oddly being categorized as an arms merchant.

Benson explains.

If two engineers are talking to each other, the assumption is they are transferring technology. So if two engineers are going to talk to each other, one a citizen of the United States and the other a foreigner, you have to have a technology transfer agreement in place or you just have to really be careful and not say anything technical. That's very difficult for engineers.

You fill out forms, and you send them in to the State Department, you make twenty phone calls over six months, you never speak to a person, you leave lots of voice mail, and that sooner or later will pop out the other end, and you go back and forth with some questions and answers and clarifications, and eventually you get the thing, but by then the business may have passed. You've lost the opportunity.

The agreement is specific for each situation. We'd like to be able to say that here's a company that we're going to be dealing with on a regular basis and doing a lot of business with them, and here are three or four categories, none of which are critical from a national security point of view, and we would like permission over the next few years to talk to them. Unfortunately it doesn't work that way.

It's all political. NASA was set up to be a foreign policy tool, and it is still used that way. First to beat the USSR to the Moon, and now to influence the Russian space industry by keeping them occupied with *Space Station Alpha*. So space is related to defense and foreign policy, and commercial endeavors suffer. The issue is, is the United States going to maintain its lead in technology and space and the high ground, and the jobs and the economy and the taxes that go along with it, or are we going to let political shortsightedness derail us and hand over the prize to our competitors? I believe I saw a report that said for the first time in history, more commercial communications satellites were built overseas than in the U.S. This is hundreds of millions lost to our competitors, making them stronger and moving revenues and jobs from the U.S. to Europe. Wrong direction.

It's very tough. In 1998 we acquired a small space company with about forty-five employees in the U.K. The problems with technology transfer caused us to reverse the acquisition a year later. They had very good electronics engineers and really nice space communications capabilities for earth orbiting, European Space Agency-type stuff. We looked at the talent of that company, and we looked at the hardware, the talent was excellent, and the hardware was typical government stuff. It was a little too big, a little too heavy, a little too expensive, but they could make it smaller and less expensive with our guidance. But months went by, and we couldn't even talk to them on how to improve our own products—even though we owned that company, and they were our employees in the country of one of our best allies—we still couldn't talk to them. So eventually we just reversed the acquisition and turned the company loose. They failed the next year—a real shame.

Needless to say, Benson is unhappy about this turn of events.

The U.S. is forming a sort of Chinese wall around itself, or trying to keep the technology in the country, but it's like the barn door's open, the horses are gone. You can't get 'em back again. We're only hurting ourselves. The satellite companies are losing sales in the hundreds of millions of dollars per year. SpaceDev could have more sales prospects—we don't even try to sell overseas today, where half the market is. We can't talk in a technical way to prospective customers overseas. The U.S. has gone from 100 percent of commercial launches down to about 40 percent in ten years—a real economic and national security disaster, and just because of petty politics.

Benson feels that his greatest strength is that he comes from outside the space arena. He knows not only how to start a business, but how to grow and exit from it. That makes him unique among space entrepreneurs. But why? Why shouldn't the space field be rife with experienced business people? "Space has a high giggle factor," he explains, using the term that one hears so often. Benson is used to working under the cloud of the giggle factor.

> I cut my teeth on mainframes in the mid-sixties. And when the first little microcomputers came out, people literally laughed about Apples. Apple? What kind of a name is that? What kind of a computer is that? Manufactured in a garage? The Radio Shack TRS-80—oh, you mean a Trash 80? Microsoft: They were a joke, except to the people who were serious about them, who were manufacturing them and improving them and programming them and writing BASIC and FORTRAN compilers, and people who were buying them. I believe my biggest contribution to the current $100-billion space industry is bringing the microcomputer mentality to an industry bogged down with an old mainframe-like mentality.

What's really keeping things back in the manned space arena, according to Benson, is safety. He thinks SpaceDev may just be able to crack that nut.

> What's holding everything back is that the space plane designers are stymied over the safety issue. Liquid propellant rockets are bombs. They will blow up. Solid propellant engines are bombs and their exhaust is toxic. They do blow up and they kill people. Like they did on the shuttle [*Challenger*]. Hybrid motors, on the other hand (which SpaceDev is working on, see page 252), are environmentally benign and can't blow up. They're storable, restartable, throttleable, nontoxic, non-explosive, transportable. They're not the highest efficiency in the world. But see, rocket motor manufacturers have gone

for ultimate precision and squeezing every last bit of performance out of rocket motors. That's why we're using liquid oxygen and liquid hydrogen. Thousand horsepower turbopumps for the shuttle. That's why it's so expensive and dangerous. With hybrid rockets it's not very sexy, not very glamorous. And most engineers like to work on the latest and greatest thing. But when you look at hybrids, they are extremely practical, only one moving part. The main safety major risk is the fuel. It's only a four- or five-minute ride. Once you're up there, the risk is much lower because you're gliding back to a landing.

But you can't go orbital now, and I don't even want to think about it. I am too busy working on practical things. There are things that I know that I want to do and that could be done if I had enough money, but there are things I just wall off because I know I can't do them. But I'm not forgetting it either. I can't make it happen now, even though from a business point of view I know how to do a reasonable launch vehicle. I just have to wait. I haven't got the credibility, and I haven't got the money. So I have to build up to the point where I can do it. In the meantime I don't torture myself by thinking about it because I've got all these other littler things I have to do in order to climb the stairway to heaven, so to speak.

THE AUTHOR'S OPINION

I think Jim Benson is going about the space business in the right way. He's taking baby steps, each of which should help build his company and make inroads into commercializing space; he's made himself and his company accountable by going public; he's researched his markets; and he's recruited a strong management team. He's succeeded in business before, and it shows. Others would do well to emulate him.

He's innovative. The fixed price list is brilliant. It's not exactly mass merchandising space, but it's making prices predictable, which everyone who buys products and services appreciates. His investigation into hybrid rockets may result in bringing down the cost of getting to orbit significantly.

Benson can be single-minded, driving, and obsessive, but those are all qualities that make a good entrepreneur. The fact that he's willing to question his assumptions is a huge asset that most people's toolboxes lack.

The one area I'm not so sure about is the property rights question. I can appreciate Benson's reasons for wanting to get the issue settled, and I can see Norris' point about holding off until we're better established in space. I think I'm in the latter camp on this one.

—14—

Denise Norris: Marketing Space

We examined space tourism markets in Chapter 5. Here we look at a company that started not with a technology or a mission, but an assessment of its markets. Applied Space Resources makes no hardware, yet plans to engage consumers with activities on the Moon. Can this business model work?

Denise Norris and I have spoken at Los Angeles space conferences and on the phone from our respective coasts. During our last interview, she was dead tired from the combination of work and travel. She'd been preparing a presentation to venture capitalists. It went well.

Denise Norris is the only space entrepreneur who employs an ethicist. She's also an ardent capitalist, admirer of libertarian philosopher Ayn Rand, and eloquent evangelist for space. She's so riveting that you can't help listening to her even if you disagree with her.

Norris' company, Applied Space Resources, plans to land a rover on the Moon, leave public and private memorabilia there, and bring back dirt for sale. She's also going to engage television and Internet viewers with video feeds, the ability to click your own picture, watch your own Moon dirt being scooped up for later delivery to you either raw or in a crystal, e-mail the spacecraft and get customized data back—all either courtesy of sponsors or for low prices. All the technical work will be contracted out. ASR is a "marketing company" whose job is to ferret out what customers want and give it to them. Not your usual rocket shop.

Norris is adamant about what her company is and isn't. She's small, specialized, and happy to be so. "ASR's not going to build a spacecraft," she says, almost delightedly. "We're going to turn to people who do these things. I would say that the *Lunar Retriever* [her lander] when built will probably have the input of 100,000 people in it. We'll buy an engine off the shelf, but think about how many thousands of people contributed to the final design of that engine throughout the years. We're riding on the shoulders of giants." Such expansiveness is quintessential Norris. We're all in this together, not for individual glory, and we each have a role to play. "The Boeings, the LockMarts [Lockheed Martin]—they're absolutely necessary, but they're not in the business of space exploration and development. However, if we're successful, Boeing will be very interested in building spacecraft frames for us."

WHAT HAPPENED TO THE MOON ROCKS?

According to Space.com, September 27, 2000, the disposition of moon rocks from the Apollo missions was as follows:

711 pounds are in NASA and military vaults
60 pounds have been studied and returned to NASA
24 pounds have been lent to museums and schools
22 pounds were destroyed in experiments
15 pounds are out for study
7 pounds were used in experiments but not destroyed
0.6 pounds became gifts to foreign heads of state
0.078 pounds were lost

As for some other space startups, Norris believes that they've got a more complicated and risky business model because they're trying to do too much. "This is where some of my competition might be missing the boat. I refer to them as micro-Boeings. They want to build the launch vehicles and the spacecraft and do the exploration missions; my company doesn't need to rebuild all that stuff." There will be no shortage of hardware, in her view. What's missing is someone who will use it, and she's happy to be that someone. "As we create more demand, companies like mine will become markets for Boeing; it's in their or LockMart's interest to make us more of a market. I don't see the need to become a vertical integrator because there will be someone out there willing to service my needs. So I don't make as much money—big deal! As Boeing develops the technology I need, someone else will also be able to use that technology. Maybe it will be one of my competitors, which means not that I should try to stop them but that I should become a better business."

Norris' unusual ideas have grown partly out of her admiration for libertarian philosopher Ayn Rand, who advocated acting in one's "enlightened self-interest." "I start sounding like a Randite, and it gets me scared sometimes," she admits. "Rand's philosophy of pursuing one's self-interest can achieve the betterment of the whole much better than collectivism, which advocates gutting yourself for everybody else. It's a fundamental truism, even if she did have some errors in it. Everybody on our team has read Ayn Rand and understands the basic concepts behind her approach."

Until recently, Norris lived her life as something far different from a space anything. She was a computer consultant till 1997, when she read an article that changed her life. "There was an article in *Scientific American* called 'Buck Rogers, CEO,'" she relates. "In it they describe that when lunar dust is sold at auction, if you extrapolate the price, one kilogram would bring one or two billion dollars. Being a good capitalist, that immediately got my ears up. Make lots of money, that's a good thing." This is revolutionary talk in the space community. Money is the last reason people become rocket engineers or astronauts.

"Some of us are going to end up being Carnegies and Rockefellers. I can tell who's not, but I can't tell who is, just from listening to how they think about doing this. I like to think it's going to be me." She won't say who's not—it wouldn't be fair.

> Projects which have inadequate business plans or technology will not thrive; they will drop by the wayside. But I'm very hesitant to point to [anyone] and say, "Your project sucks." To open up space is not about one person. We need as many participants as possible in this dance to actually get to where we want to go as a group. Some of them will succeed, some will be absorbed, and some will wither. I see my role and ASR's role in the great dance that way, but I also sit back and say, "My company is going to do it." If I have to I'll change my plans if they're wrong, but I want to be that person who gets to that great attractor, space—or who blazes the path to the great attractor.

Profits or no, space has been Norris' passion for most of her life. She just couldn't figure out how to build it into her career until now. "I remember where I was on July 20, 1969 [the date of the first Moon landing]," she recalls. "I was one of the people who was actually following *Apollo 13* even though I couldn't get television coverage. And I was gung-ho NASA. My mom took me to see *2001*. But I think after the return of the shuttle, the first landing, I kind of lost interest in the space program. There's a reason behind that. I was saddened that the shuttle was a compromise. And what NASA was doing on a lot of levels was not the excitement that I had been promised in the sixties and seventies, was not the Moon bases, was not the space station. We let *Skylab* fall simply because we were arguing politically about the space shuttle. The original goal was for the space shuttle to go up and reboost Skylab and that would become the start of our home in space," she says sadly.

> Seeing *Skylab* fall and seeing the calamities of the space program, the *Challenger* accident, just kind of made me shake my head and walk away. I still loved

space, but I didn't see NASA as the answer. The biggest disillusionment was after the *Challenger* explosion in '86, we endured a multiyear hiatus of anything in space. Yeah, shuttle missions go up and down and became routine. In a larger sense the shuttle is a nontrivial problem; it's very complex to put a shuttle up, but it ceased being exciting and glamorous. NASA continued to be increasingly bureaucratic, and it just wasn't what had excited me during the Apollo era.

It wasn't until working with Jay Manifold, my cofounder, and realizing that here we have a way to go to space and not be a government program, that I really started to regain my outward enthusiasm for space and re-fan the fires of space. That was because we realized we could do something that fundamentally changed the model about how humanity perceives itself in relation to space, which is not to need a government to do these things.

So Jay and I—Jay's background is in physics and communications—sat down and did the back-of-the-envelope calculations, just as a thought experiment. How could we go bring back Moon dirt and make a ton of money? And we first came up with all those great science fiction concepts, like a mass driver on the Moon that would launch the stuff back, not taking into account that we'd flood the market considerably and wouldn't get those prices, we also realized that there's no way we could do a mass driver because it would have to be 100 yards long. So we then looked at the Soviet Luna sample return missions of the seventies and we realized that that was a very adequate model for bringing back fresh samples.

We spent from September to December '97 doing some quiet market research, saying, "Is this a good enough idea?' Is it doable from a technical and a marketing position? What do we really think we can get for regolith [Moon dirt]?" All without really making a stir. Jay and I were incredibly paranoid that this was such an obvious way to do things that somebody else was doing it. One of the big concerns was trying to find out if anyone was doing a similar project. During that time we became reacquainted with Dave Gump's project and Lunacorp and the Artemis Project. During this time frame Jim Benson made his announcement about his asteroid mission. So we added Beth Elliott [her public relations person] to our project, and we incorporated, and started to more publicly test the market. If someone else was doing it, it wouldn't be doable. There's not enough of a market for two missions to try this at once. And we started getting involved more with the conferences and making a public statement that we're going to go back to the Moon and we're going to go get this dirt.

We're doing a very simple mission despite everything else we may talk about, media tie-ins, all these other things, the basic mission is real simple, it was done by the Soviets in the seventies with ten-year-old technology at the time, which makes it thirty to forty years old now. It's a very technically doable mission, and the market is very simple. We sell some Moon dirt, though we recognize that the media opportunities are greater than the sale of the dirt.

People have liked what they've heard from us. They've liked the fact that our company is a marketing company, not a rocket company building paper rockets [rockets that exist only as designs on paper], that our primary concern is identifying terrestrial and hopefully later space markets, and then finding the resources to address those markets. If there were no market, there'd be no project. People like hearing that. We were one of the first companies out there talking not about how great our spacecraft is—"Look at these detailed engineering diagrams, look at these orbital dynamics." We were out there saying "Look at this market." A year ago that was unheard of.

We sat down for lunch with Alan Binder [director of the *Lunar Prospector* mission, which searched for water on the Moon], and we were talking about projects and stuff, and he was dead set against logos on spacecraft or anything else, it had to be pure science. We see in the past year, though, that he's kind of like a barometer. We hear him talking about logos on the mission to help pay for it and other people starting to talk about the market, asking how you generate the money.

The "Buck Rogers, CEO" article was the catalyst. I'm absolutely proud to say that. You know a lot of people would think, "You're showing that you're only interested in yourself, isn't that greed?" You know, bottom line, at some level greed is good. It's not like Gordon Gecko's [an Ivan Boesky-like trader in Oliver Stone's movie *Wall Street*] "greed is good" motto, but we call it responsible capitalism where the capitalistic organization has a responsibility to make sure that the markets and services are kept healthy and are able to generate the capital to consume the products you're generating. A lot of times when people talk about capitalism they perceive that companies are just going to gut their market; that's classic shortsightedness. One of the lessons I learned from Rand and reading about Japanese business was "Think big." You're not just there to take money out of your market. You're there to nurture your markets and make them think of ASR as benevolent influence. Your market's going to spend money with someone; it might as well be you.

These days Norris finds herself in the undreamed-of position of being surrounded by groupies. "There are groupies out there. 'You're doing something in space? Oh, I worship you,'" teases Norris.

You get groupies. People who think it's really great what you're doing, may or may not have the courage to

> do it themselves, but who really identify with what you're doing. It's weird to have groupies. A lot of people speak disparagingly of space weenies and use that term, but these people are the salt of the space enthusiast movement. They are the grass roots. We would not be where we are without these people, willing to just hold the flame. It's weird to watch this process. When we originally announced the *Lunar Retriever*, there were people who called us wannabes, and now people are telling us that they think we have the most doable mission of all the ones being talked about. It's becoming more and more of a consensus. That's a kind of spooky thing. It's also a real kind of responsibility. These people are seeing what I'm doing, what ASR's doing, and then putting their hopes for space on our accomplishing something. I don't want to let them down. That can wake you up in the middle of the night sometimes when you realize, "My goodness, we're starting to become something of a force within this groundswell, and if we blow it, if we do something stupid, we could in theory set back the movement also."

Norris may be a reluctant icon, but she understands the power of media. Like Greg Bennett (see Chapter 10), she recognizes the entertainment value of going into space, which translates into big bucks. She also understands that there is no such thing as bad publicity.

> If half a billion people watch the live television broadcast hoping the spacecraft will crash, I'm just as well off as if half a billion people hope it lands. The mission's fully insured, so if it crashes, all my investors get their money. Even if it crashes I make money—all the advertiser revenues. Secretly I would be tickled to have the bookies in Las Vegas place odds against it landing. That'll drive people to watch it. Why do we watch NASCAR races? To see them slam into the wall. We have

a camera [on the spacecraft] that's downward looking. If it crashes, you'll see it.

Entertainment. Actually, more edutainment. We'll teach 'em stuff. We want to develop a national curriculum for schools that'll be tied with the various broadcasts so that students can do experiments, mostly aimed prior to sixth grade. Our secret agenda is to give them the space bug before they're too old, because once they hit puberty they have no interest in space; they're interested in each other. Every space activist you talk to got the bug before they hit puberty. So our secret agenda to control the world—no [she laughs]—increase the market and the potential and the enthusiasm for man being in space—we can achieve it if we will it. We just aren't willing it. Twisted, but I think it's a fairly good assessment. Some people just don't understand that. They think the benefits of being in space are self-evident. Every rocket company that makes a paper rocket and then wonders why no one wants to do anything with it is assuming that the benefits of their paper rocket are self-evident.

You have to explain. For example, take helium 3 for MRIs. Make the benefits evident by saying, "Hey, you can find and diagnose colon cancer quicker if you use helium 3 in your diagnostic process, and guess what—you can't get helium 3 except by making nuclear weapons or going to the Moon." Is that self-evident? No. The people who build the rockets don't understand how the average Joe thinks. It's evident to them, so it must be evident to everyone. That's a problem I find with a lot of very intelligent people. They're unable to empathize with people who sit at the middle of the bell curve. We have to remember that we have to talk to them, not to each other. The ideas that actually get to space will be the ones that suddenly start attracting the most attention.

And those may be completely unexpected. For example, who ever thought that cell phones would be considered fashion accessories, and that people would buy them in order to look cool? Had they been marketed that way initially, they might have flopped. Or that e-commerce would grow like Topsy because of pornography? It may be that we get into space because some fad we haven't dreamed of and never could have planned for catches the public's fancy. If that happens, it's possible that Norris will be the first to spot it and cash in on it.

Like many other space enthusiasts, Norris became interested in space because of the Apollo program and science fiction. "It's Walt Cunningham's fault," she laughs when asked how it all started. When she was in first grade, her teacher put on the TV, and she watched Cunningham (*Apollo 7* astronaut) launch. That, coupled with *Star Trek* and every other good or bad schlock sci-fi show that came along gave her the space bug.

> It was the majesty and the amazement of launching people into space. It wasn't anything intellectual, it wasn't anything about understanding the future of humanity, it was just totally awesome, magnificent. I accept that it's nothing rational—it's a totally irrational desire on some level. I live in a world where I have to apply rational thought to whatever I do. I have to, or I'll never get to space or do anything else. So when I'm confronted with something for which I have to accept an irrational explanation, I'm comfortable saying, "This is irrational and it's one of the things I accept." What adults are capable of doing—and I have a fourteen-year-old so I'm seeing it in action now as she changes—is tempering what we desire with a good dose of what is real and applying that desire to create—and I think healthy people apply that desire to create something that not only betters their own condition,

but as a result betters everyone else's condition. A little bit of Ayn Rand coming out there.

The philosophies of Ayn Rand, a Russian refugee who settled in the U.S. and contributed major works like *Atlas Shrugged* and *The Fountainhead* to the American literature canon, drive many private industry space advocates. Rand devised a philosophy called objectivism, which has become more and more popular since her initial essays and other nonfiction writings on the subject in the 1960s. Rand died in 1982, but her ideas thrive. The most important aspects of objectivism in the context of this book arc politics, ethics, and value theory. In a nutshell, Rand advocated the following:

Politics

- *Individual rights.* Since people have to deal with each other voluntarily, any action that violates the consent of any party is immoral. The use of force is legitimate only when one is protecting one's life, liberty, or property. People are by nature entitled to rights of life, liberty—including the right to justly acquire, own, and trade property—and the pursuit of happiness.

- *Limited government.* Government should do only those things that are necessary for protecting individual rights: police, courts, and national defense. Every other function of government currently in place is morally invalid if it is supported financially by involuntary means and it forbids the honest and peaceful conduct of business.

- *Laissez-faire capitalism.* The only just social system is that of a free, voluntary exchange of goods, services, and ideas. There's no such thing as a victimless crime, no centrally planned redistribution of income or delivery of education, healthcare, transportation, food, retirement income, or housing.

Objectivists are opposed to wars on drugs and other attempts to legislate behavior according to one group's tastes. People have the right to be self-destructive if it doesn't hurt others. Any governmental action that aims to attain a social objective through regulation compromises the freedom to which we are entitled.

Rand declares that we have the right to live by our own minds, in our own ways, and for our own sake. That means that we must refuse to recognize the rights of others to demand from us either unearned money or unearned time.

Ethics and Value Theory

- *Naturalistic value theory.* To make a value concept valid, we must evaluate it by asking for whom and what a particular action is valuable.
- *Rational egoism.* People should think for themselves. In fact, it is the individual's moral responsibility to "look out for number one." Relying on others for one's existence—society, government, etc.—transforms a person into a helpless parasite who demands that other people make sacrifices on his behalf. Nevertheless, people should not look out only for number one. We are morally obligated when we assume responsibilities to others. Altruism, the practice of acting for the benefit of others by sacrificing one's own values, is to be rejected. However, it's all right to do things that benefit others out of benevolence or compassion or charity. It is not okay to benefit from someone else's sacrifice.
- *The trader principle.* You cannot attain well-being by force. You must use reason and must produce and trade in order to survive. If we pursue our true interests and make production and trade central activities, we're not being

selfish. We're not trying to grab the biggest piece of pie in a zero-sum game; rather we're trying to make bigger pies for everyone. A person must uphold promises and contracts—it's in his or her best interest to do so. By being honest and just, we're advancing the smooth working of free markets and strengthening society. We also must contribute to our community.

I would say that *Atlas Shrugged* answered so many philosophical questions that I struggled with through my twenties. I kept trying to understand the idea of looking after my own self-interest as opposed to what I kept getting, the bombarding message of self-sacrifice, self-sacrifice. I wouldn't call myself a Randite and say that I'm ready to go run off and pray at the temple of Ayn, but I certainly appreciate how she contributed to clarity of thinking. The only problem is she never addressed children, and I think that's the biggest flaw in her whole theory because it falls apart when you look at the parent's willingness to jump in front of an oncoming bus to save their child. So part of what I went through when I dealt with Rand was to try to understand that we are irrational creatures struggling to become the romantic character she paints in her novels. In essence that struggle is what makes us so great—we can conceive of something better than ourselves and we can actively work to achieve that betterness.

Like Rand, who extolled industry and machines, Norris is a metal head. "I love technology," she admits. "I work in computers.

I once had a contract with Pratt and Whitney who make commercial jet engines up in Connecticut. I got to crawl inside these engines just for the fun of it—it was amazing. I love that stuff."

Norris is also a transsexual. Everyone knows it, and nobody talks about it. She's an imposing 6 feet 4 inches and speaks with male inflections in a dark brown man's voice. She wears makeup, has long, dark blond hair, and dresses like a woman, though at times with a distinctly tomboy flavor. She has a distinctly pretty, if not feminine, face. (She is definitely not in the Tony Curtis category.) In no way does she resemble a drag queen, the thought of which doesn't interest her in the least anyway.

Norris feels that her transsexuality, which is obvious, hasn't affected how people deal with her, at least not very much, and not that she's aware of. She figures that only 20 percent of people are going to have an opinion about her as a transsexual in the first place, and half of those are going to accept her without even knowing her, just because she is transsexual. "The other 80 percent of the people—I really work on making a good impression, being an intelligent person," she says. "I designed advanced computer networks for large international financial firms. If the gender stuff was an issue, I'd never get in the door. I mean if I came out like something out of the Queen Mary in L.A.—the Queen Mary, not the boat—you know, those types of places [a club for drag queens], yeah, I'd probably have trouble. But I work very hard to come across as people."

She does admit to selected unpleasant incidents, however. "There was an entrepreneur at a space conference who went haywire—how could they let someone like me do this. And the conference organizers basically said, 'Screw 'em.' The person still comes [to the conferences], I'm still friendly with him, he doesn't know I really know the full scope of what he said. I don't care."

Another time, an employee of the City of New York complained about Norris using the women's restroom during her stint as an independent contractor. "Their whatever they call it, their thought police, the people who check out biases and discrimination claims, hauled me in and then I felt like I was grilled, even though I'm postoperative. See—legal, passport, birth certificate. And I felt so trashed after talking to the people. And they're like, 'Well, as long as you're that, you're fine, but you know ...'" Norris brightens right up, like a news anchor who's just delivered bad news and is ready to move on. "That was the only problem I ever had, and it was more misunderstanding. The laws protected me right away, and the city respected the legal limits. But that was the only time."

Aside: Yes, they do indeed change the birth certificate of a transsexual, at least in New York State. "New York State feels that since transsexuality is a natal issue, it's not a lifestyle choice, that it's something that should be corrected back to birth because it's an undiagnosed issue." New York City is not quite so definite about it. "The city just leaves 'indeterminate,' or leaves it blank. The Federal government is phenomenally simple. I got a new passport. Here's my surgeon's letter, here's my new birth certificate, here's my old passport, here's my new picture. 'Oh, okay, no problem.' They have procedures for it."

Norris adamantly refuses to make herself out as a victim, though she does admit to some uneasy thoughts at times. "It always colors things when you're doing a presentation, how will someone's prejudices interfere." But she reminds herself that "if someone wants to discriminate against me because of a preconceived notion, that's not somebody that I want to even know. And it's a big wide world out there. I don't have to convince the people who don't like me to like me." She's happy with herself, she works hard, and she's got other things to think about. "I don't try to hide it, I don't try to pass. I look presentable, I comb my hair, I brush my teeth, I wear clean socks, and I change my underwear. Isn't that the criteria for a decent human being? That's what my mom taught me."

She makes it sound as though she's had a fairy-tale life. She emphasizes her past success as a computer consultant to Fortune 500 companies. She's very proud of her fourteen-year-old daughter, who's already becoming an entrepreneur selling trading cards. Custody problems? Bah. Yeah, she's had them, but all's well that ends well. "I'm a parent. I had to fight very hard in the court system for my right to be a parent. My daughter's friends know I'm transsexual, and they think, 'Yeah, no big deal.'" In fact, try to get Norris to say something bad about just about anything or anyone and you'll be disappointed. If she's got a problem with the way things are, she'll try to change things, or she'll go about her business. But complain? That isn't her style.

Norris has intrepidly chosen to compete in an almost exclusively male field. Try to find other women to profile in a book like this, and you'll practically have to turn over rocks to find them. There are a few, though not very many in leadership positions. There's Randa Milliron of InterOrbital Systems; Pat Dasch, executive director of the National Space Society; current and former astronauts Sally Ride, Shannon Lucid, Mae Jemison (who now runs her own consulting firm helping developing countries), and Eileen Collins; a number of Russians; now-retired Donna Shirley of JPL; Brenda Forman, who wrote a column for *Space News* and worked for Lockheed Martin until 1998, has now retired and has turned her attention elsewhere. You'll also find women in "soft" roles, like Gloria Bohan of Space Adventures, Colette Bevis (see Chapter 1), and space nurses/doctors like Dr. M. Theresa Verklan. Lori Garver, Associate Administrator for Policy and Plans at NASA, is that agency's most visible female representative. Andrea Seastrand works for the California Science and Technology Alliance.

However, there are no women X PRIZE contestants, no women presenters except Norris at venture capitalist seminars, almost no women in the pages of *Space News*, and none on major programs like that of the U.S. Space Foundation. Listen to presenters at space conferences (largely male), and you'll quickly see why. The language immediately deteriorates (okay, loaded word) into tech talk. Equations, jargon, and the unfortunate but common mangling of English practiced by engineers, test pilots, and military people. Most women find this not only incomprehensible, but off-putting. They don't relate to images of missiles and aircraft the way men do. They become discouraged when they see their dreams, inspired by *Star Trek* and Carl Sagan, reduced to such indignities. They want to be there, to experience all that space has to offer. The hardware for doing so is just a tool (as are computers to women, who tend to want to *use* rather than understand them), not an end in itself. And so they remain frustrated.

The fact that there are few women in the nitty gritty part of the space field, that is, on the technical side, comes as no surprise to anyone. The U.S. National Science Foundation (NSF) publishes statistics year after year quantifying the phenomenon. As of 1997, the latest year of data available at press time, the National Science Foundation found that women comprised 32 percent of American scientists and just under 10 percent of engineers. They made up just 22 percent of physical scientists, which represents just 14 percent of all scientists. (Women are much more interested in social and life science than physical science.) That means that women physical scientists make up a mere 3 percent of American scientists. Since space depends heavily on the physical sciences and engineering, those statistics show that it's pretty darn woman-poor. (Science & Engineering Indicators 2000, Table 3-9, www.nsf.gov/sbe/srs/seind00/start.htm). And it's a lot better than it used to be.

Men and women diverge in perception as well as employment. The same publication by the National Science Foundation concludes that a significant gender gap exists regarding perceptions of space exploration. American men value space exploration by almost fifteen percentage points over American women: Thirty-one percent of men and 19 percent of women think the benefits of space exploration strongly outweigh the costs.

A year 2000 study conducted by the Siena Research Institute, a branch of Siena College in Loudonville, New York, found an even wider split: 42 percent of men and 23 percent of women believed the U.S. space program had provided an "adequate economic return." (One wonders how they'd feel after hearing Patrick Collins' argument in Chapter 5.) Seventy-three percent of men and 52 percent of women felt that the U.S. space program had provided an adequate scientific return.

The NSF provides some insight as to what women are focusing on instead: new medical discoveries, environmental pollution, and local school issues, for a start. And while men find these issues important, they're far more interested in the use of new gadgets than women. An index of interest ranks men at 72 out of 100 on that point; women come in at 59.

By the same token, women don't feel very well informed about space (an index of 29 as opposed to 44 for men) or business conditions (necessary for succeeding at commercial space enterprise) or military policy or the use of new technologies (49 for men, 38 for women). What *do* they feel they know? Local school issues, followed by new medical discoveries. These findings confirm the popular wisdom that American women are largely interested in health, welfare, and family issues, while men gravitate more to business, government, and technology. No news there.

Level of public interest in selected policy issues, by sex and level of education: 1999

(Mean index scores)	**All adults**	**Male**	**Female**
New medical discoveries	82	77	87
Environmental pollution	71	69	74
Local School Issues	71	65	75
About new scientific discoveries	67	70	64
The use of new inventions and technologies	65	72	59
Issues and business conditions	65	69	61
Military & defense policy	64	68	60
International and foreign policy issues	53	59	48
Nuclear energy to generate electricity	55	58	52
Space exploration	51	59	44
Agricultural and farm issues	47	45	48

How well informed Americans think they are about selected policy issues, by sex and level of education: 1999			
(Mean index scores)	**All adults**	**Male**	**Female**
Local school issues	58	54	62
New medical discoveries	53	50	56
Issues and business conditions	50	57	43
Environmental pollution	48	48	48
About new scientific discoveries	44	50	40
Military and defense policy	44	51	38
The use of new inventions and technologies	43	49	38
International and foreign policy issues	40	47	34
Space exploration	37	44	29
Agricultural and farm issues	33	34	32
Nuclear energy to generate electricity	29	34	24

These data and anecdotal evidence (walking into an engineers' bay at an aerospace company or examining the roster of speakers at a space advocacy group conference) support the assertion that women eschew space because of its techie image. And yet lots of women like science fiction. Lots of women write science fiction, though admittedly, they gravitate more toward science fantasy—the dragons and warlocks genre. (Of course, not all science fiction is about space.) Many women are put off by the fictional images of them looking bodacious and seductive out there. More pressure to be something they're not, and denial of their interests and capabilities. Space tourism surveys too contradict the evidence. For example, Patrick Collins' Japanese survey found no significant difference in the responses of men and women. Is it inevitable that women aren't going to be interested in making space happen?

Barbara Brown, an occupational health specialist who has a master's degree in space studies from the University of North Dakota Aerospace Center, thinks it's a question of how women are portrayed in popular culture. "I believe, like anyone else, women are very interested in space; they just may not be interested in the institutions of space because they've been so excluded and can't conceive of themselves as, say, astronauts," she says. "But all the popular culture ways that space is depicted are extremely popular with both sexes [science fiction literature, *Star Trek* TV series and movies, etc.]," says Brown. "There is latent demand for space in females. What's needed is more realistic images of women going about their business in private space businesses and in space. And women in leadership positions. Women from all walks of life, not just techies."

Another problem for women, according to Brown, is that space has largely been associated with the military. Most women can't relate to the issues, the terminology, or the mindset. The military is highly regulated, conformist, and hierarchical. Women tend not to gravitate toward hierarchical arrangements when left to their own devices. Many of them don't like the idea of war and weapons and think there's a lot more pressing business to take care of.

Brown talks about her educational experience. "I recently finished a master's program in space studies from an institution that was awarded a NASA space grant. One of the space grant objectives is to recruit and train professionals, especially women, for careers in aerospace science and technology. My class was predominately military personnel. There were four women in my class of twenty-four, three of whom were civilians and one who was in the military. Conversations at social functions usually centered around military issues, rockets, and missiles, none of which have ever been part of my life's experience."

Brown cites Cornell University Professor Margaret Rossiter, who, in *Women Scientists in America,* notes that "NASA officials, who were mostly former military and Naval personnel, showed themselves to be among the most sexist and discriminatory in the federal government when they refused to let women become astronauts in 1962. Attitudes haven't changed much in the intervening forty years," says Brown. She recalls sitting in a 1997 meeting at NASA Ames in

Northern California, where she was manager of occupational health. "The participants were told that all NASA centers had to conduct 'parity studies' to compare themselves with the hiring practices of private industry. We were told in a meeting that the institution would have to hire 200 women to come into 'parity' with private industry. It seems that this issue came as a surprise to our management. I remember thinking that I was aware of the institution's masculinist culture after one week of employment."

Norris discounts the problem for herself, though she realizes its existence in general and wants to see it change. "I work in computers. It's just as bad [as space]. I'm a well-paid computer systems developer. I design heavy-duty infrastructures for large corporations. Maybe I have an 'unnatural advantage' in my 'lifestyle choice.'" She pauses to make sure that the sarcasm is noted. "Because I've seen both sides of the fence. I know how to interact with men, and I think I do a fairly good job at interacting with women. And not necessarily making too many jarringlike, [spreads legs and sits back in chair with man's attitude] 'So, Paula, tell me ...' Smoking a cigar, having a brandy, I can do that, but I'm comfortable being not necessarily, as I said, the Queen Scary transvestite image."

Any entrepreneur who wants to establish a permanent presence on the Moon, Mars, or an asteroid is sooner or later going to confront the issue of property rights. To what am I entitled, and what are my responsibilities to others who come after me? It's a complex subject (see Chapter 18)—one fraught with treaties, international and national law, precedents, and ethical issues. To debate property rights is to re-create basic political and economic ideological arguments. Private rights and economic incentives vs. the common rights of mankind. In many ways, the issue of celestial property rights mirrors those involved with Earth's oceans and seas, but there are additional analogs like homesteading, which doesn't apply to the ocean.

Here Denise Norris assumes an unexpected stance. While well versed in and willing to discuss the issues and particulars, in general she wants to wait to see how things shake out. Her view is this: Let's not claim anything for now. We're too vulnerable as an industry. Let's wait until we're better established because any lawsuits or negative publicity could stop all of us in our tracks, maybe even kill the industry. "We take our space property rights position because we try to think that far ahead and in that large a picture," she explains.

Like Greg Bennett of the Artemis Society, who's also trying to get to the Moon, Norris doesn't feel compelled to go herself—yet. "My daughter's 14. It won't be until she's on her own and on her feet. Before she was conceived, yeah, I would have gone into space. But being a parent certainly tempers the way you look at the world. And I need to know that my daughter's on her own. Certainly I couldn't get life insurance if I went into space right now. I don't necessarily have to be the next person to walk on the Moon. I'd love to, but the other side of the coin is a lot of neat things are going on here on Earth too."

THE AUTHOR'S OPINION

I like the way Norris has gone about putting together a business plan. She's done her research, realizes she has to be market-driven, and has adjusted her plan and regrouped when necessary. It's refreshing to hear her say that ASR is a marketing company, not something other space companies do. She likes technology, but she's not obsessed with it. She's gotten into her business because she sees a market, not because she likes to fool around with gadgets.

However, I'm not entirely convinced that her plan is going to work. I fear that the public will become jaded too quickly, especially if some other company gets to the Moon first. I recognize that Norris herself acknowledges this fact and knows that she has only one shot at making a stir before she has to move on to something else.

A word about the "women" issue. I've had readers tell me that although my analysis of women's attitudes toward space is interesting, it doesn't belong in this chapter because Norris wasn't born a woman. I disagree. She's a woman now—a woman working in a field dominated by men, facing issues that women in such situations face. I would be remiss in failing to explore those issues.

—15—

Charles Miller: Legislating Space

Boeing and Lockheed Martin employ lobbyists in Washington to influence legislation so that it's favorable to them. But who speaks for the space-interested public? ProSpace, that's who.

In a way, ProSpace was responsible for this book. At a Space Frontier Foundation Conference in Los Angeles, ProSpace board member Jeff Krukin invited me to come to Washington to lobby Congress with the organization. I immediately dismissed the idea. Too much trouble, not my thing, journalistic objectivity—every excuse I could find. But Mr. Krukin is nothing if not persuasive, and I was soon immersed in the legislative side of space. That led to my meeting with ProSpace's then-president, Charles Miller, and the rest is history. Miller and I spoke over the phone for much of this interview even though we first met in person at a Foundation conference. When we first spoke he was living in Ohio. He has since relocated to the San Fernando Valley, my old stomping grounds right over the hill from Los Angeles.

In the late 1970s, two senior L5 Society leaders attended a U.S. congressional hearing in hippie garb. One of them stunned the attendees when she lifted up her blouse and began breastfeeding her baby. A short while later, another space advocate lobbied Congress while wearing *Star Trek* Mr. Spock ears. Many years later, people on Capitol Hill remember these incidents. The ones who were there were not amused.

That's one reason ProSpace, a citizens' space lobby founded formally by Charles Miller in 1996, insists on a strict dress code when its members talk to legislators and their staffs. Men must wear suits or blazers, women suits or business dresses. They even tell you what kind of shoes to wear and how much jewelry is okay. The group, which astonished itself and just about everyone else with its initial success in creating a $25 million program back in 1995 before it was a legal entity, isn't taking any chances.

March Storm, ProSpace's annual trek to Washington to brief and persuade legislators on matters relating to commercial space, grew out of political intrigue and infighting among space advocacy organizations, of which there is no shortage to this day.

The space advocacy community, which is relatively small, consists of several organized groups, the most visible of which are the National Space Society, the Planetary Society, the Space Frontier Foundation, the U.S. Space Foundation, and ProSpace, a very select group that requires its members to participate in Washington lobbying. Each group adheres to a particular philosophy and performs a specific function; though hedging their bets, many people belong to more than one of them.

The Planetary Society was founded by the late Carl Sagan. It supports robotic exploration of the solar system and international cooperation in doing so. It also likes the idea of a humans to Mars program based on cooperation between the United States and Russia.

One of the National Space Society's predecessors, the National Space Institute, was founded by rocket pioneer Wernher von Braun. The other, the L5 Society, was a grass-roots organization that grew out of the ideas of the late Princeton University professor Gerard O'Neill, advocate of floating space colonies. The National Space Institute focused on the government as the source of space activity. It received money from the aerospace community, but it lacked excitement. The L5 Society was more radical. Founded essentially by hippies, it had lots of energy and vision but a flaky image with the establishment. In the mid-eighties, the two organizations got together and decided to take the best from each—relationships with the establishment and money from the NSI, and grass-roots activism from L5—and merge them. And so the National Space Society was born. The NSS promotes the human settlement of space, and it doesn't care if that happens via a government program or privately, through international partnerships or unilaterally.

The Space Frontier Foundation was founded by disgruntled National Space Society members Rick Tumlinson, Jim Muncy, and Bob Werb. Its members are extremely active and believe in commercializing near space (everything between the Earth and the Moon) through private efforts, not government. On the other hand, they believe NASA, the dominant government space agency, should be out on the fringes exploring and making way for human settlement and private enterprise as it spreads outward.

The U.S. Space Foundation is a staid organization whose members hail primarily from the establishment: large aerospace companies and government. The group's purpose is to bring together the sectors of the space community and serve as a credible source of information about the practical use of space.

And then there's ProSpace, an organization that proudly asks, "Is one week of your life worth $3 million?" Even Charles Miller, who founded the organization, has been astonished at ProSpace's successes, and Miller is no stranger to political success.

In 1990, Miller, whose mother is one-eighth Native American, was approached by a Native American friend of his mother's from high school. Her tribe needed help with computers, which the former Cal Tech science-student-turned-business-major agreed to provide. It turned out that the tribe needed more help than that. They were having trouble obtaining government funds to help them survive, and difficulty in negotiating contracts with the U.S. Bureau of Indian Affairs. They had planned to hire an ex-staffer from California Senator Alan Cranston's office, but that didn't work out. The ex-staffer's loss was Miller's gain, as well as that of the seventeen tribes he ended up working for. Miller was hired as a planner/coordinator; the result of this collaboration was that in 1993, Congress awarded the seventeen tribes $1.7 million per year rather than the $1 million they were seeking—$100,000 per year per tribe in perpetuity. For tribes that were used to receiving $20,000–$40,000 per year, this victory was heady stuff. It completely changed the future of tribal communities throughout northern and central California.

How did Miller pull off such a coup? He told the tribes that rather than relying on others, they should brief Congresspeople themselves. He put together a position paper for them stating why they should be funded. He told their story—one of dastardly deeds done to them—and suggested a course of action to Congress that included generous grants. The win garnered the notice of tribal leaders all over California, who said, "How did you do this? What happened?" Their elders had been trying to do something like this for decades and had failed.

But Miller's success was not without controversy. Even though his mother is part Native American, Miller is white (he is adopted). His being so as well as being prominent didn't sit well with some tribal leaders; as a result, Miller became uncomfortable with his position. So when the tribes held an organizing meeting at which they nominated him to be chairman of a California task force for tribal leaders, he declined. Sure, he'd help out, but he didn't want to be in a leadership position. They insisted. Miller agreed to be vice chairman, but that wasn't good enough for the tribal leaders; he and a tribal leader ended up being co-chairmen of the "California Tribal Base Level Funding Committee."

They spent two years lobbying and succeeded in changing U.S. federal Indian policy. They got the Bureau of Indian Affairs and Congress to agree that there should be base-level funding for small tribes across the United States—many on postage-stamp pieces of land tucked away and forgotten. More than 300 tribes would benefit. The result? Tens of millions of dollars per year for these impoverished tribes and Miller's adoption by the Buena Vista band of Mewuk Indians, who call him "Peach" for his figurative yellow outside and red inside. And Miller learned about lobbying.

Miller's involvement with space began early. Growing up in Durham, a small rural town in Northern California, the boy would lie in his backyard and stare up at stars. He felt wonderment and a certainty that with so many billions of stars and galaxies, there had to be other life out there. Maybe there was even an alien "boy" staring up at his own stars and thinking the same thing. It was on a night like this that Miller decided that at least part of his life had to be about space. He concluded that humanity's future depends on our expanding into the universe. And he wanted to help make it happen.

So he went to Cal Tech with the goal of becoming an astronaut. There he discovered that he didn't like science and math as a lifestyle. He's a gregarious fellow who likes being around people, and he found that science and engineering are like being locked in a room by oneself. "They slide a problem under the door," he says, "and when you've solved it, you open up a trap door and slide back the solution." After three years of the best and worst of times at Cal Tech, Miller switched to Chico State University, also in California, and got his bachelor's of science in finance and business administration.

During that time—the early 1980s—he became connected with the L5 Society through a friend. He started L5 chapters at both Chico State and Cal Tech. He immersed himself in grass-roots space activism, first by getting involved in and then becoming a leader of a coalition of state chapters called the California Space Development Council. He was noticed by people at the national level, and when he graduated, the National Space Society hired him as its number two staff person.

Miller, who goes by the name Chaz, stayed for one year. Never one to be shy about his opinions, he disagreed with the number one staff person about how things should be done, and their relationship suffered. Miller laments that back then he didn't have the skills to change the relationship (he does today, he says). Miserable, he left and went back to California with the intent of finding a job in Silicon Valley. He never got there. His mother's Native American friend called, and he was off on a journey that would last five years.

In 1994, Miller returned to space activism. After his experience with the California tribes, he realized that citizens had been expecting and waiting for the government to solve their problems for them, and that wasn't going to happen. His conviction deepened when George Bush the elder's call to return to the Moon failed to inspire action. When the Republicans took over the U.S. Congress in 1994, this new way of thinking began to permeate American politics. It was time to act. Miller would not even try to convince government to undertake another Apollo-type program. Instead, he would attempt to persuade government to facilitate the private sector's efforts to push outward.

And so Miller came to D.C. the week after the Republicans took over, met with space staffers, and announced that he wanted to organize a week-long lobbying event. He'd invite citizens from all over who care about opening space. This group would lobby for cheap access to space and commercial space legislation. His idea met with some enthusiasm. Nick Fuhrman, an aide to the new chair of the House Science Committee, Bob Walker, loved the idea and suggested that Miller organize a seminar on Capitol Hill for members and staffers as well as the would-be lobbyists. Miller went off and got the National Space Society, the Space Transportation Association (a group of aerospace and defense executives), the California Space Development Council, and the Space Frontier Foundation to organize a "21st Century Space Policy from the People" seminar. Then he invited people to come to D.C. to lobby for the X-33—a reusable launch vehicle prototype—cheap access to space, and Bob Walker's Omnibus Space Commercialization Act, which was first introduced in 1991. The bill would do several things:

- Require the federal government to purchase commercial launch services
- Create a tax credit for purchase of a commercial space company's stock
- Eliminate capital gains on commercial space common stock (and other things)

Little did Miller realize the political intrigue that was about to explode.

When the National Space Institute and the L5 Society merged, the fledgling National Space Society ended up with a humongous board of directors—thirty-two people. Miller recalls sadly that their primary activity was fighting over who would control the organization. Part of the problem was that two organizations with conflicting visions had merged. Another had to do with the huge board, which Miller believes acts as an impediment rather than a facilitator. But another factor, charges Miller (and a number of others), is the culture of the organization, which directs that you really aren't anyone unless you're on the board. "Treating a director spot as a privilege is not what you want to do," says Miller. "You want privilege to come from accomplishment." The organization isn't a meritocracy, and it should be. That's why a lot of good people started leaving, he says. Member Henry Vanderbilt started the Space Access Society, a small group of rocket scientists and entrepreneurs who write and phone their Congressmen on issues related to space transportation and bringing its cost down. Tumlinson, Muncy, and Werb started the Space Frontier Foundation, with which Miller later became associated. Miller clashed with the NSS culture because, he says, he wasn't willing to give his vote over to a faction; he wanted to do things, and the leadership was just looking for people who would support its faction in the ongoing internal political fight.

So it was with some negative history that Miller invited the National Space Society to work with the other groups on the seminar. Miller, who was working independently rather than as a representative of any one of them, declared that all the organizations would get equal billing, but two weeks before the event, he got a call from an apologetic staff director of the House Space Subcommittee. She had been given orders that the seminar would be managed by the National Space Society and only the National Space Society. It turned out that Bob Walker, the Chair of the House Science Committee, was scared by "radical" organizations like the Space Frontier Foundation. He was afraid that these "grass-roots revolutionaries" would insult someone and "tip over the apple cart." After all, the Space Frontier Foundation had tried to kill *Space Station Alpha* because it felt it was anticommercial.

At the same time, Miller and the NSS had an agreement for NSS' role in the week-long lobbying event. NSS interns were to set up appointments for the volunteer lobbyists to brief members. Fearing that somehow something could go wrong and the NSS might be blamed, the organization's executive vice president got cold feet and insisted on retaining complete control of the project. Miller explained that no, that wasn't possible. The NSS pulled out, figuring it would collapse without NSS support. With less than two weeks to go and no meeting arranged, this was a likely scenario.

Miller was determined to proceed with the event anyway, and decided to make all the calls himself. He and his secretary set up about fifty congressional meetings from Miller's California office. The seminar was held: six members of Congress attended, plus many staffers. One of the Congressmen was Newt Gingrich, who told the seminar attendees they weren't thinking big enough—Mars and the Moon were just the tip of the iceberg. Gingrich also backed the privatization of space, starting with the space shuttle. This kind of thinking hadn't been heard from Congressmen before, but it started a trend. Within six months of Gingrich's pronouncement, the United Space Alliance—a joint venture between Boeing and Lockheed Martin—was created to operate and maintain the shuttle. Miller had found a powerful ally in Congress.

Nine private citizens showed up to brief fifty-two congressional offices, and Miller's little band achieved some major victories. Despite initial opposition from the chief staffer of the House Defense Appropriations Subcommittee, Miller was able to effect a compromise, call in a political favor from an influential subcommittee member, and end up with $25 million for reusable launch vehicle development for the Air Force, which hadn't even requested it. Miller was ecstatic. He and his ad hoc assembly had created a brand new $25 million program. Is it any wonder that in March of the next year, 1996, forty people showed up to lobby Congress under Miller's aegis?

Miller realized that he needed to reorganize to handle events on this scale, so he approached a national organization, the Space Frontier Foundation. They responded positively. With Miller now on their board, the Foundation sponsored the second March Storm. But because lobbying wasn't the main function of the

Foundation, whose purpose was to change the broad public conversation about space, a new organization was needed—one that would change inside-the-beltway conversations about space, a goal that required using different words and methods to reach a different audience. Miller and some people from the Foundation incorporated ProSpace in July of 1996.

Miller reserves high praise for the Space Frontier Foundation. "The Space Frontier Foundation has had a big impact on how we think about space," he explains.

> Most space groups have thought of themselves as space supporters, which puts you in a box because you're supporting someone else's agenda, either that of NASA, the White House, or aerospace companies. The whole thing tends to have a strong overtone of "more money for NASA." They're not vision-focused, they're not taking responsibility for promoting their own agenda because they don't have enough power. They feel that they need more power and influence before they can become really influential, so they put themselves in a box—the same box that most of the American public has been in. Many of these space organizations are dominated by money from aerospace companies and executives, so no big change can occur. We wanted to open the frontier for all people, and what was being done would not get us there, but most people didn't want to hear it. The Foundation would say, "Apollo never again. We're going back to stay." I thought that was great, but other people got hung up on "Apollo never again" and got upset.

It's not surprising. The Foundation's ideas are complex and counterintuitive to many people. The least misapplication of language, and they can be seen as against space altogether. They're not. And they've learned from their early experiences. Now they choose their words much more carefully. Miller continues:

> People were focused on the past. The Foundation was focused on the future and what it was going to take to get there. The organization has a very good understanding of the power of words and language. What they do is called "cultural cruise missiles." It's a phrase for something that sticks in people's minds, that grows, makes sense, and they pass on. The Foundation put the phrase "cheap access to space" back in the popular lexicon. The space community has agreed that this is the number one priority, and then even Dan Goldin came to say it's his number one near-term development priority.

On the subject of NASA, Miller offers not only opinions, but insider insights. He honestly feels that the agency's shrinking budget is more a reflection on who's president than on what NASA is doing. The issue of NASA's poor return on its investments is a separate one.

> Tom Rogers (see Chapter 2) says, "Take all the money taxpayers have put into space—over a trillion dollars—and show me the one Nobel Prize winner that's come out of it." NASA argues about spinoffs, but the bottom line is that their projects are a very ineffective way to invest money to create useful technology. We'd get much more bang for the buck giving money to the National Science Foundation or the National Institutes of Health. The space program didn't invent Velcro or Teflon or the cordless power drill, as some people assert. Myth is larger than truth.

It certainly is. The fact is that NASA didn't invent Tang either. Miller understands how government works.

> NASA lowballs the cost to sell the program. You either tell a white lie or you don't get what you want. This is one reason Americans don't trust government. Goldin

> has done some great things in overcoming some of this with faster, better, cheaper. Also, NASA has a policy that any program that exceeds estimates by more than 15 percent is cancelled, and he's done this a couple of times. [The exception is *Space Station Alpha*.] Many space activists refer to Goldin as the Gorbachev of NASA. He's bringing revolution but can only take it so far. He's never worked in a real commercial industry. His background is with TRW, with the government as a customer. He's doing really well considering his background and experience. We're expecting him to understand commercial industry, but no one who works at NASA has ever worked in commercial industry. Some of them acknowledge this.

It doesn't matter much within Miller's world view. "Commercial space is growing by double digits each year. If advocates put all their energy into increasing NASA's budget, the best they can hope for is a 3 to 4 percent per year increase. Which one of those makes sense? Commercial space revenues are now as large as world government space activity. We should put our efforts into a place where growth compounds. Goldin is beginning to realize this."

Once ProSpace was official, Miller began to rely on an e-mail list he had compiled of 20,000 to 30,000 people interested in space. The group sent out Internet alerts asking recipients to write or call their legislators about an issue. A specific opportunity soon arose. Late in 1996 the House passed the Commercial Space Act, a revised version of Bob Walker's 1991 Omnibus Space Commercialization Act. The Science Committee came to Miller and asked if ProSpace could help them get the bill through the Senate—within four to five days. Miller hand-delivered three separate flyers to all 100 Senate offices: one to environmental staffers (about commercial remote sensing), one to telecommunications staffers (about GPS), and one to space staffers (about reusable launch vehicles). It was close. They almost got the bill through the Senate that year, the year Bob Walker retired. Walker had fought tirelessly, but ultimately he failed to build sustained support. It took a public-interest group, ProSpace, to get it passed. In 1998, the bill became law.

Anyone can attend ProSpace events. You don't have to be particularly knowledgeable. You don't have to work in the space industry. In fact, it's better if you don't, because you'll be lobbying to get specific items funded, and if your company benefits from those funds, you'll be involved in a conflict of interest. No, ProSpace is a citizens' lobby, a vehicle through which ordinary people who care about space can make themselves heard on the subject of space policy in the U.S.

When you sign up to attend March Storm in Washington, D.C., ProSpace will begin to prepare you immediately. Weeks ahead of time, you'll receive a number of backgrounders explaining the issues as the group's leadership sees them. According to the organization, if you read these materials you'll be better informed about space policy than nine out of ten congressional staffers. ProSpace tends to libertarian views that support the vigorous involvement of private enterprise in space, but there's plenty of room for NASA and other government activity as well. Nevertheless, you'll be urged to think carefully about what you're reading and ask yourself how the material relates to your personal vision of the future. ProSpace will encourage you strongly not to accept its philosophies blindly, but to come to your own conclusions. In fact, at your in-person training on the first day of March Storm, you're supposed to hand in 100 words communicating your personal vision of the future and "why we need to open the space frontier for all people." This personal vision will help you clarify and crystallize your own thinking so that you can understand yourself and communicate your rationales effectively to legislators.

Backgrounders present and past include the following, among others:

- ProSpace's previous year in review. The list of successes is designed to inspire and motivate. The failures help to set out a new agenda.

- A reprint of an article on space commerce and space entrepreneurs from *Fortune* magazine. This piece serves to acquaint ProSpace volunteers with the players and some of their activities. There's a lot about financing and the role of government spending in fostering private space activity. (Basically, the small companies aren't taking money from the Feds, but the large companies are.)

- Excerpts from "The Significance of the Frontier in American History" by Frederick Jackson Turner. This short paper helps to acquaint the novice with the concept and importance of the frontier as ProSpace leadership sees it.

- A preview of the House version of the 2001 ProSpace-drafted tax credits bill, not yet introduced in the Congress. The group drafted the bill, then took it to Capitol Hill to find a sponsor, which they did, in the person of Congressman Ken Calvert of California.

And more. You get the idea. The point of the papers is one, to inspire, and two, to educate.

When you get to Washington, you'll receive a full day of rigorous training that includes role-playing as well as more education on the issues. ProSpace will have set its agenda for the week, which will comprise only a few items to keep its message simple. In 2001, the agenda was:

1) Lower the cost of space transportation by stimulating demand and helping companies developing space transportation systems by:

   a) Providing investment tax credits for developers.

   b) Ending funding for NASA's Space Launch Initiative (SLI), which ProSpace feels will not work. SLI calls for meeting NASA's routine space operations needs, including Earth-to-orbit launch, by giving the business to the private sector. ProSpace sees SLI's contractor selection process as unfair and unworkable. By focusing on NASA's needs alone rather than on those of general space transportation, the program could exert a chilling effect on promising approaches before they're even off the ground (because funding will go only to those companies that support NASA's objectives). In addition, NASA has already made choices that have eliminated many companies—choices that were not supposed to have been made for several years.

c) Stimulating the launch market by allowing private companies to resupply Space Station Alpha and setting up reasonable procurement policies that give all companies an equal chance at securing contracts and by government investment in precompetitive transportation component technologies rather than in specific vehicle architectures like the X-33.

2) Privatize space shuttle operations.

3) Increase funding for space solar power research and authorize a space-to-Earth power transmission demonstration.

4) Help mitigate the threat from Earth-crossing asteroids by increasing funding for asteroid searching and studying human and economic repercussions of impacts from near-Earth objects.

As you might expect from people who care passionately, the issues are hotly debated via e-mail and computer bulletin boards ahead of time and during the event. Sometimes participants disagree with the official stands. At other times, they simply fail to understand the issues completely. Such disagreements can lead to potential participants dropping out, but then, there's nothing new about any of this. That's politics.

During each March Storm, ProSpace arranges a meeting with key NASA and/or congressional officials, often at dinner. This event allows participants to exchange ideas and questions directly with people to whom they ordinarily would have little or no access.

WHAT HAS PROSPACE ACCOMPLISHED?

Today, ProSpace enjoys a high level of credibility. Participants who brief Congress now get warm receptions. Legislators and staff hear from the aerospace community and NASA all the time, but not from the people, and not about what's really best for the people. So they welcome hearing from citizens. There was some resistance at the beginning, but now legislators have started coming to ProSpace for information, and they're taking action based on that information. The organization also wrote the 1999 Commercial Space Transportation Investment Incentives Act, which would grant tax credits to investors in new space transportation technologies. The bill didn't pass that year; it was reintroduced in 2001 but as of this writing has yet to be passed. However, the ProSpace-backed Commercial Space Act of 1998 did pass.

The 1998 legislation streamlines regulations and institutes policy to promote a stable business environment for the commercial space industry. For example, it directs NASA to study commercial possibilities for *Space Station Alpha* and to further privatize the shuttle, which is already being operated to some extent by United Space Alliance, a joint venture between Boeing and Lockheed Martin. It allows commercial space vehicles to reenter Earth's atmosphere and return payloads to Earth—something that, believe it or not, previously was outlawed. It requires the U.S. government to purchase space science data from the private sector if such is available. And it requires the government to procure commercial transportation services for government payloads.

The group has enjoyed additional successes, such as:

- Funding for space solar-power research: $15 million in 1999, $8 million in 2000, and $8 million in 2001. This funding was acquired despite opposition from Dan Goldin.

- Future-X: $37 million in 1999, which was used to create the X-37 space vehicle.

In 1998, Miller stepped down as ProSpace's president and eventually left the organization entirely. When he founded a commercial space venture that year, he knew that if he stayed, he'd face a conflict of interest. He's now CEO of Constellation Services International, Inc., a company that plans to repair satellites in low-Earth orbit at a cost far lower than that of replacing them.

THE AUTHOR'S OPINION

Miller has come up with a winner—a citizens' organization that's effective in advocating on behalf of commercial space. It's too bad that space legislation is so complicated, though. The heavy issues that people need to weigh through in order to participate in ProSpace can be daunting.

It's difficult pulling together a lobbying organization in which all participants agree on the issues and the approaches. ProSpace has to be commended for doing such a good job of uniting its members and hearing out diverse points of view.

—16—

Rick Tumlinson and James George: Advocating Space

When you first heard about the Russian space station Mir being refurbished and made a private tourist destination, did you wonder how such a thing came about? Wonder no more. The Space Frontier Foundation, profiled in this chapter, thought up and executed the whole thing. Who are these guys, and what else are they up to?

At this point I've spent an awful lot of time around Rick Tumlinson, but James George is actually the one who'll sit still long enough to talk to you. Some of George's interview here was conducted in my favorite Los Angeles area Indian restaurant on Ventura Boulevard in Sherman Oaks, some at Jerry's Deli in Studio City, and some via e-mail.

The small but sassy Space Frontier Foundation was endowed in 1988 through a grant from Walt Anderson, the reclusive telecommunications billionaire who is also behind the founding of MirCorp. An official imprimatur was stamped on the organization with its first membership check, which came from Robert Heinlein, the famous science fiction author. Even so, the organization is seen as a bit of a rogue in the space community, mostly because of its pro-business stance and the strong unforgettable personalities of its leaders.

If Anderson is the body, Rick Tumlinson is the face of the Foundation. Both are quirky, but both are changing the world. At forty-five, Foundation president Tumlinson is so intense that you can almost see the calories burning up in his body. He favors black clothing, often punctuated by a white shirt. Bearded and ponytailed, he invariably looks neat and professional. He's half English, but he grew up in West Texas, where it's almost as dry as Mars.

Even though the Foundation was started by three people (Tumlinson, Bob Werb, and James Muncy, a consultant in Washington and former aide to California Republican Congressman Dana Rohrabacher), Tumlinson is the one you most often see and the one you remember. His speeches can make you cry. He's so skillful with words that he could persuade a fuchsia to bloom in winter. At the same time, he can so anger people that shouting matches ensue. At the Foundation's 1998 conference, Tumlinson and Alan Ladwig, formerly NASA Administrator Dan Goldin's right-hand man, got into it following one of Tumlinson's many attacks on NASA for stonewalling humans-to-space projects and monopolizing space vehicle development. Ladwig, who is no shrinking violet himself, told Tumlinson to cut the criticism. Tumlinson said he would when NASA changed its ways. Amazingly, Dan Goldin showed up to address the conference the following year, announcing a change of heart and direction for NASA: The agency would support space commercialization by turning over *Space Station Alpha* to a nongovernmental organization when it's ready and by forging a public/private partnership in the exploration and development of space, though with qualifications. The two sides had begun to come to an uneasy truce.

ANATOMY OF A SPACE ADVOCACY ORGANIZATION: THE SPACE FRONTIER FOUNDATION (SFF)

Founded 1988
President, Rick Tumlinson
Executive Director, James George
www.space-frontier.org

SISTER ORGANIZATIONS:

ProSpace (lobbying)

FINDS (Foundation for the International Nongovernmental Development of Space)

Annual conference attracts 500 space businesspeople

BELIEFS:

Private enterprise should develop near space (everything in Earth's orbit, including the Moon), while NASA should explore far space (everything beyond the Moon).

GOVERNMENT SHOULD:

Be a good customer for services offered by private companies.

Explore; people should settle.

Set laws, create tax and investment incentives, streamline regulations to support safe and swift development.

Act as a cheerleader.

Create opportunities from which the private sector can bootstrap (transport, service, fuel and data purchases, prizes).

Invest in long-lead technologies beyond the scope and abilities of the private sector.

GOVERNMENT SHOULD NOT:

Finance private space projects.

CHARACTERISTICS:

Dominated by Libertarians.

Angel funded.

Loud, messianic, effective, practical.

Ambivalent relationship with NASA and current space industrial paradigm.

Pro-frontier rationale.

Funds astronomers worldwide who search for and track near-Earth objects.

STRATEGY:

To wage a war of ideas in the popular culture for a new American space agenda.

ORGANIZED AROUND PROJECTS:

- Cheap access to space
- Alpha town (the first human town in space, on Space Station Alpha)
- Space Station Alpha Authority (project to establish an independent authority to operate Alpha)
- Keep Mir Alive (a completed project with the deorbiting of Mir)
- The Watch (tracking near-Earth objects)
- Return to the Moon
- Space solar power
- Space shuttle external tanks (reclaiming and using them on-orbit for other purposes)
- Space Enterprise Project to train entrepreneurs and match them up with investors

- Arthur C. Clarke awards for best portrayal of private space efforts in feature films and digital short films, and for best screenplay depicting same

ACCOMPLISHMENTS:

- Successfully negotiated with the Russian Space Agency and Russian space firm Energia for the lease of Mir by MirCorp. Unfortunately, the Russians ultimately decided to deorbit the station, and MirCorp has since turned its attention to other projects.
- Brokered the deal by which private citizen Dennis Tito was to spend a week on Mir via MirCorp. (Once Mir was deorbited, Tito switched his destination to Space Station Alpha, an arrangement he made with the Russian company RSC Energia, which was 60 percent owner of Mir.)
- Convinced NASA Administrator Dan Goldin that Space Station Alpha should be privatized and commercialized and that NASA and the private sector should partner to explore and develop space.
- Popularized the ideas of cheap access to space and space as the frontier.

A skillful politician, Tumlinson never strays from his message: that space is a place, not a program, that NASA should emulate Lewis and Clark and pave the way for settlers, and that near space is the province of people, not governments. He writes article after article, gives speech after interview, and he always, always says the same thing. If you've heard him two or three times, you've got it down pat. Frontier, Lewis and Clark, the near frontier should be handed over to universities and private firms to explore and develop for human use, "no more flags and footprints," NASA should get out of the operations business, we can have a future without limits, welcome to the revolution. Then the specifics: *Space Station Alpha* should be handed over to a private landlord (or a consortium, or a public-private partnership, or even a private international consortium) and commercialized to further opportunities for the private sector in near space; space should be declared an enterprise zone; the government should become a good customer and avoid handing out sweetheart deals and picking winners and losers (by awarding loan guarantees); the shuttle fleet should be turned over to private operators who are allowed to carry anything or anyone they want; NASA should explore Mars and the rest of the solar system in such a way as to set the stage for human settlement; NASA should support research that's beyond the financial reach of the private sector, including reusable launch vehicles and technology demonstrations; the NASA centers should be privatized so that they can act as true research centers and incubators for new space industries. Each pitch contains language that conveys a sense of urgency while at the same time praising his adversaries and whipping up his audience into a frenzy of inspiration. Tumlinson used to be a marketing executive, and it shows.

For example, take the following excerpt from his opening address to the Foundation's "Return to the Moon Conference" in the summer of 2000:

> The reason I had to come up with this division [Near vs. Far Frontier] was so that I could sleep at night. I was, like many of you, in constant conflict and facing continuing disappointments as I watched the failed and confused interaction of the public and private sectors in space. It was making me nuts. It seemed that all of space belonged to the government. Belonged to NASA. I love NASA like everybody else, and I want to see them do great things in space. But there is a problem with them. I do love NASA, they are part of the American family, but if I had a brother or sister or son or daughter like NASA it would be like having a heroin addict living in your house. NASA is addicted to the control and power they get from ruling space. Now this didn't come about because they did it to themselves, but as a result of a

> lack of visionary leadership at the top of our government here in the U.S.A. NASA is a bureaucracy, and like any bureaucracy without leadership it hunkers down and begins to live for itself, expanding and defending its turf, and for NASA that meant all of space.

Tumlinson once worked for the late Space Studies Institute scientist Dr. Gerard K. O'Neill, author of *The High Frontier*, which sets forth how humanity can establish orbital colonies. He also worked with preeminent physicist Freeman Dyson, who once proposed "hijacking" comets and using them for commercial purposes. He helped found FINDS—Foundation for the International Nongovernmental Development of Space—and Lunacorp (see Chapter 10), where he served as both marketing executive and director. As early as 1992, he was calling for the U.S. to buy or lease *Mir*, likening it to the purchase of Alaska, which was also pooh-poohed in its day.

Tumlinson thrives on confrontation. He loves shocking people, as he did in a 1997 *Space News* piece where he screeched that only three major bastions of socialism remain: China, Cuba, and *Space Station Alpha*. He flies a pirate flag that has been to *Mir* in his office. He also wrote an outrageous e-mail to Mars Society members in August of 1999, right before the organizations' second conference. Summary: "Mr. Zubrin (see Chapter 8) has asked me to resign and is not allowing me to speak at the conference." Tumlinson appreciates the beauty and promise of space so intensely that he feels his pain, frustration, and anguish over the obstacles that keep him and the rest of humankind from reaching and "enjoying" it physically. This physical passion, one that may be felt in the gut, shoulders, or other parts of the body, is palpable among space advocates. Most are not as eloquent in expressing it as Tumlinson. Like the verbally facile but less polished Zubrin, Tumlinson knows his mind, and wants others to know it as well.

But disputes like the one between Tumlinson and Zubrin are common within the space community, where egos swell and personal philosophies are embraced as part of one's identity. Even polished practitioners like Tumlinson fall off their horses now and then. Like it or not, space is intensely political, whether private enterprise- or government-based.

The Foundation has its fingers in so many pies it'd be the envy of Jack Horner. It's no wonder that Executive Director James George works till 1:00 A.M. most days. It holds lunar development conferences (together with other organizations like the International Space University) and a yearly conference that attracts space businesspeople, runs a public outreach project aimed at establishing a dialog with environmentalists about solar power from space, oversees a program (The Watch) that aims to mitigate the threat from Earth-crossing objects, lobbies to establish an international quasigovernmental authority to manage *Space Station Alpha* so that NASA and the other Station partners can be freed up to explore far space, sends telescopes to Africa, and more. One of the Foundation's new projects is its Enterprise Program. This badly needed brainchild of James George helps train and educate new entrepreneurs and put them in front of investors. Corporate sponsors like aerospace giant Boeing pony up funds to help make it happen. You'll often find George and his colleague Bob Hillhouse of the California Space Development Council networking at organizations like the Los Angeles Venture Association (LAVA) and Vojo, a technology networking group also in Los Angeles.

A new Foundation program involves a bit of Hollywood glamour. The group has introduced a series of awards named for science fiction author and space visionary Arthur C. Clarke. All of the awards are meant to recognize filmmakers and screenwriters who depict space as a place where private citizens and private enterprise go. There's an award for best feature film, one for best digital short film, and one for best screenplay. Films and scripts that focus on governmentally controlled space and government space activities do not qualify. The first awards were presented at an event at Hugh Hefner's Playboy Mansion in Los Angeles.

"NASA should not be a landlord." Foundation Executive Director James George is referring to the idea of NASA operating *Space Station Alpha*. "They should not be a trucking company." He's referring to the shuttle.

> They should be the explorer. They should be out there pushing the limits. They should be doing probes. What they really want to do is go to Mars. Well, they don't have the budget for Mars and no politician is going to say, "We should do a mission to Mars. Here's $500 billion. Go do it, guys." It's not going to happen. But if NASA is able to give up those two things, suddenly they have about $9 or $10 billion a year, easy. They could start doing these things. In stages they could do interesting cool stuff. Doing habitats and testing them out in orbit, on camera. They could be doing a lot of different things with a budget of $8, $9, $10 billion a year that would be freed up. Plus they could do a lot better probes.

This kind of talk is quintessential Space Frontier Foundation. They worry about NASA's monopoly, about its being subject to political moods, and about lack of opportunity for everyone else. Meanwhile, the aerospace establishment sees the Space Frontier Foundation as fringe. Wendell Mendell of NASA (see Chapter 11) comments on the organization: "The SFF is a very small group of people. They have an inordinate amount of influence because they're hooked into Dana Rohrabacher. They are sort of libertarian in the sense that they are looking for a pure capitalist economy and they argue that very forcefully, and normally you wouldn't pay any attention to them except for the fact that they can influence certain kinds of legislation."

And they put *Mir* back on the map, at least for a moment, and their conference attendee lists read like a who's who of the private space movement, and they get a lot of press, and Dan Goldin spoke at their conference in 1999, and other important NASA officials turn up at every conference, usually as speakers, and they presented the Senate Space Roundtable for 2000 in partnership with ProSpace and the staid Space Transportation Association. They brokered the deal to get now-famous civilian space traveler Dennis Tito to *Mir* and are working with both Russia and the U.S. to make sure other private citizens can go into space as well. Fringe? I think not.

THE AUTHOR'S OPINION

Okay, so I really like these guys. The fact that their politics and mine don't mesh is irrelevant. I like their ideas, I find their passion infectious, and I admire them for being doers.

I also happen to like NASA a lot. I won an award from it once, and it means a lot to me. I get upset when I see the Foundation and NASA yelling at each other. I want to play Rodney King and ask them if we can't all just get along.

There's one thing about the Foundation—well, two things—that I have a problem with. First, I think it should moderate its divisive rhetoric. Actually, I have noticed it toning it down in the last year or two; it doesn't seem so angry any more. This is good. It seems to me that you don't make progress by alienating people. The other thing, and this is something that will not change because it's at the core of its philosophy, is this frontier thing. I have to agree with Kim Stanley Robinson and Rene Schaad here (see Chapter 19). The idea of the frontier doesn't sit well with non-Americans; not even with some Americans. Perhaps the idea of "expanding our horizons" would work better.

Two artist's renditions of luxury resort architecture firm Wimberly Allison Tong & Goo's space resort design. The structure will simulate gravity in some portions and retain zero gravity in others. Courtesy Wimberly Allison Tong & Goo.

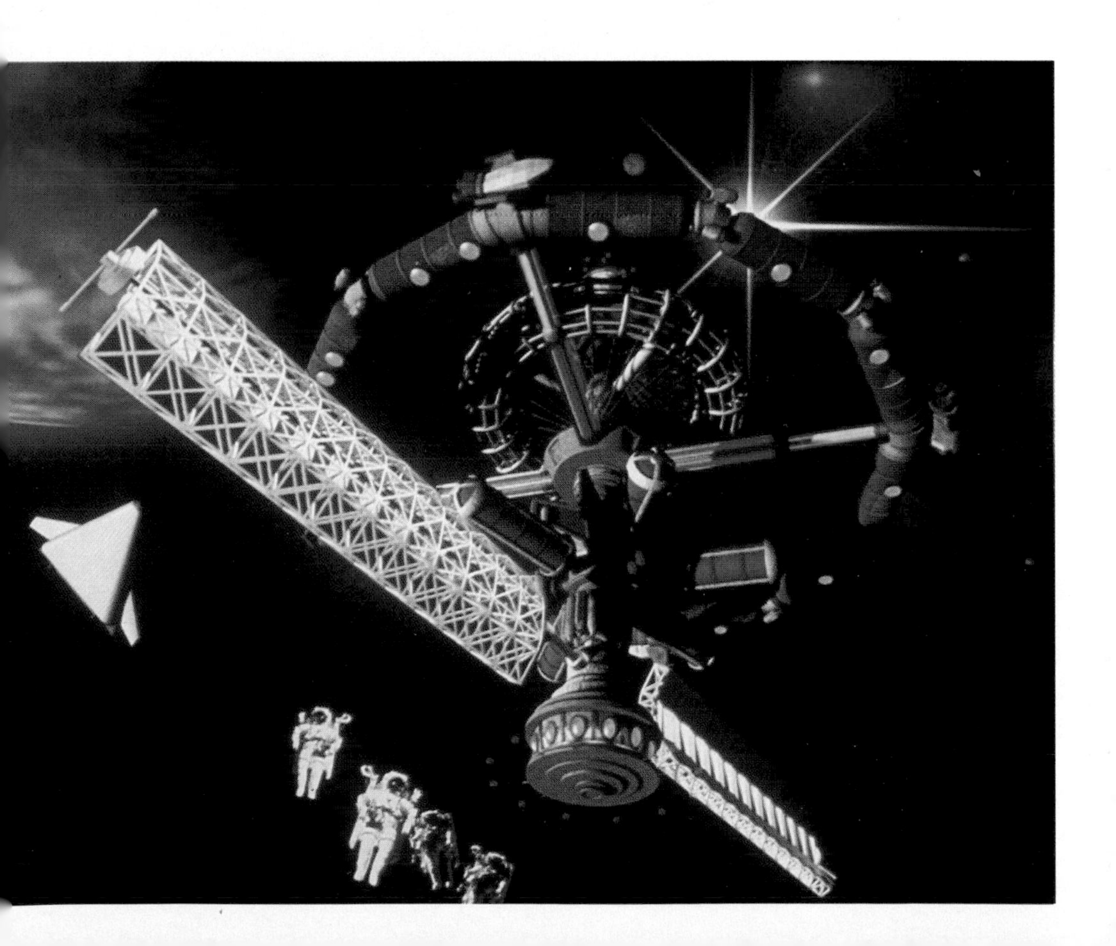

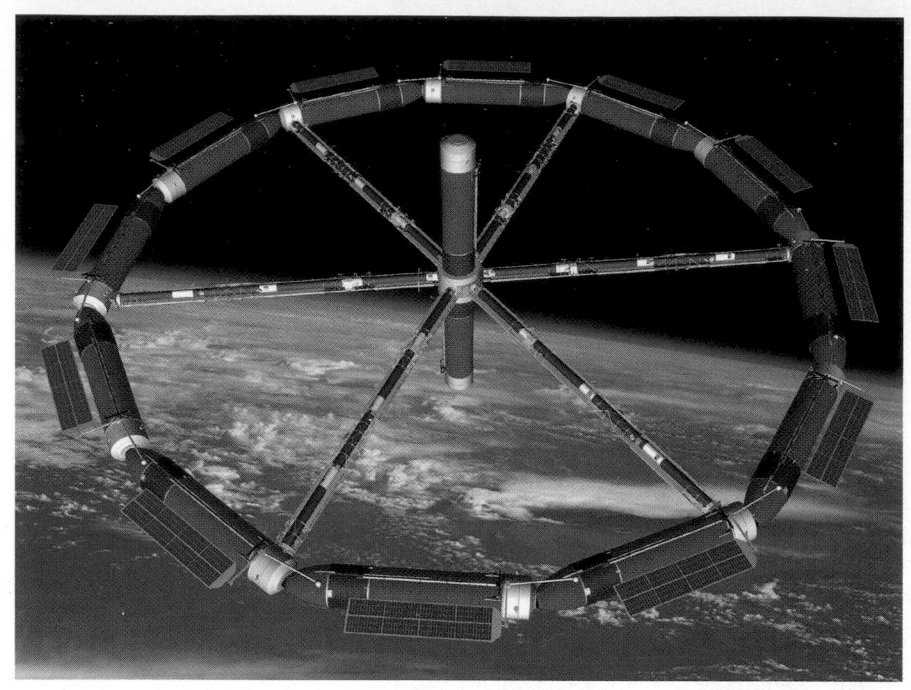

Artist's conception of a 12-ring station. Courtesy Space Island Group, www.spaceislandgroup.com.

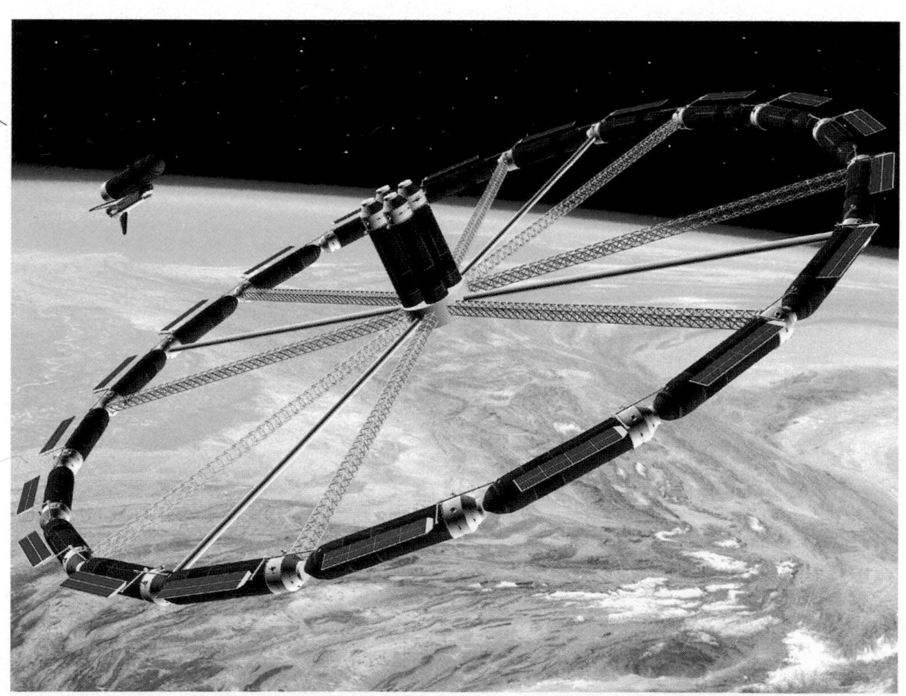

Artist's conception of a 20-ring station. Courtesy Space Island Group, www.spaceislandgroup.com.

PRIZE contender Ascender from Bristol Spaceplanes. Courtesy Bristol Spaceplanes.

ɔace tourism and entertainment company Space Adventures is planning to send passengers into suborbital space efore 2005 on vehicles like this. The company is working to build a spaceport in Oklahoma to accommodate ıch flights as well as jet training and zero gravity flights. Courtesy Space Adventures Ltd.

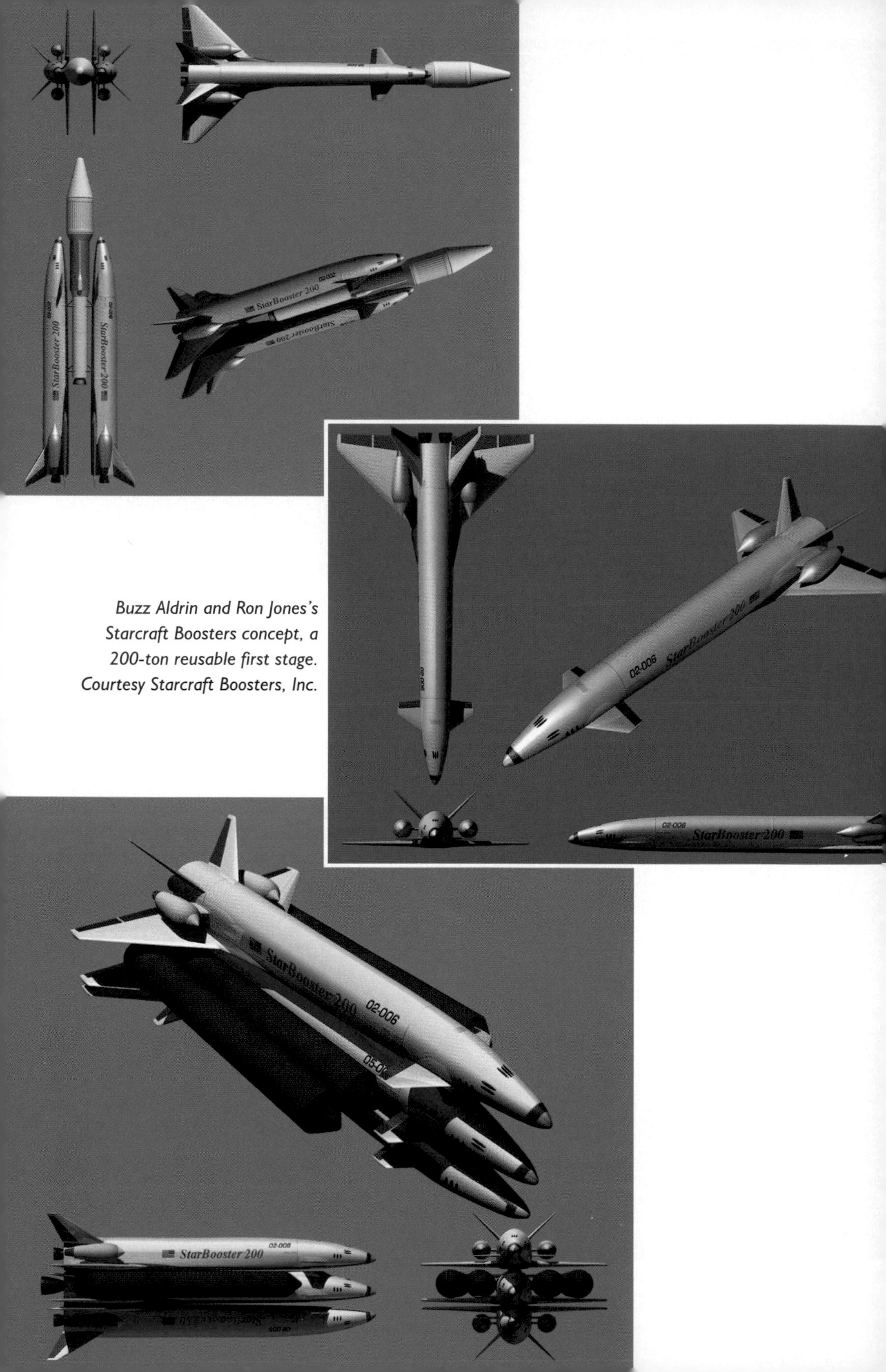

Buzz Aldrin and Ron Jones's Starcraft Boosters concept, a 200-ton reusable first stage. Courtesy Starcraft Boosters, Inc.

Myhome.com's interior design of a space station suite. Courtesy Space Island Group, www.spaceislandgroup.com.

Artist's conception of a space station suite. Courtesy Space Island Group, www.spaceislandgroup.com.

Hans-Jurgen Rombaut's lunar hotel soars 160 meters (520 feet) over the Moon's surface, offering guests dramatic views of both the landscape and the Earth. The hotel extends an equal distance below the surface, housing staff and protecting them from radiation. Courtesy Hans-Jurgen Rombaut, www.rombaut.nl.

X PRIZE contender Canadian Arrow's full-scale engineering mockup. Courtesy Canadian Arrow.

You can fly to the edge of space (over 85,000 feet above Earth's surface) in this MiG-25 sponsored by Space Adventures. Courtesy Space Adventures Ltd.

U.S. company Pioneer Rocketplane has an exclusive right with Space Adventures to sell seats on this reusable vehicle for suborbital passenger flights. Courtesy Space Adventures Ltd.

Neutral buoyancy tank at the Gagarin Cosmonaut Training Center in Star City, Russia. Space trainees simulate space walks and repairs here. Courtesy Space Adventures Ltd.

oSpace March Storm volunteers in 1997. Front row from left: Charles Miller, former NASA Administrator Dan ldin, U.S. House Space Subcommittee Chair Dana Rohrabacher. Credit: Michael Anthony Hall.

X PRIZE co-founder Peter Diamandis floats on a zero g flight. Courtesy Peter Diamandis.

Tom Olson, Colony Fund co-founder. Courtesy Tom Olson.

Jim Benson of SpaceDev.
Credit: Paula Berinstein.

Economist Patrick Collins at a Los Angeles space conference.
Credit: Paula Berinstein.

Denise Norris of Applied Space Resources. Courtesy Applied Space Resources.

The Artemis Society's Greg Bennett in his adopted city, Las Vegas. Credit: Paula Berinstein.

'an Wasser with his own bit of Martian real estate: He owns a piece of the Martian meteorite that nded in Zagami, Nigeria in 1962. Courtesy Alan Wasser.

Psychology professor and space ergonomics specialist Harvey Wichman. Credit: Paula Berinstein.

Former astronaut and moonwalker Buzz Aldrin believes that space should be open to everyone. He and colleague Ron Jones have proposed a stepping stone approach to developing space tourism. Courtesy Buzz Aldrin.

Share Space Foundation Executive Director Ron Jones is working with Buzz Aldrin to forge a public/private space tourism and Mars exploration partnership. Courtesy Ron Jones.

ark Shuttleworth, the world's second self-financed space traveler, training at Russia's Star City for his flight to ace Station Alpha. Courtesy Space Adventures Ltd.

Dennis Tito on a Space Adventures zero gravity flight. Courtesy Space Adventures Ltd.

Dr. John Lewis advocates "mining the sky" for asteroid metals, an endeavor that could bring universal prosperity to Earth and support a solar system population of many billions. Credit: Paula Berinstein.

X PRIZE contestant David Ashford of Bristol Spaceplanes thinks that winning the prize will be much more about business than technology. Courtesy David Ashford.

Tom Rogers, grandaddy of space tourism in the U.S., says that what he'd do if he went to space is nobody's business. Credit: Paula Berinstein.

—17—

Alonzo Fyfe: But Is It the Right Thing to Do?

A whole new arena of human activity brings with it ethical questions never before faced. What are they, and how do we riddle them out?

Fyfe and I spoke by phone for this interview, although he's happiest communicating by e-mail. I know what he means. It's easier to say what you mean in writing than in speech.

The ethical issues surrounding venturing into space are numerous and complex. Many are less than obvious. Alonzo Fyfe riddles these out for Denise Norris's company, Applied Space Resources (see Chapter 14). An ardent environmentalist trained in philosophy and ethics, Fyfe contributes insights on issues like multiculturalism, use and abuse of property, and environmental concerns. However, he eschews the term "ethicist," preferring to refer to himself as a marketer.

When you listen to Fyfe, you can understand how such two seemingly disparate subjects, ethics and marketing, fit together. He practices his profession according to the precept of "value for value." The Golden Rule. Give value to your customers, get something of value back. This philosophy, gleaned from his study of value theory and human decision making, profoundly influences how he evaluates which products should be sold and how best to market them. It also leads to a socially responsible outlook on the use of space, which is only fitting for someone who became involved in the field because of his environmental concerns. Fyfe believes that if space isn't rationally and carefully developed, the Earth will suffer.

Pose some hard questions to Fyfe and his value-for-value approach will lead to answers you hadn't thought of. For example, ask him what he thinks of the frontier rationale for going into space, and he won't say, "I like it" or "I hate it." Instead he'll come up with something like the following:

> I have two responses going along different routes. One route is marketing, using the frontier as a way of selling space. As a marketing slogan, I would treat it like any other type of marketing slogan; that is, I would do research, I would create some focus groups, and if the frontier is something the subjects value, then I see nothing wrong with it. I don't see it as a cure for what ails us. Some people value living on a frontier. They romanticize it. ASR's job is to provide something of value, and if people value a frontier and we can provide it, that's what we'll try to do. But that's on a personal level. There is another level. People allege that the frontier historically has provided a relief valve and inspired technological growth. I don't think that's very well proved. I would have trouble as an ethicist marketing a lie. I would say to customers, "If you value the frontier, ASR will try to provide it for you." But I would not go to customers and say that the frontier provides all these benefits if the argument is unsubstantiated.

Here's another one. "We have a policy that we call value for value. That is, if anybody makes a contribution to ASR, then ASR is going to make sure that they get something back. A corollary of that is that ASR won't accept any subsidies from the government because we see a subsidy as violating the value-for-value principle."

Fyfe elaborates.

> A subsidy means that the government's giving us something of value but isn't getting anything of value in

> return. An agency in California wanted to hand out grants to what it thought were worthy space companies and didn't ask for anything in return. We were considering applying for one of those grants, but when we were in the debate process I brought up the fact that such a grant counts as a subsidy. This violates the value-for-value principle. If the government of California wants to buy something from us, we'll sell it. If they want to buy payload space to send something to the Moon, we'll sell that to them. But we're not going to accept a grant that isn't based on them getting something of value in return. We decided not to apply for one.

Isn't the State of California getting overall business development, however, especially if the company ends up opening the Moon? Fyfe explains.

> That's one of the things that was brought up in debate. Don't the grants advance a certain political end, and isn't that the value that California would be obtaining? In this specific case, that would be a weak argument. The state is interested in economic development, but if it were to purchase payload space or buy 100,000 pages of the Millennial Archive [ASR's lunar time capsule] to put down some sort of state record, that would accomplish the same end without violating our value-for-value principle. What we didn't like about the grant was that it was open ended and seemed too one way. It may accomplish some politician's end, but it's the public's money. The state may want to spend it to promote a space-developing situation, but nothing prevents it from asking for something of value in return.
>
> We may not have made the wisest business decision, but we'll take a cut because to forego the grant is the best thing to do. In that respect, I'm certain that ASR is in the minority, and we may even be unique. Most companies would accept that type of contribution. But ASR decided at the beginning that part of its corporate

> philosophy would be value for value. One reason we did that is that there are a lot of enthusiasts out there who want to go into space. They end up making a lot of free contributions to organizations and companies simply because they like to get involved. We don't like the idea of milking these enthusiasts for free ideas and not allowing them to benefit from their own genius, as it were. Having California giving us some money without getting anything of value in return is a lot like some engineer doing a free mission profile for us. That engineer also deserves something of value in return. It's just an extension of that basic philosophy where we know, yes you're interested in space, but we're not going to exploit you like that.

Unlike most space activists, Fyfe doesn't try to force his opinions on you. Even more unusual, he doesn't naturally assume that he's right and everyone else is wrong. He's very, very careful about such things.

> When I was a junior in high school, I had this concern to make a difference in my life, except I noticed that everybody disagreed with what making a difference consisted of. On any issue you want to name—abortion, capital punishment, minimum wage, free trade—you have people on both sides, each equally convinced that people on the other side are wrong. I had a problem with how you decide who's right and who's wrong. When I went to college, I took a history course which was actually history of philosophy. There I discovered a branch where people studied these issues, where they didn't presume to know the right answers but actually tried to figure out what the right answers were. So I got into that. I went to graduate school at the University of

> Maryland, College Park, in philosophy studying ethics and philosophy of law.
>
> Another group of issues that ASR has had to deal with recently is the simple fact that different people have different views. Even if you think something is the best idea, if people aren't going to adopt it, then it's worthless. You have to make sure that you're working in the real world and not in some ideal ethical utopia. One of the facts that you have to consider is that there are people out there who disagree with you, and you need to respect their views as much as your own. Don't arrogantly assume that everything you believe is correct. Respect the fact that there are people on the other side who are at least as intelligent as you are who think you're wrong. And if there are people who are at least as intelligent as you are who think you're wrong, then there's a possibility that you're wrong. You should give that view some respect. So, for example, even though ASR believes that property rights in space are important, we also believe that it's important to develop a system for working in space which compromises with people who have different views. And ultimately it's probably the compromised view that's closer to the truth.

It's a centrist approach, and a practical one. "If you're pursuing a policy that nobody's ever going to adopt, you're wasting your time. But I think it's important to go one step beyond that," says Fyfe. "Not 'I'll work with you because I'm forced to,' but 'I'm working with you because I'm not going to arrogantly presume that I have the sum total of all the wisdom of the universe at my fingertips.' Not 'Your view is wrong but you happen to be powerful enough to force me to consider it.' I think it's necessary to go that extra step from an ethical point of view, not just a practical one. Give those different views the respect that they deserve."

How does the ethically minded person decide what's right and avoid picking the answer that's a mere personal preference? Fyfe explains.

> In professional ethics there are procedures and criteria used to evaluate different positions. It's not really much different than science. For example, you'll find one scientist who says that dinosaurs were warm-blooded and another who says that dinosaurs were cold-blooded. They will disagree, but they have methods at their disposal for resolving that disagreement. They might not in fact resolve it, they could shout at each other across a table for millennia with their conflicting views, but there still is a set of criteria, and ultimately the situation in ethics isn't that much different. There are people in ASR who disagree with me on some issues. They don't give me carte blanche to decide everything for the company. But when we disagree, I present my arguments, this is my evidence, these are the rules of logic that are supposed to apply. One of the things that I have from my twelve years of college is a background in the formal rules of logic. I present my case and they get to decide if it's strong enough. It isn't a matter of personal preference. In fact that's one of the main things that one learns when one studies academic ethics, that the idea of ethics as a personal preference doesn't get very far. Because there's also the personal preference of a mass murderer. If you adopt a view that ethical positions are merely a matter of preference then those become as ethical as anything else. And that seems to be an absurd result.

Ethicists can disagree with each other. "Just like scientists aren't always going to agree with each other," says Fyfe. "Ultimately the person who's responsible for making the decision will listen to the evidence presented on various sides like a court case and make a decision based on the evidence. Given that we're human beings, people will make flawed decisions or present bad arguments and accept them as good arguments and vice versa. They'll reject good arguments thinking they're bad for some reason. That's just part of the human condition. There's no perfect solution. But there are ways of proceeding and reducing the possibility of error."

To the lay person, this kind of precision in such a fuzzy area may boggle the mind. In fact, it isn't as precise as all that. At least, not so you can put it in black and white. "Even the methods themselves are under debate," Fyfe explains.

> There's a branch of ethics called metaethics, which is an examination of the rules and procedures by which ethical decisions can best be made. Even at that level people disagree. Again it is much like the type of debate that goes on in science. Scientists debate the evidence and attempt to determine which theory best fits the evidence. But there is a branch of science, actually philosophy of science, which discusses what are the principles by which science is and should be done. Even the principles of science are under debate.

Then there's the question of religion and the role it plays in professional ethics. Are ethicists' conclusions colored by theology and religious teachings?

> In academic philosophy where I learned my ethics, there tends not to be much influence from religion. Academic ethics tends to separate religion and ethics. But even in metaethics there's a debate over what role religion should play, and that's a metaethical question that academic philosophers do debate. It involves other things like whether there is even a God. If one answers that in the negative, obviously that would imply that religion shouldn't have much of an impact on ethics. Religion does play a large role in ethics with the public. One has to respect that.

He does. If the public holds a religious view that conflicts with the ethics that ASR subscribes to, "The public wins. Ultimately that's going to be the short answer. In a business sense the public is our customers. The customer wins. ASR's job as a company is to provide its customers with value. Whatever our customers value, that determines the course that the company will take. Unless, of course, the officers of the company feel that an activity would effectively be evil. And it is possible for a large number of people in a society to adopt a course of action that is ultimately evil. There are plenty of cases in history, such as Nazi Germany and slavery. In that case ASR's only option would be to cease as a company if we cannot provide our customers with values that we share with them. If the customers are asking us to do things that we think are wrong, then we will cease to exist as a company."

One might ask, as in politics, who leads: the company/politician or the customer/voter? And what if the customer doesn't know what he wants, or doesn't realize that he wants something? "It's very expensive to engage in an education campaign," says Fyfe. "There is some flexibility, particularly with respect to needs in providing what customers value. That is, how does one go about it? For one thing you can say that customers always value a lower price. But lowering your prices by robbing somebody else is not an ethical option." It's unethical to undercut the competition? In fact, it's illegal in the U.S. if the result is a monopoly, but beating your competitors' price is a key tenet of American business. Fyfe's unusual approach is either going to start a trend or lead to his company's demise. But then, those who are ahead of their time often become martyrs.

The common assumption is that this combination of free-marketeering and social responsibility does not represent the norm, at least not in U.S. business practice. Fyfe is quick to question that and to point out that he may not remain an anomaly—an ethicist on a corporate staff.

> Heads of companies are people. And though there's a strong tendency in popular media to depict everybody who runs a major corporation as a selfish, inconsiderate

> ogre, in fact they represent as diverse a group of human beings as any other group with their own values and interests. The major officers of Microsoft, for example, contribute significantly to environmental causes. There's a strong correlation between devotion to environmental causes and income, if you look at the statistics. That is, wealthy people are more concerned about the environment than poor people, who on the whole are caring more about where their next meal is coming from. It's the people who have more disposable income who can actually afford to visit a national park. Somebody living in the ghetto of New York isn't going to get to Yellowstone, will tend not to care much about what happens in Yellowstone. So I do think there's a place for people like me in corporations. And I think space development is going to be a place where people like me are going to find more of a home because it's a new industry. New industries tend to break molds. They don't do things the way they were done in the past.

Right now, the closest thing to ethicists in the space community is public policy analysts. "But there are no organized professional-type ethicists," says Fyfe. "Policy analysis involves a fair amount of ethical consideration. You can't analyze a policy without analyzing the ethical implications of it. And for that you'll find people in places like the Cato Institute [a Libertarian think tank]. Any major political think tank has people in it who look at space issues for policy analysis."

Where does this analysis go? Who uses it? "They publish a report," says Fyfe. "People who publish their reports hope that space activists will read them, and they do. The space activist community is actually fairly small, which means it doesn't have a huge lobbying arm with which to influence public policy. The think tanks advise policy people in government. For example, there are a lot of conservative politicians who pay attention to what the Cato Institute has to say on a specific issue. And these reports have a fairly strong influence through that route. I think they've contributed significantly to legislation like the Commercial Space Act."

For now, though, Fyfe is alone. His boss Denise Norris says, "I don't know of any other space project with an ethicist on board." When asked how this unusual situation came about, she says, "We found each other. Alonzo had a midlife crisis and decided he wanted to get involved with something that helps humanity, and he chose space. He was watching us, and we started watching some of his Internet posts. We liked what he was saying because we felt that it was in line with what we had already come up with. He does our strategic planning and market research. And we say to him, 'If you don't think it's ethical, speak up.'"

Fyfe expands.

> I've always been interested in space. In February of '98 I decided to have my midlife crisis. I saved for it, I planned for it, it was about time. What I wanted to do, what I've wanted to do ever since I was young, was to make sure that I made a difference, that I did something important. I believed that space fits that bill. I believed that the future of the human race depends on getting at least some people off this planet. One of these days the planet's going to be a barren hunk of rock. The future will rest entirely in the hands of those who've made it off the planet, whenever that might happen to be. It might be five billion years from now, but it may be in just a few years. Nobody knows for sure. So in my concern to do something important, space was it, and I decided to get involved in some type of private activity. I believed private activity would be better than government activity, mostly because government activity seemed to be stalled. I looked at the companies, and I thought that ASR was the best company around, so I approached them with the possibility of working for them. The way that I did that was by sending them a loss leader. They wanted to launch to the Moon to collect some moon rocks. I had some ideas for how to get more money for

their moon rocks, so I sent that to them for free, largely with the idea that if you like this, there's more where that came from. I set myself a goal. I started this process in February. I said by the first of August I would be involved in one of these private space projects. On July 16, 1998, one day after my birthday, ASR offered me a position.

If Fyfe is on the leading edge of a trend, then space may indeed break the mold. And with all the previously unencountered ethical issues it raises, it will have to. Here's Fyfe's take on some of them:

- *Common heritage of mankind vs. private property.* "One of the issues we're concerned with now is the issue of private property in space. There are principles like common heritage of mankind that effectively hold that anything in space is owned by everybody—effectively communal property. As an environmentalist, I tend not to like that idea because things that are commonly held also tend to be either destroyed or unused. There's no middle ground. There's no rational use if it's owned by everybody. And that's because nobody can prevent anybody else from doing something with it. There's a famous paper by Garrett Hardin written in economics called 'The Tragedy of the Common' that demonstrates how common property is destroyed or misused. One of the issues ASR's involved in is private property. What should our position be? Not only what's good for the company, but what's good for space? Good for the Earth? My view is that if space isn't rationally developed, the Earth will suffer. And we need space to relieve some of the burden that we're placing on this planet. If we go with the communal property view of space, then it's going to stay like it has over the last five billion years: unused, The earth is going to be increasingly overused, and we will as a race suffer and perhaps perish as a consequence."

Here's another Fyfe spin on property rights. "There's a possibility that somebody may interfere with our mission or that somebody will try to confiscate some material by saying it doesn't belong to us. That's a property rights issue. Part of the mission involves bringing a capsule down to Earth and retrieving the contents. This means that there will be material things in our possession which could generate some conflict."

- *Space debris and other pollution.* "When ASR was considering its Millennial Archive, one option for putting messages into space is putting the archive in orbit. But if you put something in orbit, then that's going to be a navigation hazard for how many thousands or tens of thousands or hundreds of thousands of years that thing's floating around in space. You are asking future generations to spend their resources to plot whatever they do in space to make sure they avoid colliding with your message, and that seems, at least according to ASR, to be an unfair burden to place on future generations. Our Millennial Archive will be on the Moon, where maybe the worst thing that will happen 100 years from now is an astronaut will trip over one of the legs, but we're not asking future generations to suffer that type of imposition. We also selected a means we think is ethically responsive to the needs of future generations, not to dodge our spacecraft."
- *Multiculturalism.* "We want to make sure that the human race is adequately represented. That led to two decisions. One is what are we going to put on the archive. Are we going to censor information? And a decision we made is that it would be arrogant for us to presume what future generations will be interested in knowing about our culture, and also knowing about our culture means knowing the good and the bad. So we decided not to censor except to the degree required by law. We recognize that some customers will not value having their text placed next to certain other kinds of text. We allow ways for customers to choose the type of disk, the general content of the disk

that their message will appear on, but by and large we're not going to censor except where it violates the law. Another thing that we want to work on is something that represents cultures that would not otherwise be represented. Think about a small aboriginal culture in South America. They're not going to send us a page at $39.95 to put in the Millennial Archive. But we still want them represented, so we're trying to work out forms of sponsorship where cultural organizations—and we'll offer as much of a discount for this as we can—can get these unrepresented or underrepresented groups onto the Millennial Archive. I'm afraid we are probably going to miss people. It's like the 2000 census. You can't count everybody. But we'll try our best to insure that those who can't afford to be represented are in some way represented on the Millennial Archive.

"We will include anything that's legal, but different nations have different rules with respect to what's legally permissible. So no matter what it is that you're talking about, you can probably find some society somewhere that prohibits it. For example, you can't have Nazi material in Germany. You can't have material that's critical of Islamic values in some Middle Eastern countries. So what do you do with different sets of cultural values that don't always go along with each other?"

- *Governance.* "We haven't discussed specifics with respect to how colonists would get along or what rules or laws should govern colonies. ASR supports the idea that whoever finances the operation gets to call the shots, with certain limits. That is they can't do harm to others, and even others within the same company. So, for example, slavery is ruled out in space even if someone thinks they can get away with it. But by and large we tend to support a policy that whoever creates the enterprise gets to determine how it's run." That includes people outside the U.S. "If the Chinese decide that they want to fly something to Mars or put something in orbit, then they get to determine the rules for operating according to their culture and beliefs and principles. Unless of course they're going to use it to attack somebody. That creates a whole new set of issues."

- *Whether to develop space at all.* "ASR's interested in the development of space. In looking at the future of the company, one of the things we need to consider is what's going to be the future of space development. And what would be good and what wouldn't be good. To some extent these issues are purely on an academic level at this point. And the difference here is that we don't have to make any choices. We don't have to decide that something is going to be one way or another at this point. Although we can start to throw out our various ideas so that when we do have to make a decision, we have some background behind us."

Fyfe has a novel idea for not only making space happen, but for doing it right from the beginning. He and author Kim Stanley Robinson want to enlist the help of the environmental community. To them, space is all about saving the Earth, and they both feel that we can learn a lot about Earth by exploring space. They have their work cut out for them. Robinson admits, "The environmental community has never thought about space as anything except part of the enemy, the high-tech solution to all our problems, a kind of arrogant military-industrial complex." But he's positive nevertheless. "I think you can make a case that what we've learned from going out into space, everything from the public relations value of that photo of the Earth rising over the Moon to actual data about how tall is the canopy of the forests and why is it that Mars changed, the hole in the ozone story, all of these kind of comparative planetology things and direct study of Earth from space."

"I'm involved in space mostly because of my environmental concerns," says Fyfe. "I see space as a great way of helping the Earth. Ultimately, we have a choice as to where future generations can get their mineral resources, for example. We can either further cut into the living biosystem of Earth, further scar it, or get our resources from dead planets. The latter seems to me to be the best option. That's the way the business should be marketed." Targets should include large groups with specific needs and interests, like environmentalists. "The largest space advocacy group, the Planetary Society, has approximately 100,000 members. The largest environmental organization—the National Wildlife Federation—has 4.5 million members. If you're looking for customers for your space project, then space activists really aren't a sizable market compared to some alternatives. So you sell space by selling space to environmentalists."

Robinson echoes this sentiment.

> It seems like the space program does not have much of a constituency. The people who support it are highly skilled and enthusiastic and vocal and well-educated, but it's a small group, and I think that fact reduces the interest of the people who control the purse strings of government. I think what will get us into space and off to Mars is a public relations push that makes it clear that going into space is not just a hobby or a specialized interest of people who happen to be interested in space. It's a useful tool for Earth management. If that were made clear, then all of the people who are interested in their children's welfare might become more inclined to support going into space. It'll become obvious that we need to be up there in order to keep the Earth in balance over the long haul. There are a lot more people interested in their children and grandchildren's welfare than there are people interested in the stars or whatnot. This approach makes space an environmentalist cause, part of the Green program. There are huge agencies devoted to protecting the environment, and if space becomes part of that project, I think its case is strengthened.

This may be news to environmentalists. Fyfe admits that piquing their interest might take some doing.

> They're not up on the technological facts. I think that's a flaw of the space activist community, not of the environmental community. The space community is almost entirely made up of engineers and scientists. They like technology, but ultimately one needs to do engineering for a purpose. Engineering for its own sake doesn't really produce anything of value. And environmentalism is a purpose. Environmentalists probably don't know so many of the facts, but they don't need to. They know what we're doing by using carbon-based fuels in warming up the atmosphere. They know what happens when you strip-mine a mountain and the copper tailings get into the water. And they know that none of that would happen if you were mining an asteroid. They'll leave it to the space geeks to figure out how.

But won't environmentalists insist that we keep space pristine?

> The environmentalist community is quite large, and as such, represents a lot of different views. It's a mistake to look at any group of that size and say, "This is what defines them, this is true of everybody." Most of the people in the environmental community have regular jobs; they're not giving up driving. They don't give up many of the conveniences of life. There are a few on the fringe who are going to have any objections to even setting foot on the Moon or an asteroid. But by and large if you present them with a case that makes common sense, that is, "Here's our choice: We cut into the living ecosystems of the Earth or we mine a dead asteroid," then the bulk of the people will see the sense in that.

So far, interest from the environmental community has been tepid. What worries Fyfe more is the space community's view of its potential allies. "The space community tends to be hostile to the idea of environmentalism," he says. "My theory is that it's a case of 'politics makes strange bedfellows.' The space community tends to be very conservative and a limited-government type of crowd, at least the parts that I've had much contact with. And environmentalism tends to be associated with big government. So, just by making that surface connection, a lot of space people view environmentalism with a bit of hostility."

It's not that space activists are in favor of despoiling things. "Mostly they don't like the methods of environmentalists. They don't like the idea of huge government programs aiming to clean up or protect the environment. They tend to think that a better enforcement of private property rights will do a better job. They're not out for the wanton destruction of the universe. But they do tend to have some latent hostility to any organized environmentalism."

How, then, to get the two communities together? "That's one of the main things that I've been puzzling over," admits Fyfe.

> I think an alliance between the two would be a good thing. I do think people are basically rational. These antagonisms may get in the way at the surface, but ultimately when enough people look at the issues seriously that they see the sense behind things. I do believe that there has been progress in the last 10,000 years of human existence. We have by now practically wiped out slavery. We're reducing most forms of bigotry and prejudice. I believe these trends will continue. They demonstrate that humans are basically rational. All it will take ultimately is just getting the facts out. I'm starting to work with a group of space activists who are going to plan a conference on space and environmentalism. We've gotten some contacts among the heads of environmental organizations and are arranging some meetings. So far, it looks like they'll be interested in attending.

Robinson and Fyfe have both suggested that the environmental angle may be the way to whip up support for space. When asked about whether this approach might be the way to sell space to Europeans, who don't like the frontier argument and who tend to be less gung-ho about space than the Americans and the Japanese, Fyfe again turns the question on its head.

> First thing, you have to find out what Europeans value. They tend to value the environment more than Americans do. That probably has to do with the fact that things are a bit more crowded over there and they have less nature to deal with. Also, they've been using their resources more than the rest of the world, so scarcity of resources is more of an issue for them. But the question "How do we sell space to the Europeans?" seems to presume that one ought to sell space to the Europeans, and that tends not to be the way that I approach it. That is, if you don't value the product, then it's not my job to force you to value it. My job is to find a product that you do value.

But if certain cultures don't exhibit an interest in space up front, isn't it possible that once it's developed they may feel that they're being left out? "That question dovetails with some ethical concerns we've discussed at ASR, and it ties in with including other cultures in the Millennial Archive. One of the things we're looking at is making sure that the *Lunar Retriever* mission is an international mission, not only with respect to the Millennial Archive, but also in manufacturing the flight hardware and operating the mission. Because we don't want people to be left out. We want it to be a global adventure."

THE AUTHOR'S OPINION

I find it fascinating that any company has an ethicist on its staff, space-related or no. That one of the early private space companies does may set a good precedent for an industry in its infancy. Fyfe is a careful, conscientious thinker who raises questions that would escape most of us. Whether his value-for-value approach will lead to his company's success remains to be seen. He may be so principled that the business will fail. I hope not.

—18—

Alan Wasser: Who Owns Space?

If we are ever to have space hotels, or asteroid or Moon mining, or settlements, someone is going to have to be able to claim ownership of property out there. People aren't going to want to take financial and physical risks for nothing. The ability to claim property, however, will provide an incentive for them to take those risks. Or will it? And if it does, how do we ensure that a property rights regimen is fair?

I first heard Alan Wasser speak at a Space Frontier Foundation conference in Los Angeles, but we didn't actually meet until I called him about an interview. We were supposed to speak for a maximum of one hour. However, the interview went so well that we ended up talking for about two. Or was it two and a half? Later, we communicated via e-mail.

No issue inflames passions so ardently within the space community—other than that of government vs. private industry—as that of property rights in space. The rest of the world has barely heard of the matter. But one day, the question of who owns space, or parts of it, will determine fortunes at the least and the fate of humanity at the most.

There are two basic reasons people get so riled up over property rights. One is that land can be very valuable. Proponents of private property rights see them as an incentive for developing space and a way of funding exploration and establishment of an infrastructure.

Opponents worry about rich corporations and individuals owning all the resources. The other reason is that some members of the space community fear that we shouldn't be addressing the issue yet, that it's premature and could lead to major setbacks for the industry should things go wrong (see Chapter 13).

Some people take the idea to extremes and talk about claiming an entire planet or moon. Most of us can't conceive of such a thing. But if you start at the beginning and ask if it's okay to claim territory on a body, then if it's okay to claim a tiny body like an asteroid (of which there are hundreds of thousands), you begin to see that the line between what's okay to claim and what is not gets fuzzy. Is it okay with you if I claim 100 acres on the Moon? How about 100,000? How about 50 percent of the Moon? No? Where do you draw the line and how do you justify your decision?

Here are some other questions:

- Can property rights in space work the way they did in the American West? Does homesteading entitle you to claim the land? How can you homestead land where there's no air, where living communally is the best way to survive?

- What constitutes a claim on a piece of "land?" Must you land a human on an asteroid to claim it or is a robotic probe enough? How long does the human have to stay there in order for the claim to be valid? Is getting out of the vehicle and planting a flag sufficient? Should someone be required to develop the land in order to own it? What constitutes acceptable development?

- What if I establish a rule that rewards the first person or company who takes the risk and actually gets there? How do I decide how much of a reward to bestow? It has to be substantial enough to motivate, yet leave room for others who come afterwards. How do I make sure that it isn't just people from rich countries who get to claim property?

In the past, the issue of property rights boiled down to the Cold War. During the years following *Sputnik*, Americans worried that the Russians would claim territory on the Moon;

Russians feared that Americans would do so. These fears led to the 1967 Outer Space Treaty, a United Nations document ratified by ninety-one countries including the U.S. and the USSR, that bars claims of national sovereignty in space and forbids countries from claiming land or objects there. (The full name is Treaty on Principles Governing the Activities of States in the Exploration and Use of Outer Space, Including the Moon and Other Celestial Bodies.) The Treaty stipulates that a country is internationally responsible for its "national activities" even if they are conducted by nongovernmental entities. No nuclear weapons or other weapons of mass destruction are to be placed in space, and countries are not to interfere with other countries' exploration and/or use of space.

The treaty doesn't address the issue of private property at all—back in the sixties no one thought such a thing could happen—so the idea that private individuals or corporations can or can't make claims is hotly debated. There are different property rights systems in effect around the world. The countries that negotiated the Outer Space Treaty subscribe to different systems, which has contributed to the confusion. Another treaty, the Moon Treaty, dating from the early 1980s, was meant to remedy that omission in the Outer Space Treaty by banning private individuals from claiming private property on the Moon and making it illegal for any Earth country to claim sovereignty over a celestial body. The problem is that no major spacefaring nation actually signed the Moon Treaty, and in fact, only seven nations did so: Australia, Austria, Chile, the Netherlands, Pakistan, the Philippines and Uruguay. The U.S. and the USSR wouldn't sign it because of an inability to agree on the theory of property that would apply to resources deemed the common heritage of mankind. Third-world countries were keen to define resources as common property, the U.S. favored a capitalist free-market approach, and the USSR fell somewhere in between. The developed countries feared a restriction against mineral exploitation, while developing countries were worried about a failure to achieve equity and the establishment of colonialism in the new environment.

Alan Wasser is one of the fathers of the idea of property rights in space. A retired journalist, Wasser has been working on his scheme for incentivizing the settlement of space for over ten years. "I've got what I believe is an economic incentive to interest private industry in establishing privately funded space settlements," he explains. "The history of expansion has tended to be for economic reasons. I personally have spent a great deal of time trying to figure out what could possibly pay for privately funded space settlement. The idea of land ownership, private property, is not easy for people to accept. It's a pretty far-out idea, and it's been a tough sell to convince people that it could be the answer. But nobody's pointed out to me yet why it isn't. I still believe in it and I still try and sell it."

Wasser relates the story of his attempts. "One of the first things people gave me a hard time about was, 'Why would anybody want to own land on the Moon—it's all the same?' Any one piece of land on the Moon is no more valuable than any other. I pondered this one for a while and came to realize that actually there was an area that was more valuable than the rest, which was the poles." The reason is that the poles are in full-time sunlight, which would solve one of the major problems of colonizing the Moon: two weeks of dark followed by two weeks of brilliant sunshine. 'It's obviously not a way humans like to live," Wasser reminds us.

> The heat gets too hot and the cold gets too cold. But if there were a mountain at the pole of the Moon which you could live on top of, you could have sunlight all the time. And because there's no atmosphere on the Moon, I realized that that midnight sun would be every bit as warm and warming as a noon sun at the equator would be because there's no atmosphere to weaken it. So I published an article saying that the first settlements on the Moon might well be on a mountain top at the poles where people could have sunlight whenever they wanted it. They could block it out whenever they wanted to, but they could have it whenever they wanted to. And not only that, but if you build a big tower on the mountain top, you can have unlimited electrical power from solar power. And everybody thought that was a nutty

> idea until the *Clementine* established there really was a mountain top on which you could do that and which really does get full-time sun.
>
> In fact I was very flattered. Ben Bova [a well-known science fiction author] wrote a book in which much of the action takes place on that mountain top at the pole, and he actually called the mountain Mount Wasser. I had no idea he was going to do this until somebody called me up and said, "You've go to see Ben Bova's latest book." He had taken huge portions out of my original article. I mean pages that were straight out of my article talking about why the mountain top was so valuable and why that piece of land would be different in value from the rest.

In his article, Wasser also pointed out that there might be a lake of frozen water at the bottom of the mountain top, "which means now you take your unlimited solar power from the tower on top of the mountain and use it to heat the water at the bottom, and lo and behold, you have a water supply, which is a major problem on the Moon. And that [water at the poles] turns out to be true, too." (Actually, that's not yet confirmed 100 percent.)

Now that some of Wasser's ideas have been proven true, he's gained credibility. "When I first started the whole idea of property rights in space was regarded as ridiculous. I used to get very good laughter for that suggestion. Then they stopped laughing," he says. "Suddenly the idea of property rights in space is not quite so silly. However, they're not yet ready to accept the idea of land grants as a way of inducing private investment into putting up a settlement."

But he's got a proposal ready just the same.

> I've actually got a draft law that I would like Congress to pass that says the U.S. government would recognize a claim to a certain amount of land put forward by someone who has actually established a permanent human settlement on the Moon and met certain other conditions, including offering to sell tickets to ride on the ship. In order to have a settlement, you have to have

a ship going back and forth. The settlement should be required to offer tickets for sale at a profit, to go back and forth to the Moon on that ship for anybody from any nationality. Any peaceful person, I suppose. And that if you did that the U.S. government would recognize your claim to a piece of land.

Actually, on the Moon my [proposed] piece of land is about the size of Alaska, which ends up being only about 4 percent of the Moon's surface, but it's a huge piece of land, obviously. And you could start selling off pieces of it immediately and recouping your expenses for putting up a settlement.

Part of the problem is that there are space treaties in force which were passed during the years when socialism was much more popular. These treaties tried to discourage private ownership, and so they have to be worked around. The treaty we signed and agreed to is called the Outer Space Treaty, and that one merely bans nations from claiming land, from claiming objects in space. It doesn't speak to private property at all. It does not say anything about private individuals or private corporations making claims. The reason being that when that one was passed, nobody thought that such a thing was conceivable. It was only the U.S. or Russia. When they realized a little later that private property was a possibility, they tried to pass another treaty called the Moon Treaty which literally bans private individuals from claiming it as well. Fortunately, the organization I was already involved in by that point—the L5 Society—mounted a major campaign against it in the U.S. Senate and got it killed. And so the U.S. government never ratified it. In fact, no major spacefaring nation ratified it, and while it technically exists still, it has about five or six minor signatories, one of which is currently in the process of trying to get out of it, so I figure we can safely ignore the Moon Treaty, but we certainly do have to deal with the Outer Space Treaty, which means the U.S. government can't claim the Moon and award it to you.

> But my way around it, which is also now finally moving from the category of "You've got to be kidding" to "Well, maybe" is that the U.S. government would recognize your claim. It wouldn't claim anything itself. In fact, it would specifically say, "We don't have any rights to the Moon as a nation, but we do recognize that the XYZ Settlement Company that has in fact established a permanent settlement is claiming 600,000 square miles around that, and the U.S. government recognizes that claim and U.S. courts will consider a sale of land from that claim a valid legal transaction." Suddenly XYZ Corporation can recoup the cost of putting up a settlement by selling off pieces of land just the same way the transcontinental railroad companies recouped the cost, or at least were supposed to have, recouped the cost of building the transcontinental railroads by selling off pieces of the land grant they were given in return.

Wasser sighs. "I've made enough progress to be pleased but nowhere near enough to be satisfied. And I'm often frustrated and disappointed because people say, 'Gee, what a great idea,' and then they forget about it."

Wasser has always been interested in space. He graduated from the Bronx High School of Science the year *Sputnik* was launched (1957). But it didn't take that to pique his interest. "The year before *Sputnik* was ever launched, I was able to figure out correctly that there were going to be astronauts, there was going to be a space program. The United States was working on rockets and it wasn't too hard to figure out what was coming. And so I went to MIT hoping to be an astronaut, and everybody thought I was stark raving mad until *Sputnik* went up, and suddenly I wasn't so crazy any more."

Wasser was prescient. There weren't any astronauts in those days, and there wasn't any astronaut training, so he had no guidelines by which to plan his future.

> I guessed that it would be scientists with pilot training, and so I went to MIT in order to get the science. So there I am at MIT studying physics, which doesn't really fascinate me, and marching around in my little blue Air Force uniform looking forward to being a pilot, and everything's fine until my sophomore year when the Air Force finally got around to giving me an eye test, which had never occurred to me. I didn't need glasses, but I don't have 20/20 vision, and they informed me I was just a little too nearsighted to be a pilot and farsighted to be an observer, which was the second seat in the plane. And I wasn't going to get to fly jets. Well, it wasn't too hard to figure out, if I wasn't going to get to fly jets, I wasn't going to get to fly rockets.
>
> So I left MIT and decided I'd go be a journalist and write about it if I couldn't fly them myself. I got a job with the *New York Times* and *CBS News* and indeed I did cover—actually mostly for *ABC News*—I covered all the *Gemini* flights and that sort of stuff. Incidentally, I was wrong in one respect. They didn't take scientists as astronauts at that point, they took plain ordinary military test pilots, so I probably wouldn't have made it anyway, at least until the second group when they started to take scientists. But, in any case, I had been right that there would be guys flying in space, just wrong on a couple of minor details.

Later on, Wasser was again ahead of the crowd.

> When after Apollo everything fell apart in terms of space flight development, I was one of the first to realize we needed political activism. I got involved in what was then the L5 Society, just beginning to be formed. I was active in that and merging with the National Space

> Institute which Wernher von Braun had started, and I've been involved ever since in trying to promote, well, the phrase they use is "promoting a spacefaring civilization." There's still an element of hoping somehow they'll take a tourist and I can go, but I just believe that it's the most important thing I can do for my species, for humanity, is to help expand its habitat beyond the Earth. And so I figure this is my good deed on Earth.
>
> I don't assume that humanity is a given forever. We're a relatively recent species, we've been expanding like mad and growing like mad. We've certainly filled our Earth pretty near to capacity—not quite yet but we're getting there. The potential risks to humanity if we're all trapped down here are the same as true for any other species with a limited habitat. Things go wrong in habitats, whether it's an asteroid hitting you, or a disease, or a weather change, or a calamity no one could have expected, the weather changes. If you've only got one island that you all live on, you're vulnerable. Or you're more vulnerable—I guess you're always vulnerable—you're more vulnerable than if you've got two islands. And the next island we should move out to is beyond the Earth. The numbers of threats that we solve or at least reduce the risk of by expanding are tremendous.

Going into space provides breathing room too. "You can have an infinite variety of different societies suited for different people, for whatever your tastes in life are," says Wasser. "The sedentary ones get to stay home without being harassed by the adventurous ones. So it's good for everybody. My favorite analogy is, 'Would the poor of today's Madrid be better off if Queen Isabella had given her jewels to buy food for the poor of 1492 Madrid rather than for sending Columbus out to find the Americas?' Everybody benefits by expanding our habitat."

Wasser believes that his approach is sound, but he laments its being met with so much apathy.

> I keep waiting for someone to say, "Here's the fatal flaw," and tell me why it's no good. That hasn't happened yet. I get a lot of people saying, "Gee, it sounds like a great idea." I got an e-mail from somebody who told me what a great idea it was and how much he was in favor of it, and a few months later I happened to see an e-mail discussion he was involved in with somebody else about how space settlement might take place, and he hasn't even mentioned the possibility of land grants as even a factor in the consideration. I sent him a little note, and he came back and said, "Oh, I'm so sorry, you're absolutely right and da da da," but watch what happens next time. I don't know if you remember Everett Dirksen [U.S. senator from Illinois], but at the time when integration finally passed, his comment was it was "an idea whose time has come." And I feel like this may be an idea whose time has not yet come. So there are times when I get very frustrated and say, "Why am I wasting my time? No way I'm going to get any benefit out of it and I put all this time and effort into selling this idea," and then I say, "On the other hand, look how far you've come." So it depends on what day of the week you get me on as to how I'm feeling about it.

Nevertheless, as far as the substance of his ideas is concerned, Wasser feels that he's pretty much on the right track.

> It's a funny thing. No two people ever raise the same objection. The guy I was dealing with—one of the officers of the National Space Society—insists on hard figures on how much lunar land will be worth. "How many dollars an acre will you get for it?" I'm working with a minimum estimate of one hundred dollars an acre simply because that seems to be the lowest anybody ever can think of paying for land with absolutely no ability to

reach it and no availability. So I figure that's the least it's going to be. And with that you still have enough money to pay for a settlement. But I can't prove it's going to be that, and I think it will be a lot higher. How do you prove how many dollars an acre's going to be worth? And keep in mind you're only going to own this land after you've built a ship capable of going back and forth at an economic cost and have established a permanent settlement and are offering tickets for sale. So now this is accessible. Obviously, nobody is going to regard this as ridiculous any more because it's going to exist, and now it's for sale and the U.S. government endorses your title to the land and your right to sell it. Which is obviously going to change everything from the current situation. How on Earth can I give him a hard-and-fast dollar price for what land will be worth at that point? But he wants that, and he isn't about to go for all-out supporting this idea until I can prove to him what the market price for land will be at that point. Well, I can't.

It's a chicken-and-egg situation. "Everybody is waiting for everybody else to accept the idea publicly before they do. And you know, unusual ideas have that problem." Part of the problem is the up-front cost of designing and building the vehicles and settlement needed to win the land in the first place—land that isn't going to repay the investment until after the settlement is established. But Wasser isn't addressing that issue directly yet; he's focusing on the prerequisites.

What I'm pushing for at the moment is the low-cost part of this activity, which is passing the law. If the U.S. Congress passes a law now saying when somebody establishes a settlement on the Moon or Mars that meets these conditions, we will recognize their claim to so much land, that's not going to cost a great deal of money to do. And if it doesn't produce anything, so what, so the *Congressional Record* is three pages longer than it would have been. But my hope is that this will suddenly justify the cost of research on low-cost access to space. On bringing down the cost of taking humans to the Moon. Because somebody will say, "Look, if we can bring it down below this cost, and we get this land for building the settlement, we can recoup our costs." At that point it won't be that hard to raise the money to do the research. But at the moment, absent either a law that says you will get the land or the existence of a low-cost ship, there's no point in pursuing it. It's too hard.

I've had other people say, "Will Boeing swear that if we pass a law that they will then put up a settlement?" [He laughs.] Well, no. But obviously they're going to start looking into it. And if they can find a low-cost way to get there, then they will do it of course. Or somebody will because there's a huge profit available. I'm assuming the settlement once there and paid for by the land speculators will be able to meet its own expenses through contracts with governments and scientists and researchers and selling souvenirs, and selling tickets to tourists, and dozens of things that nobody really can think of now. One of my analogies is saying, "Could you ask the Wright Brothers the day before Kitty Hawk who's going to be flying American Airlines in April of 2002, and why would they pay to get on an airplane?" Obviously the reasons people are paying to get on an airplane, some of them you could have guessed, but most of them you couldn't possibly have guessed.

People say, "What's lunar land going to be used for?" Well, the most lucrative usages of the land grants for the transcontinental railroads were totally unpredictable in advance. Sun Valley ski resort. I mean, skis hadn't been invented. You couldn't have asked the land grant buyer to tell you that the land he's buying will someday be a ski resort. I'm assuming that ways will be found to bring in money that are currently unpredictable, but they aren't enough to pay for putting up the settlement in the first place. My hope is that the land sales will pay the initial costs and then once that's done, maybe give it a year or two to figure out how to make money. Frankly, even if they establish a settlement and then go bankrupt, which happened to plenty of the land-grant railroads, somebody else buys them for pennies on the dollar, and he gets rich. And we still have the land-grant railroads, even the ones that went bankrupt. The railroad existed and continued to exist, and I assume the settlement would exist and continue to exist once somebody's paid the money to develop and build a low-cost ship. And his incentive to develop and build the low-cost ship was to get the land grant and the ability to sell that off.

Wasser admits that he's fighting an ideological battle.

I've had some people object because they simply don't like the idea of injecting commercialism and private enterprise and private property and profit into what should be holy and pure, such as space exploration. It should be only for altruistic motives. And the idea that somebody might make a profit doing that offends them. This was a major problem when I first started this. It's gotten less so as people have come to realize that "from each according to his means, to each according to his needs" doesn't work in practice, as beautiful as it may be in theory. And that making a profit is a very good way to get people to do good things—things that will benefit everybody if you give them a profit for doing it. You can't reason with them. There was a letter to the editor about one of my articles in one of the magazines berating me for proposing something so crass as private enterprise.

When we, that is, the world, wanted to stop settlement and exploitation and development of Antarctica, the solution was to ban private ownership, or to ban ownership, period. We all signed an international treaty—the nations of the world got together and promulgated an international agreement that prohibits ownership in Antarctica. The reason was, if you couldn't own it, you weren't going to develop it, and the world didn't want it developed.

I think the truth is there was an element of that [the Antarctica approach] in the Outer Space Treaty, which banned nations claiming it. I've got articles from newspapers saying that by the time [Lyndon] Johnson signed the treaty, actually proposed the treaty, the Russians were ahead. Not much, but still by a few months. And there were articles like, "Will the Russians claim the Moon when they get there first?" The Moon race was on, and I think a large part of it was to either claim it ourselves or at least stop the other guy from claiming it. So when they signed this treaty that banned claiming the Moon, suddenly there was less of a risk, the prize wasn't so great, and in fact the budget we spent on space every year until the day we ratified that treaty has gone down every year since then. By banning ownership of the Moon, by banning ownership of objects in space, we prevent settlement and development. In Antarctica, we did it deliberately. Maybe in the '67 treaty we did it deliberately. By now, everybody's used to that idea that nobody can claim the Moon. It seems like it was handed down by God on Mount Sinai. But it really wasn't. The world changed in '67 when Johnson and the Russians negotiated this treaty. It pretends to be a U.N. treaty. It's actually a bilateral treaty that the U.N. adopted. Johnson sent Arthur Goldberg to the Russians with this idea that they signed off, and the Russians pretty soon dropped out of the Moon race, although we didn't know it. We kept going at least as far as the Apollo program, but that's it. Because there was no more danger of them claiming the Moon there was no incentive to us because we couldn't claim the Moon, and so in fact banning ownership of the land has stopped development and settlement of the Moon, and I'd like to reverse that by finding a way to reverse the effects of that treaty.

Everybody tells me there's no way I'm going to get the '67 treaty repealed or even amended. That's where I started ten years ago, proposing to get rid of the treaty. But it's been brought home to me that there are other provisions of that treaty that everybody loves so much, they're just not going to tamper with it. So I came to this idea of going around it by having private property without national sovereignty.

We already have Earthly precedents governing how common areas are to be used: ocean law, for one. Antarctica, for another. The Moon Treaty, to which only seven nations are signatories, is derived from the Law of the Sea.

Under ocean law, no one can own the High Seas. You can go there, but you can't claim territory past twelve miles from your shores. Within the twelve miles and then again within 200 miles, you're allowed a certain progression of rights, such as the ability to regulate environmental protection, but you can't claim sovereignty. If you sail within somebody else's twelve miles bearing the flag of another country, you're under the jurisdiction of the flag country unless you meet "special circumstances" such as smuggling, bearing weapons of mass destruction, broadcasting on unlicensed frequencies, or engaging in some other dubious activity. If you engage in one of these activities, the coastal state can take action against you.

Past the 200 miles, that's it. You're on the High Seas, which means that neither you nor anyone else can claim any sovereignty whatsoever. Furthermore, you can't sail the High Seas unless you're flying under the flag of one nation. No flag? You can be seized as a pirate ship. More than one flag? You can be denied the protection of both or all countries involved. Flying a flag to which you're not entitled? The nation owning the flag can board you.

As far as using the resources of the high seas are concerned, a number of special cases apply. Usually, no one attempts to restrict fishing or whaling. However, sea-bed mining in international waters is another story. Citing such resources as the "common heritage" of people, the Law of the Sea states that more than 30 percent of any revenues from such activity can be taxed in order to spread the wealth. (That is, to allow nations that can't engage in sea-bed mining to share in the benefits of a resource that's considered "common heritage," i.e., the province of all.) Likewise, nations that are capable of exploiting sea floor resources can be forced to share their technology with those that can't. As a result, investors are less than keen to back sea-bed mining ventures.

Wasser has run his idea past a number of international law specialists without finding consensus on its legality and effectiveness.

> They all have different takes on it. There's one guy who says the only way to do this is to auction off the land. He's actually a friend of mine. He supports the idea of private property. He agrees that private property could be the incentive that provides space settlement. But he doesn't like the idea of land grants, especially given out by the U.S. Congress. He wants some international organization to auction off the land. The problem is how do you get an international organization? What international organization is going to do this? And my problem is that if you auction it off, you cost people money, and they own the land but they have no particular incentive to run out and develop it. I want a race in which it's who can get there first and establish the first settlement to win the prize.
>
> Every part of my proposal has had the endorsement of at least one space-law expert, if not several, but there are some parts that at least one person disagrees with, someone who says, "I can support everything but that."

> At one point I had a congressional aide who was saying, "I'm very excited about this idea. I think I'm going to get my Congressman to introduce your law." And then nothing at all. He said give me two weeks. And two weeks goes by and I call and nothing. And I call back a month later, and nothing. He's swamped, and he's too busy, and whatever, and maybe his Congressman thought it was too far out or whatever. At one point NASA was forming a task force to look at this, and I was called down to Washington to talk to the woman who was setting up the task force. We had a nice conversation. "Oh yes, I'm going to get Dan Goldin to endorse it." And it also dropped into limbo.

But Wasser is optimistic. He thinks he's planted a few seeds. "One of the seeds I planted is a guy named Jim Benson of SpaceDev." (See Chapter 13.)

> I think of him as one of my children in the sense that when he first started, I started pushing space property, and now he's a big supporter of space property. But of course he also has his own scenario as to how it should happen. SpaceDev is just going to go drop its rocket on an asteroid and claim it. I'm sure we planted a lot of seeds. And they grow into many different weeds. The fine detailed plan that I was pushing, I never had any illusions that that was exactly the way it was going to come out and they were all going to say, "Oh, Alan Wasser suggested this." I'm sure by the time it happens they will have long forgotten I ever suggested anything of this sort and it will have been shaped and reshaped and re-reshaped by the people who actually did make it happen, and I'll be lucky if somebody says, "Yeah, I kind of vaguely remember, yeah, Wasser." A guy named David Anderman [a space activist who works for Charles Miller's company, see Chapter 15] keeps saying, "Wasser, you're going to end up winning the Nobel prize for this,

> but not yet. It's all going to happen after people are ready to go there, and then suddenly it's going to happen."

* * *

Could a private company or individual technically go out and, privately, set up a small settlement and claim an asteroid or part of the Moon because no one could stop them? "It's absolutely technologically possible," he admits. "It could be done within relatively few years. It certainly would not be technologically prohibitive, it would be financially prohibitive. The problem is, how do you make a profit? Why do it? You're going to have to raise a few billion dollars in capital, although that's not so much these days. The current estimates are $40 billion to establish a lunar settlement. It's financially doable." Even the life support wouldn't be that big a deal.

> We've had a few years to advance science since Apollo, and so it would be a whole lot easier today than it would then. We also didn't know about the deposits of water at the poles then. Now it would be much easier. Boeing could do it out of their current resources. But why? There's no way to make a profit. You certainly can't start selling the land if nobody recognizes your right to claim it.
>
> Let's assume I do what you're talking about. I raise the money, I do the technology, and I go establish a settlement, or Jim Benson does, he claims the Moon and he starts selling land. Well, somebody's going to haul him into court for fraud. He's going to spend three times on much on lawyers as he did on putting up the settlement trying to justify his claim. So there's no way he can make a profit. He can bring back Moon rocks and probably sell them, although even that's open to question. But you can't make the money on that. You certainly can't justify the cost of putting up a settlement on that. He can sell space to astronomers who want to do astronomy on the far side where they're blocked from

> light from the Earth. But again, those are income sources, not justifications for putting up the settlement in the first place. So it is not at all science fiction.

But that wouldn't be the only problem if someone were bent on flouting international law. "You've obviously got to maintain a base on Earth to do it from. And that's where he's vulnerable," Wasser explains.

> If you say he could go and move to the Moon permanently and never come back and what are they going to do about it, yeah. You've got to have a ship going back and forth. You've got to have permits to take off and land. And more to the point, you got to be able to raise the money to do this. You've got to buy fuel every time you want to take off. You've got to answer the insurance companies and the licensing boards, and more to the point, your investors are going to expect a return. They're not going to be willing to wait twenty-five years for a return. They're going to want a return in a few years. With the current legal system and economic system, there's no way to do those things.
>
> My problems with Jim Benson's drop-the-rocket-on-the-asteroid approach are it doesn't get us human settlement, it violates the tradition that land claims should be based on "use and occupation," and it won't do him much good. It's technologically doable, nobody's doubting that. He wants to claim an asteroid, which everybody agrees is probably filled with trillions of dollars of minerals. All the titanium, iron, nickel, whatever, you could ever want is sitting out there waiting for you to mine it and bring it back, and he's going to claim ownership of it by dropping his rocket on it. He could make a fortune if he could turn around and start selling that stuff. Suddenly he would have a huge profit, and knowing that that would be possible, he'd have no trouble raising the money to go do it. But in the current environment when he claims it, everybody's going to

laugh at him. And nobody's going to start buying it from him because there are no laws granting him the right to claim it.

He could sell the iron. And in fact the treaties allow you to do that. As currently written, the treaties say that if you go to the Moon and extract Moon rocks, and bring them back, you do legally own the Moon rocks, which of course the U.S. was quite concerned about since we were planning to do that and, in fact, we did. Nobody has turned around and said, "Hey, the Moon is common heritage of man, therefore we all have equal ownership of those Moon rocks you brought back." Nobody has made that claim because the treaties clearly say if you go and take a resource from the Moon and bring it back, you own it. So, yes, you could go to the asteroid and mine it and bring back the minerals and sell the minerals, but there's no way currently to make enough of a profit at that to justify doing it in the first place. And that's where I get to my land ownership. If you can sell the mine ... the prospectors didn't literally get rich on the gold they brought out. The ones who got rich got rich selling the mine to mining companies. You want to be able to establish ownership and then use that for collateral for bank loans and whatever to establish a mine. If you've got to wait till you've brought back the minerals before you get your first dollar back, it doesn't work economically. On the other hand if you could start selling parts of your mine, that changes things. And that's what you can't do.

As far as support from the big guns like the aerospace and tourism industries, that hasn't happened yet. Wasser hasn't even approached anyone in those areas, though he acknowledges that some of them have heard of his ideas. Resolving the issue of property rights will be in their interest if space tourism is ever to get off the ground. But until the right vehicle exists, it's less than urgent.

Wasser elaborates.

> If someday, a Boeing scientist, like Seth Potter, discovers the secret of cheap access to space and goes to management and says, "Look what we could build," suddenly I suspect Boeing is going to discover Alan Wasser's crazy idea and be out there supporting it. But they haven't got that vehicle yet. So from their point of view, to waste time and money on supporting the law when it could end up being Lockheed that invents the vehicle is not really a logical use of their stockholders' money. It'll be a great benefit for whoever invents it, but not for anybody else. I want to pass the law first because then there's a pressure on everybody involved in aerospace to be the first one who develops the ship. Suddenly you're in a race. If you don't work on this ship and your competitor does, he's going to claim that first colony. And the law is set up incidentally to make first-come much more valuable than second-come. Deliberately. It creates a real strong reason to work on that ship.
>
> Now at the moment there's no real reason to do research on a cheap ship to the Moon. What are you going to do with it if you had it? You go to Boeing and say, "Hey, I got this great idca for how to bring down the cost of bringing humans to and from the Moon. I need $100 million a year for research, and that'll go up to a billion pretty soon." Do you know what they're going to say to you? "Why should we do this? What are we going to do with it if we develop it? How are we going to make a profit with it if we develop it? So why should we put this research money into this rather than into finding better ways to pack more people into a 747, where we can make a nice quick profit on our research?" So in some ways they're not going to be enthusiastic about it because it suddenly means they have to devote a certain amount of their resources into researching an inexpensive way of getting humans to and from the Moon. And they have to

> do it in a hurry before somebody else gets there first. Well, yeah, that's a huge potential profit for them if they succeed, and in fact if the law passes, my assumption is that finance people will see this potential, and will put together a pool of money and will go shopping for that ship, and in fact at that point Boeing will be real interested because here's KKR or Mitsubishi or whoever else that's looking to buy that ship and willing to throw money at them to develop it. But on the other hand at this point, you'd have to be pretty far-sighted to see why you should push this law rather than a loan guarantee for doing NASA's current project.

Wasser believes that even though his law would affect countries outside the U.S., the American locus would be critical.

> If the law passed I would assume that there would be international consortia that would pursue these things. But these days if you want to start selling your property, you'd really like to be able to sell it in the U.S. market. It's a whole lot less valuable if you can only sell it in Japan and it isn't recognized as valid title in the U.S. courts. Even if the money that comes in and buys it is Japanese or European. My assumption is if the U.S. Congress passes this law, suddenly every major space-faring nation going will want a piece of this thing and will be looking for reciprocity, and say, "We'll pass the same law and recognize your people's claim to the Moon if they get there first if you'll do it for our people." Aerospatiale [a European space consortium] and Energia and the other international aerospace companies will all be very active in this race, and in fact, Aerospatiale might be the one that invents the inexpensive Moon rocket. But the impetus could come out of the other nations, and in fact Australia is the kind of place where it might happen.

That might scare some people in Congress so much that they'd rather not pass the law.

> I've had people who said, "I'd support your idea only if you agree that only an American can do it. It can only be Americans who own the property." Well obviously you can't do that. The rest of the world would not be amused. There was a guy who started out all enthusiastic about supporting me, until he realized that I was proposing to allow "furriners" on the Moon. Not only that, I was going to allow them to buy tickets on a rocket, and that was the end of his support.
>
> Part of my problem is what they call "the giggle factor." Basically the whole idea of space activism has always dealt with the giggle factor. I remember the giggle factor when in 1956 I said I wanted to go to MIT and learn—I didn't say "astronaut"; the word hadn't been invented—but to be one of those guys who was going to go into space, and everybody obviously knew I was a nut and laughed hysterically at this idea. Now the idea of land grants, or private property rights in space, is no longer giggled at, at least in Congress and in those centers that pay attention to this. But it sure was before.

If we don't get private property rights for celestial bodies, what will space development look like? "It isn't a given that it will occur," says Wasser.

> There is a theory that says we have a relatively narrow economic window on Earth which we will either expand or not. There will come a point in time when the Earth's population has grown so large, and resources have been so depleted, the oil has been used up and there are now forty billion human beings crowded on Earth. At that point, we simply will not

> have the economic and industrial capability of expanding, and humanity will retrogress. I obviously don't like that scenario. I don't really believe it's going to happen, but it certainly is possible that we would pass through a window in which humanity had the ability to go into space and it will never happen.
>
> If we don't have private property, well, okay. It could happen if there is a Steve Jobs and Steve Wozniak working in their backyard garage/laboratory, and they come up with the really inexpensive way to get into space.

If such a thing could happen, which seems highly unlikely, "You get a couple of possibilities," says Wasser.

> One of them is you could end up with "Stop me if you dare." It could also be that they go ahead and establish de facto ownership, don't attempt to sell the land on Earth but just keep going back and forth and occupying it. Eventually they declare their own state with their own rules. This requires some nation on Earth to cooperate with them and let them land and take off, and presumably you could find somebody who'd be willing to do it, especially if you weren't trying to sell stuff or you were paying them. Another possibility is some nation's leadership becomes so fixated on this that they're willing to spend the nation's treasury on doing it on a government basis. The U.S. taxpayers don't seem inclined to do that, but it is conceivable that there would be a charismatic leader so powerful and effective that he could talk the taxpayers into anything. Or a dictator in some wealthy country. If the new prince of Saudi Arabia decided to devote the entire wealth of Saudi Arabia to establishing a Saudi Arabian colony on the Moon. Seems improbable. You know, these are all longshots. But sure, there are ways it could happen.

There are a number of other property rights proposals floating around. One comes from John Lewis (see Chapter 12), whose concern is mining. Lewis suggests the following:

- The United Nations or the World Court could serve as a registry for mineral claims on extraterrestrial bodies. This does not mean that either body would exercise control or authority over the activity—it just would administer the records.

- Claims could be registered by individuals and corporations of any nation with their own government, with full mutual recognition of claims. This alternative involves executing reciprocal agreements as the need arose and as each nation entered the mining arena. The World Court would adjudicate disputes. Enforcement would occur on Earth.

Lewis also asks what level of activity would represent the threshold for making a claim. Should it be the mere discovery of an asteroid? How about remote characterization by examining the asteroid's spectrum—a characterization that reveals the presence of an economically attractive resource? Would the physical presence of an unmanned vehicle that documents the presence of an ore or other resource be the appropriate point, or perhaps the presence of an unmanned vehicle that documents the presence of the resource by returning samples to Earth? What about the presence of a human crew that proclaims mineral rights or ownership, or finally, the presence of an established human settlement?

Lewis points out that almost a thousand asteroids have been subjected to photometric or spectroscopic study, and not one claim has been made on that basis. He believes that this evidence proves that there is no consensus that the threshold should lie with such activity. However, maritime law has provided a precedent for claiming mining rights to shipwrecks that have been scouted by unmanned vehicles that document the presence of valuable material. Lewis deems this level appropriate, arguing that the need for human presence puts a huge economic barrier in the way of asteroid resource use, preventing the participation of private corporations and small nations. He'd rather see a more egalitarian opportunity.

Lewis roundly criticizes Wasser's idea that Congress should pass a law authorizing an "extraterrestrial land claim made by any private entity that has established a true space settlement." This requirement, says Lewis, is way too onerous and amounts to one's having to build and occupy a house before being allowed to own the land the house sits on.

Finally, Lewis asserts that the abundance of resources from near-Earth asteroids and in the Asteroid Belt suggests that competition for mining claims won't be a common problem. However, the discovery of a particularly rich and accessible asteroid might present such a situation, which would create a strong incentive to get there first and claim the whole thing because it's so small. Claiming a large area on the Moon, say, the entire Mare Imbrium, would be ridiculous, however—tantamount to claiming all of Kansas or Switzerland as a mine site.

THE AUTHOR'S OPINION

There is a difference between the right to use something and ownership of it. I think the former should be made fairly easy; the latter should carry a respectable threshold. The point is to effect a balance between the risk-takers, who should be rewarded for sticking their necks out, and the rest of us, who may not have the resources with which to take those risks.

One idea might be to set aside certain amounts of public land for each X number of acres claimed by a private entity. Eventually, the Moon or Mars will be composed of private and public lands rather than just the former. The public lands might be tantamount to the high seas on Earth. As far as asteroids go, that's another matter. No one is going to settle permanently on an asteroid, so we probably don't have to worry about public lands there.

I think the threshold should involve human presence—at least for Mars and the Moon. A permanent settlement isn't necessary, although if you build one, you're entitled to more land than if you don't. The more effort, the greater the reward. As far as asteroids are concerned, the more effort you put into the endeavor, the more of the asteroid you're entitled to claim. Landing a probe entitles you to less of the body than actually mining it does. Speculators can't get as much land as those who actually occupy it or work it.

I showed Wasser these conclusions, and he asked for the chance to rebut, to wit: "A permanent settlement should and must be necessary for space property rights to be granted." Wasser objects to anything less than that for several reasons: First, that is the only legal way under traditional international law.

- Where the U.S. has sovereignty and is the source of ownership, the government can give ownership of land, or limited rights to its use, for whatever reasons it chooses. But since no nation can claim sovereignty on the Moon and Mars, the only thing governments can do is to recognize, or not recognize, a claim made by a private entity.

- In fact, in countries like France that follow what is called "civil law" (as opposed to "common law," which the U.S. inherited from the U.K.), property rights are based not on territorial sovereignty but on the natural law theory that individuals mix their labor with the soil and create property rights independent of government, which merely recognizes those rights.

- Throughout history, actual settlement—"occupation and use"—has been the traditional basis for a claim of ownership of land that had no sovereign. Space claims, even limited ones, based on less than settlement would be much harder to justify to the courts and the world.

Second, human settlement of space is the real goal. We are a lot more likely to see it happen if it is the required condition to win anything.

- Giving limited ownership for less could reduce the incentive, for the winners and losers of the first round, to keep going full out toward settlement.

- The last thing we want is to have given people a land grant for some small step and find that they never do anything more with the claim for the next twenty years, then stop others who are ready to settle and develop the land.

- The existence of a permanently inhabited settlement is the point of no return for development. Only when there is a live human being waiting on the Moon for the return flight can we be really sure that there will be a return flight, even if the accountants say, "Put it off for a few years, or more."

But the most important reason to reserve grants for actual settlement is that it is the only way lunar land can really be worth enough to make a real difference.

- The dollar value of an acre of lunar land goes up exponentially the day buyers can actually buy a ticket and go there, or send a representative or a customer. That means the value of the grant goes up exponentially if we hold it back until there is a space line going back and forth.

- The value of the land claim can be similarly increased if we capitalize on the media coverage of a space ship taking off to try to win the race to establish the first human settlement on the Moon. The day people land on the Moon, set up permanent habitation, and stay there while the ship goes back for more people, they will be the whole world's heroes. At that very moment, back on Earth, their representatives will finally be free to start selling, at huge prices, seats on subsequent trips, and, at prices people all over the earth can afford, valid deeds for land around their base. People will even buy land as a way to support, or just feel part of, the project.

Then, and only then, according to Wasser, will a lunar land claim reach the multibillion dollar value that would make a real difference, enough to justify even the billions it took to win it.

—19—

Rene Schaad: Space Around the World—This One, That Is

Where will space development come from? It's in no way a given that it will be from the U.S. In fact, the Russians have already shown themselves to be pretty resourceful space capitalists, leasing out Mir *and ferrying Dennis Tito to* Space Station Alpha.

What are people around the world thinking about space, if they are, and what are they doing about it? I was at a Mars Society conference in Boulder when Rene Schaad got up to speak after hearing an address by science fiction author Kim Stanley Robinson. Robinson had mentioned that the idea of space as a frontier doesn't exactly resonate around the world. Schaad got up to thank him for his comment. This "other side of the story" was just what I was looking for, so I asked Schaad if I could talk to him at some length. And that's just what we did.

Although most of the people interviewed for this book are American, space activity around the world is not exclusive to the U.S. and Russia, which has long had a manned space program and offers commercial launch services and perhaps even tourist services now. So when you're talking about humans in space, you're often talking about the two countries that have been most willing and able to spend money on that endeavor. The reasons for the divide between those nations and the rest of the world are several. Europe supports proresearch, prorobotic, proenvironmental missions. Japan has a commercial attitude toward space, but insufficient funds to make it happen. China is interested in establishing a major presence, but it too lacks funds. Australia is becoming a hotbed of space activity but has a way to go toward developing its infrastructure. Therefore, space development involving a human presence will likely come from the U.S. and Russia.

Asia, China, Japan, and India are already space powers, although only China has been successful in selling rockets commercially. India and Japan have set the creation of such a service as a national technological priority. Japan has been one of the most active countries in the area of space, and perhaps the most active regarding space tourism. Starting in 1993, drawing on studies published over the last thirty years, the Transportation Research Committee of the Japanese Rocket Society began work on a preliminary design of a fully reusable one-stage vertical take-off and landing launch vehicle for passenger service, the *Kankoh-maru*. The Japanese estimate that *Kankoh-maru* could bring launch costs below $180 per kilogram (about $82 per pound). Veteran rocket designers in the U.S. consider the design realistic. The Japanese Rocket Society, commissioner of the market research studies cited in Chapter 5 consists of senior Japanese airline representatives and related associations. These are the people who will shape space tourism in Japan. In the U.S., airline involvement in building a space tourism industry is virtually nil. But not even a prototype has yet been constructed. Japan also offers a new commercial rocket, the H 2-A, for launching satellites.

As for South Korea, in January of 2000, President Kim Dae Jung set a five-year initiative to design, build, and launch a commercial space cargo rocket and to use it to create a commercial launch business in his country. The country plans to use only domestic industries and expertise. If South Korea succeeds in creating a commercial rocket system, its military strength could be enhanced. Meanwhile, North Korea claimed to have tested a satellite launch rocket in August of 1998, but western governments couldn't find anything in orbit.

Although China worked on a manned space flight program for several years and launched an unmanned capsule in 1999, no specifics were known until the end of 2000. Then, officials from the Chinese Space Agency announced that they will send an astronaut into space within the next five years and launch thirty commercial satellites.

Australia wants to be a major commercial space player. In 1998, the country passed legislation that established a regulatory framework for commercial launchers. It also covers requirements for liability insurance. Kistler Aerospace, an American company, is establishing a spaceport in Woomera, South Australia. Australia also provides satellite remote sensing data.

EUROPE

Even though the European Space Agency is planning a number of solar system science missions like a Mercury orbiter and a solar orbiter, and European companies are deeply involved in the commercial satellite launch business, Europeans don't seem to share the "humans-in-space" bug with Americans and Japanese and Russians. While European countries have sent people into space, all of them have gone on American or Russian missions. The European Space Agency and the various national space agencies in Europe and the U.K. have yet to focus on human space flight.

One of the reasons for the difference is that Europeans are preoccupied with what's going on on Earth. They're concerned with jobs, largely because unemployment rates are so high. They're also worried about social security, fairness, and the environment. Politics in Europe lie farther to the left than they do in the U.S. The Green movement is huge in Europe, especially in Germany. Everywhere you go in the United Kingdom, you see disclaimers having to do with whether a particular restaurant serves genetically modified food.

Rene Schaad, a Swiss scientist who is uncharacteristically enthusiastic about humans in space, explains.

> Some countries in Europe, even rich ones like Finland and Germany, have neighborhoods in which a third of the workforce is unemployed. These voters don't care much about Mars, as you can imagine. You have unemployed kids. They don't know what to do with their lives, so they hang around and form gangs. This activity worries the parents and society and keeps people preoccupied with Earthly concerns. If you ask people, "What do you think about going to Mars?" they say "Huh?" And when you start talking to them, they come up with many good reasons not to go.

Schaad cites the reaction of a co-worker who seemed surprised to discover that an intelligent person like Schaad could fall for such a "weird and obviously bad idea." When asked about his reasons for not going, the co-worker answered, "What the hell for? Why go through all that pain, suffering, and expense to be far away from Mother Earth?"

Europeans think Americans' frontier argument is off base.

> The frontier doesn't have a good reputation in Europe. And we don't know frontiers from our own history, so we can't relate. We sympathize with the Native Americans. At one point that was a very hip thing—to want to be like an Indian. Even though that romantic time has passed—from the sixties through the eighties—the feeling that imperialism should not be repeated is still strong. Many people in Europe still view America's foreign policy as imperialistic. So when you talk to people about conquering Mars and use words like "frontier," people form images of Palestine or the old West with slaughtered Indians and buffaloes. Europeans think, "Why would you want to do that again? We've grown so much wiser since then." What people these days want to achieve is some kind of balance, sustainability, ecological viability. We're inward-focused. We tend to be cautious and slow. Ethics are a big driver of public opinion, and ethics in Europe

have a lot to do with social justice and attitudes toward nature.

As far as the search for knowledge goes, that argument for going into space, that's certainly something that Europeans would like to do, but it's really a matter of cost benefit, and robotic exploration is so much cheaper than human exploration. The same goes for the search for life. Why not spend $40 billion directly on Earth sciences? How about $14,000 a day for a submarine that explores the ocean bottom and helps to preserve our marine life—a steal by space program standards, but a fortune for a marine biologist—a project that much more directly influences our daily lives because that's where we get our fish.

If the discussion is about any kind of space program, it's really just about money. People say "Why do we spend $250 million on this?" Ultimately the authorities always justify doing so by saying, "It's cheaper than it used to be." That cost cutting helps to convince people. But the amounts of money involved are mind-boggling for most Europeans. They don't understand why so much money is spent on space.

Since Europeans have never been to space themselves—they've ridden on other rides—there's no history there. There's no questioning, "Why did we ever stop our space program?" like there is in the U.S. We don't have to save face; we don't have to regain our self-esteem.

Reasons for apathy notwithstanding, Schaad is crazy about the idea of humans going to space. "It's just fascinating," he says. "The joy, the ability to play." He thinks part of the reason he feels that way may be that he's spent time in the U.S. His friends think that influenced him. "Ten years ago, I came over on a regular basis for studies and vacation."

The other part of the reason is just who he is. "I think the true reason for my enthusiasm is that everybody needs something to look forward to—something that's bigger than yourself, that makes a difference—and I want to be part of that thing. And that's the thing I picked, I guess, for various reasons. I've always been interested in space flight and space in general, and I like the idea of having a frontier that you're trying to conquer. The idea of a cause, that space creates a dynamic environment for everybody and unites people, that idea is just wonderful."

But when it comes back to the question of funding for manned space projects, Schaad resumes the European point of view.

> I don't believe that initially you can get by with private funding. The more I look at these proposals, they sound so ridiculous. How can you sponsor a Mars flight by selling merchandise? Or viewing rights? I mean, look at *Apollo 11*. After one flight, nobody was interested in that stuff any more. You can't build your financial foundation on something like that. It doesn't work—at least not initially. I believe governments should be heavily involved at the beginning, playing an enabling role and limiting their involvement to those actions that are really necessary. Then they should slowly move out of the way.
>
> I don't think the U.S. is going to do this alone. The U.S. could have put up a space station all by itself, but it wasn't done. Politically it wasn't doable. They needed international cooperation to justify their expenses. They needed to be tied into a network so they could say "We can't duck out now."

Other Attitudes Toward Space

American's attitudes toward space keep changing. Interest peaks when something dramatic happens. According to the Pew

Research Center for the People and the Press, ten of the 689 most closely followed news stories between 1986 and 1999 (1.5 percent) dealt with space exploration, with the space shuttle *Challenger* explosion leading the list. The others were:

- The October, 1988 flight of the space shuttle (first time since Challenger)
- John Glenn's 1998 flight on the shuttle
- Deployment of the Hubble Space Telescope in 1990
- The exploration of Mars by the Mars Pathfinder probe in 1997
- Discoveries made by Voyager 2 in 1989
- The troubles aboard Mir in August and September of 1997
- Astronaut Shannon Lucid's return from Mir in October of 1996
- Discovery of the scientific beginnings of the universe, May 1992
- Discovery of possible life on Mars, September 1996

(Weather was the subject of twelve stories; earthquake stories accounted for four; two were about problems at nuclear plants; six were about health; three were about cloning.)

A poll taken by the U.S. Space Foundation found the following:

Percent with favorable opinion (out of 1,000 randomly selected registered American voters):

	1999	2000
Space exploration	65%*	75%
NASA	75%	82%
GPS system	44%	56%
Space Station Alpha	57%	64%

Believe NASA's portion of the U.S. budget should increase	33%	43%
Believe NASA's $14 billion budget should stay the same	N/A	46%
Believe NASA's budget should be decreased	N/A	6%
*Percentages have been rounded		

On balance, Americans have a favorable view of space exploration, except when it comes to their wallets. No surprise there.

In the United States, Europe, and Canada, approximately one in ten adults can be classified as attentive to science and technology policy, with the U.S. most attentive; the proportion is smaller—about 7 percent—in Japan. Despite their space-related activity, the Japanese seem to be less interested in science and technology in general than Europeans or North Americans (according to the U.S. National Science Foundation's *Science & Engineering Indicators 2000,* Chapter 8). In surveys, Japanese adults express relatively more interest in economic matters and local issues—for example, land use—than in new scientific discoveries and the use of new inventions and technologies. A significantly higher percentage of college-educated respondents in Japan (compared with the percentage of those with less formal education) reports substantial interest in scientific and technological issues, which is also the case in Europe and in North America.

THE AUTHOR'S OPINION

I won't make predictions about who will establish a permanent human presence in space first. However, I do want to stress that space should be an international, multicultural endeavor. And I think it will be. The fact that the X PRIZE contestants come from several countries, that we're building a sixteen-nation space station, that I see people from various nationalities at space conferences—these facts prove that we're moving in the right direction.

—20—

Whither Humans in Space?

What does all this mean? I went out to find out what's happening, heard the news and the views, and developed my own viewpoint. Here it is, dear reader.

And there you have it—a story of colorful people tackling the same basic problem in a variety of ways. People holding wildly divergent views, adopting radically different approaches regarding cost, whether we should do these things in the public or the private sector, whether the health hazards are prohibitive for tourists, whether we should push the property rights issue now, whether the Moon is worthwhile as a destination, how to fund space projects, what you need to make a space business succeed, whether the frontier image is a good one, and on and on.

A unified view of private space efforts is as elusive as Einstein's unified theory. But there are some things we do know:

- Money is the key.
- The Earth's gravity well poses the greatest technical challenge.
- Lots of people would like to go to space; at the same time, there isn't much of a constituency for space.
- Governments have monopolized space.
- Few people in the space field have business experience.

- Space activists are largely cut off from the rest of us.
- Space is an unforgiving environment.
- Doable space has nothing to do with *Star Trek* or *The Jetsons*.

Where these knowns leave us is this:

- *The Reason to Go*. There has to be a driving reason to go into space or it won't happen. My opinion is: A driving reason will emerge out of military involvement, another international space race for political reasons, the threat of extinction, or as the result of commercial bootstrapping. Space commerce or going to the Moon or Mars will grow out of these efforts.
- *The Money*. The money will follow from these involvements.
- *Space and the Public*. Governments will no longer monopolize space, but public involvement will be a long time coming, at least on a greater scale.
- *Space Businesses*. Most space businesses will fail until the industry really gets going. That's normal and okay. Some of the people who fail may start over and succeed the next time.
- *Technology*. If a new propulsion technology comes along, everything could change. But don't bet on it happening any time soon.
- *The Body*. Lots of work needs to be done on the health issues. That will take years.
- *The Cost*. Space is going to continue to be very expensive for a long time.
- *Making Space Universal*. Space needs to be marketed better in order to encourage everyone to go. Word of mouth based on personal experience will help a lot, but it won't be

the only channel we'll need. We also need to change the image from a techie, exclusive one to a user-friendly one.

- *Feasibility.* Space is doable. *Star Trek, The Jetsons,* and *Babylon 5* are fun fiction; that's all. They have nothing to do with what we can achieve in the next couple of decades.

As for when we'll get there, I'm betting it won't be soon. Predictions of massive, imminent change often fail to be realized. When the USSR dissolved, we were supposed to see Russia with a vibrant healthy economy within a few years. We haven't. We were supposed to have a settlement on the Moon by now. We don't. We were all supposed to be talking on videophones by now. We aren't. The internal combustion engine was supposed to have been replaced by now. It hasn't been.

We're only just now emerging from government monopoly of space. It took the airline industry sixty years to get going on a mass scale after the Wright Brothers flew, maybe longer, depending on how you measure the market. It took forty years for computers to be popularized after ENIAC's debut during World War II. (ENIAC was the first operational general-purpose electronic digital computer. It was used by the Allies to produce artillery firing tables.) It's been more than thirty years since the first Earth Day and more than twenty-five since the 1973 oil embargo, and we're still dependent on fossil fuels. We're not going to jump into a fully formed commercial space industry without taking a lot of small steps first.

I wonder about the health issues. Artificial gravity will be an absolute necessity for people who spend time in space. Even so, we don't know how they'll be affected. And what will happen to people who spend long times on the Moon, or Mars? Once their bodies have adjusted to low gravity, what will happen if they want to return to Earth? What will happen to any children born there who have never experienced Earth gravity?

What if we find signs of past or present life out there? Now, I don't think the latter is at all likely, and I don't know about the former, but just in case .... Life on Mars will present huge ethical and biological questions. We've just never dealt with such situations before.

But none of this is to say we shouldn't or won't go. I think we should. With apologies to Kim Stanley Robinson, I think one of the best reasons for going is to ensure the survival of humans. If the Earth is destroyed or becomes uninhabitable, that will be the end of this adventure, of 35,000 years of homo sapiens. No more Shakespeare plays, or Mozart symphonies, or Michelangelo frescoes, or any of the other achievements of humankind. Everything gone—all our history, all our discoveries, all our art. I find the thought of that painful. What a waste that would be. Don't we owe it to ourselves and to all who came before us to preserve our achievements? I think we do.

SPACE TOURISM

Along with Patrick Collins, Ron Jones, Buzz Aldrin, and others, I think the key to opening up space is tourism. Tourism would provide the mass market we need to support space transportation development, which could then be used for other endeavors such as resource acquisition. How do we get there?

We've looked at the bit of market research that's been done, and we've seen that many people claim they'd be willing to spend a month's or year's salary to go into space. We've examined how people spend their disposable income on Earth, and how much they spend on travel. And yet, we must remain open minded about how the space tourism market will really develop.

While I believe that some market is there, I don't know how big or robust it is, and I have a lot of specific questions about it. I've put these, more data, and more analysis in Appendix A. If I were writing a business plan for a space tourism business, the information there is what I'd be looking at in order to size up my market and evaluate feasibility. It's not technical; it's just a bit long. Hence the appendix.

PROSPECTS FOR SPACE VENTURES BEYOND TOURISM

As for the other possibilities—people on the Moon and Mars, space-based mining, and the rest—they're an even harder proposition than tourism. Getting to orbit is one thing. Getting to Mars is quite another, and actually being able to survive there for extended periods is way beyond that. If tourism becomes a reality, then these endeavors are far more likely to succeed because some infrastructure will exist, and there will at last be significant human-inhabited nonterrestrial places to go. If not, then the only way I see these big projects becoming reality is through politics. If there's another international space race, then yes, governments will be happily willing to spend the money, and Robert Zubrin, John Lewis, and Greg Bennett may see their dreams fulfilled. If not, it will be far more difficult to make any of the rest happen. You can't just move a few more miles out, increment by increment, the way pioneers did in the American West, until you've got continuous settlements. You need lots of people and the possibility of making billions and billions of dollars.

Which doesn't mean it's not a good idea to do these things, simply that it's impractical right now. Kim Stanley Robinson has given us all kinds of reasons why it's not a good idea to go to Mars, but he's still a fierce advocate in favor of our going. If we don't go, we'll never know what we might be missing—places and resources that might help us survive. Yes, the risk is that we'll export our negative collective personality traits and our imperfect ways of doing things. But despite the utopias we see in science fiction, we're never going to be perfect. At some point you have to take what you have and just do it.

If we don't go into space—just send robotic probes ad infinitum—what will happen? We've never reached this stage in civilization before, and we don't know anyone who has. We might outgrow the Earth and ultimately die off, or, we might learn to live within our means and go on for a long time.

—Epilogue—

Dennis Tito: The Breakthrough That Almost Occurred ... and the One That Did

I first came across Dennis Tito at a meeting of the Space Tourism Society in Los Angeles in 1999. I was to make a presentation about the Society's upcoming space tourism magazine. I noticed this smallish bald fellow I had never seen before in the audience and wondered who he was. Later I found out that Dennis Tito was a very wealthy man who was interested in space tourism. The next thing I knew, Tito was negotiating for a trip to Mir. *The rest, of course, is history. The magazine didn't take off, but Dennis Tito did. As I look back, I am convinced that Dennis Tito has done more for space tourism than any magazine could ever do.*

By now, everybody knows the name Dennis Tito. They also know that the same Mr. Tito, the world's first self-financed civilian space traveler, had the best time of his life in space, even better than he anticipated. After spending a weightless week aboard the Russian section of *Space Station Alpha* doing routine tasks to help the cosmonauts, looking at the Earth, taking pictures, listening to opera, an exuberant Tito returned safely to Earth on May 6, 2001. "One does not have to be superhuman to adapt to space," he told the world. "It's very doable." He got a little space sick, and his face puffed up a little, but most of the time he felt great.

Dennis Tito is a reluctant hero. In fact, he doesn't think of himself as a hero at all. His reasons for wanting to go into space have always been intensely personal—he just wanted to experience it. But that desire, and its ultimate fulfillment by his visiting *Space Station Alpha* in May of 2001, have transformed him from "some rich guy" into the poster boy for public space travel.

Before his announcement in June of 2000 that he intended to fly to the Russian space station *Mir*, Dennis Tito was virtually unknown in the space community. He had turned up at a meeting of the Space Tourism Society in Los Angeles, and there was noise about an X PRIZE fundraiser at his Pacific Palisades, California home. But he wasn't an activist.

He did, however, carry one credential that made people sit up and notice: He had money, and he was willing to spend it. As we all know, money, a commodity sorely lacking in the private space industry, conveys clout. And that, in a nutshell, is how Dennis Tito has become an international celebrity overnight.

Dennis Tito is almost everything that your typical astronaut or cosmonaut is not. He's much older, for one thing—sixty. He's a private citizen, for another. He has no ties with the military or any national space agency. He's really just a guy off the street—okay, a guy who used to be a NASA Jet Propulsion Laboratory engineer (he has two degrees in aerospace engineering), but that was a long time ago (he was in his twenties then), and he no longer has connections or pull inside the agency.

There have been two other "guys off the street" who have flown in space: the late Christa McAuliffe, who died in the space shuttle *Challenger* explosion in 1986, and a British woman named Helen Sharman who went to *Mir* to do experiments under Project Juno, a Soviet space-mission project. There was also a Japanese Tokyo Broadcasting System journalist who flew to *Mir* at his employer's expense, but he did so to cover a story, not just to have fun or to realize a dream. So Dennis Tito really is unique. He's the first private citizen to fly into space by paying his own way.

It turns out that Dennis Tito has become unique in other ways as well. His personal desire virtually caused an international incident between the U.S. and Russia. When he heard about MirCorp, which was formed in 1999 to commercialize the Russian space station *Mir*, Tito approached the company about flying to *Mir* as a private citizen. MirCorp was delighted, and flew into action. But controversy was already afoot. The Russian space agency wasn't sure it could afford to keep *Mir* in orbit, and NASA felt that the station was too dangerous to remain. The U.S. protested the proposed flight vehemently, even though it held no title to the station. (The U.S. had sent several astronauts to live on *Mir*, and during their residence, multiple physical disasters occurred, including a devastating fire and a collision with a Progress cargo vehicle.) In the end, the Russians decided to deorbit the station for financial reasons. They simply couldn't afford, even with assistance from MirCorp, to keep it going. It was old, worn out, and a bit iffy. On March 23, 2001, the fifteen-year-old station crashed into the Pacific Ocean.

But that wasn't the end of Dennis Tito's flight. *Mir*'s demise was preceded by much debate and lobbying of the Russians by NASA. Realizing that his flight to *Mir* might not materialize, Tito got to work on an alternate plan: flying to the Russian portion of *Space Station Alpha*. At first, the Russians weren't keen on the idea. The U.S., too, nixed the idea from the beginning, as did its European and Canadian partners. The space station was no place for civilians. It was unfinished, dangerous, and sensitive. Only highly trained astronauts and cosmonauts could go there, and only ones who could speak common languages.

Space activists, who knew that the Commercial Space Act mandated that NASA bring commercial business into the station whenever possible, were livid. Here was a chance for NASA to show the public that it meant business. That its iron grip on space was loosening, that it was serious about involving private industry and private citizens in space. Yes, there were issues, Tito's training being one of them. The Russians, who gradually had changed their minds, assured the U.S. that Tito would receive rigorous health screening and training. The U.S. continued to protest.

Tito embarked on eight months—750 hours—of training at the Gagarin Cosmonauts Training Center in Star City, Russia. He learned about the systems he'd be using, like the food and waste management system and the photographic equipment. He also learned 2,000 technical terms in Russian. (He did not learn any of the American systems.) In November 2000, the Russians informed NASA that they intended to fly Tito to *Space Station Alpha*. Yuri Koptev, head of the Russian Space Agency, says that he doesn't remember a negative reaction from NASA at that time.

In March of 2001, Tito and two Russian crewmates arrived in Houston. They were to be trained in the specifics of the space station. NASA refused to allow Tito to be trained, and told him he couldn't go on the trip, even though the Russians were providing transportation in the form of a *Soyuz* spacecraft.

NASA's argument was that Tito's presence would distract the station crew and endanger everyone aboard. The fact that Tito did not speak Russian was only one reason for their protest. They simply didn't want anyone who wasn't a trained official space worker to be there.

The flight was nevertheless scheduled for April. Tito insisted that he was going. The U.S. said no. There were meetings between the Russian Space Agency and NASA. There was an ongoing standoff. At the last minute, NASA relented, partly because if it didn't, it would be risking the necessary participation of one of its major partners. Even so, the agency issued strong caveats. Tito signed an agreement that neither he nor any of his family would sue NASA if he were injured or killed and stipulated that he would pay for anything he broke on the station. They would also hold the other space station partners harmless from anything that went wrong.

Equipped with an insurance policy from the Russian company Avikos, Tito did go (on April 28, 2001), as we all know, and he returned safely. There were reports from Tito's *Soyuz* companions that the NASA astronauts aboard the station were less than cordial to him, and rumors that he was received coldly at NASA's request. At least one of the NASA astronauts has rebutted this claim.

But the U.S. has not been gracious about Tito's flight. Dan Goldin reportedly criticized Tito for stressing NASA personnel, astronauts, and ground workers, and told Congress that the agency would bill Tito for any delays he caused in the construction of the station. And the ranking Democrat on the Senate subcommittee that determines NASA's budget, Senator Barbara Mikulski, whose

Maryland district is heavy on NASA facilities and employees, likened the role of the Russians to that of pimps. Mikulski and the committee's chairman, Senator Christopher Bond (R-Mo.) questioned the U.S.'s ability to cooperate scientifically with the Russians. Goldin said that he could no longer count on the Russians to provide an emergency crew-return capability from the station. In so doing, Goldin essentially held the Russians hostage. As of this writing, there is always a three-person Russian *Soyuz* vehicle docked outside the station to evacuate the crew. Since the station can hold seven people, which it has yet to do, this craft ultimately will prove inadequate. While NASA considers building a seven-person lifeboat, there was talk of keeping a second *Soyuz* vehicle on hand to increase capacity. Goldin nixed that, saying that the Russians have behaved so badly over the Tito issue that he couldn't recommend such a step.

The Europeans claim that flying Tito to the station violated agreed-upon crew selection requirements. They consider Tito's flight an exemption to the rules. But the dispute led to the partners adopting new standards.

Tito reported that individual NASA employees were supportive, despite the agency's public stance. He also said that the Russians treated him very well and were supportive and firm in their commitment to his flight. He said that Russia is actually a freer environment than the U.S., which is characterized by "a lot of effort put into protecting people from themselves" and a paternalistic approach to its citizens.

Certainly the Russians took much risk with this endeavor. After the flight, the head of the Russian Space Agency, Yuri Koptev, admitted that had anything gone wrong with Tito's flight, the Russian space program might have faced bankruptcy. If the station had been damaged, NASA could have presented the Russians with a bill totaling more than the Russian agency's budget, and almost certainly more than the $20 million Tito paid for the flight. But nothing did go wrong.

Let's get one thing clear right away: MirCorp didn't fail. The company that was founded to develop commercial opportunities for the Russian space station did nothing if not think outside the lines. They saw a piece of space hardware that had served well, saw nothing else on the near horizon, and went for it. (While there is *Space Station Alpha*, the company knew that that was for all practical purposes a government-run facility; there might be some commercial opportunity down the line, but *Mir* was available now, and a lot easier to get.)

After much to-ing and fro-ing, MirCorp, which is backed by Walt Anderson and headed by Jeffrey Manber, struck a deal with the Russian Space Agency and Energia, the Russian company that operated the space station, to lease the station. Soon we heard about Dennis Tito's plans to visit. The American television show *Survivor* was planning to site a season there. There were rumors that *Titanic* director James Cameron might also go.

The old (by space standards, vintage 1985) station enabled the longest in-space stays in history and more than 20,000 experiments. One cosmonaut, Valery Polyakov, lived there for 438 days; seven Americans inhabited *Mir* at one time or another, each for longer than any other American who's been in space (longest stay: Shannon Lucid at 188 days). Without this long-duration experience, we'd be much the poorer for information about how space affects the body.

There were problems, including a fire, broken-down equipment, and a collision with a Russian *Progress* cargo craft. The station wasn't in good shape in 1998 and 1999, but many felt that it was fixable. In fact, Space Frontier Foundation activists (see Chapter 16) felt so strongly that *Mir* was fixable and useful that they came up with a longshot plan to keep the station in use. Their scheme almost succeeded. Foundation Executive Director James George describes what happened. (George said the following *before Mir* was deorbited.)

> It's our belief that *Mir* represented very vital property in space. If you take the frontier mentality, which is what we keep pushing, imagine you're a settler in Kansas in

1870. If you have a wagon wheel that breaks, and the rim, which is metal, is broken, you're not going to throw away that metal rim because that's useful. You're going to heat it up, you're going to reshape that into a tool or something. I remember reading the book *Little House on the Prairie* where the little girls, part of their job was to catch nails that fell when Dad was building the house, and straighten them back out because they were too precious to waste. Well, if you've got a completely manufactured and operable space station on the frontier of low-Earth space, all of a sudden that's a very valuable piece of property. The Foundation believed that it was worthwhile to keep it alive.

Certain members of the Foundation, including Rick Tumlinson, got together and hatched this crazy idea: What if we looked at *Mir* not just from a way of getting money for it and getting customers for it, but also a way of reducing its operating costs so it doesn't cost as much to keep in orbit? The man with the money that's also a member of the Foundation is Walter Anderson. He's a very reclusive individual, doesn't like a lot of limelight, likes to do his things and let other people advertise, which of course is exactly why all the news people want his picture. So Walt Anderson and Rick—Rick is Walt's space advisor—they've known each other for years and years, and when Rick came up with this crazy idea, Walt said, yeah, it might work. They went into the whole concept feeling that they were going to fail, that someone was going to swat them down. NASA was going to stop it, or the Department of Defense (DoD) was going to stop it because NASA hates competition and DoD is paranoid where space and proliferation are concerned. So they approached the whole issue as "This is crazy, it's never going to succeed, but we may as well try it."

They got a guy named Joe Carroll from Tether Applications down in San Diego. And they drafted Jeff Manber in as the CEO [of MirCorp]. Jeff Manber was at the time working on another Foundation project called

> the space business archives. So here these guys are going merrily along, having meetings with the Russians, trying to see if they could actually pull this off. They made a trip to Russia, and that's when the first memorandum of understanding between Energia [the primary Russian space company] and the new company called MirCorp was signed.

Press reports almost invariably omit Rick Tumlinson's name from their accounts of the wheeling and dealing. George explains why.

> Rick is the director of FINDS [Foundation for the International Nongovernmental Development of Space]. That's a nonprofit. He's also president of the Foundation. That's a nonprofit. He cannot be seen to and cannot have an active role in the promotion of a for-profit venture or work on behalf of that for-profit venture because that is a conflict of interest. So did he have a lot to do with it? Sure. He was the one who had one of the original ideas. But was he the guy that runs MirCorp? No. Is he an officer with MirCorp? Is he a stakeholder in MirCorp in any way? Not to my knowledge: not as shareholder or anything like that. He's no more of a stakeholder in MirCorp than any other human on Earth. All humans are stakeholders in that because a successful MirCorp will change our space history. And even an unsuccessful one will have an impact. So Rick has no involvement per se in that sense. He cannot.

There's another factor in Tumlinson's keeping a low profile over this issue. "Another reason that could be why you're not hearing too much about Rick is the fact that NASA is livid over this," says George. "They wanted the Russians to use that *Progress* vehicle to push *Mir* into the ocean, not keep it orbiting Earth." Tumlinson already has incurred NASA's wrath over the years. It's not always in his or the Foundation's best interest to constantly irk the agency.

Dennis Tito reportedly paid the Russian Space Agency $20 million for the privilege of going into space for about a week. That's a pretty high price for a trip, but because it's the first, the price can only come down from there. By paying $20 million for his experience, Dennis Tito has set several precedents and standards: the opening price for a trip into space by a private citizen, minimum training requirements, health standards, and space travel as something that private citizens can do.

After the flight, the Russian company RSC Energia began to seek additional private space travelers to occupy seats on a week-long orbital voyage in its three-person *Soyuz* vehicle. Avoidance of the space station would keep the trip simpler on both political and technical levels. The company specified that such travelers must know Russian and have an engineering background. (Not a bad incentive for people to learn the language and pursue careers in engineering, which the U.S. has long been saying it wants its kids to do.) A medical examination will also be necessary. But there is also talk of other Russian-sponsored "tourist" flights to the space station, including those put together by German television marketing company Brainpool, which wants to stage contests to select space travelers and produce shows anchored around their experiences; some involving TV show *Survivor* participants, and possibly also another reality-TV show; one by Mark Shuttleworth, a South African Internet security tycoon; and possibly *Titanic* director James Cameron, long a supporter of space exploration and tourism.

At the same time, Dennis Tito began to investigate the commercial possibilities of space tourism from a sound market and business perspective. His first task is to assess demand. Then he will decide whether or not to set up a fund to invest in space ventures. If he does, chances are it will back winners. Dennis Tito is such a good businessman that he almost can't lose.

—— Appendix A ——

Space Tourism Market Issues Expanded

Assessing the feasibility of a mass space tourism market requires much research and financial analysis. This book isn't a business plan, but we can raise some questions and look at some preliminary data here. In the section that follows, I've raised some important questions and presented some data that might begin to reveal how people approach spending in general and travel habits in particular. There is no analog to space travel; looking at somewhat similar pursuits on Earth is all we have to go on. But that's true for many new products and services. You examine everything you can, you think carefully about what you've got, and you either go for it or you don't, making sure that you've devised contingency plans.

Possible Barriers to Going on a Space Trip

People say that they would spend a lot of money to go into space, but will they really do it when push comes to shove? For one thing, will they actually have the money? It takes a long time to save a year's salary, or to pay back a loan for that amount. If the price of a ticket is $30,000, it might take ten years to save the money. Or, enthusiasts might have to liquidate some assets—take out a second mortgage, sell some stock, etc.—which could pose a hardship. If a couple wants to go together, they have to save twice as much money, and to take the kids means saving even more.

What will the prospective travelers have to give up in order to save that money? What if some necessity comes along in the meantime that prevents them from accumulating the funds? What if they're presented with some great deal that's hard to refuse, like a discount on some other vacation they'd like to take, or the opportunity to get in on an attractive investment, or the possibility of sending their kids to an Ivy League school rather than a state one? What if an earthquake strikes, or a health emergency arises? Factoring in all these possibilities, how many people will actually have the money to take the trip? Enough to support Patrick Collins' projections of a million people in the first ten years?

Preferences change over time. For example, Stanley Plog of Plog Travel Research reports that in 1967, consumers favored the purchase of material goods over travel. However, in the late nineties, the situation reversed, with many people preferring to accumulate memories rather than material goods, which wear out. As of 1998, leisure travel was growing at twice the rate of business travel, and a Gallup poll commissioned by the Virginia Tourism Corporation in 1993 confirmed the oxymoron that travel is a "necessary luxury" for Americans. However, that situation could flip-flop at any time.

There's another question about money. Are five to ten million people in thirty years likely to have the amounts of disposable income needed to take these trips? The Japanese Rocket Society says that those numbers represent 2 percent of the middle-class population of the world at that time. Realistically, those who will have enough will most likely reside in G-7 countries, that is, the U.S., Canada, the U.K. and Europe, and Japan. Let's look at some population figures (as of 2001):

- U.S.: 281 million
- Canada: 31 million
- Europe: 512 million
- Japan: 130 million
- Total: 954 million

While world population is estimated to increase 40 percent by 2030, the cited countries gain population more slowly. The U.S. will have 30 percent more people, Europe only 6 percent more. Let's say that the population of the G-7 countries will increase by 20 percent by 2030, which will bring it to about 1.1 billion.

The JRS's figure of ten million tourists represents 0.9 percent of the total populations of the G-7 countries. Is that a reasonable figure? Would 0.9 percent of the population of the same countries today (8.6 million) have the money, time, and inclination to take a space vacation? Certainly 8.6 million people today don't spend $25,000 on a vacation. Will people be relatively wealthier in 2030, and if they are, how will they spend their money?

Economist Collins is optimistic. He asks us to consider the Engel coefficient, the proportion of people's income used to purchase necessities. The Japanese government estimates the figure at between 30 and 40 percent. The percentages are similar in other rich countries, and they're falling steadily. If that's so, a family could reduce its expenditures by more than 50 percent without lacking any necessities. In developed countries, it's taking less effort to satisfy human needs, which means that people have more resources for doing what they want to.

HOW PEOPLE SPEND THEIR MONEY

Is Collins right about this? Look at how people spend their money overall, not just on travel.

In 1999, Americans spent about $37,000 per year, not including taxes. Housing, at $12,057, represented nearly a third of that. Pensions and social security represented another 8.4 percent. Transportation took up 19 percent. That's already a total of 60 percent, and it doesn't count food, healthcare, clothing, or taxes. I'm skeptical about this generous estimate of discretionary income.

On the other hand, nearly three and-a-half million of us plop down somewhere in the neighborhood of $30,000 for sport utility vehicles, which, while transportation, are far from necessary or even practical. If we all bought economy cars instead, we'd have half the price of a ticket into space left over. And with energy prices going up and SUVs declining in popularity, that could happen.

But economic fortunes rise and fall. Recessions, or fear of them, can erode consumer confidence to the point where people reduce their spending on all sorts of things. My guess is that space vacations would be hard hit under such circumstances, but that may depend on where in the tourism cycle we are. At the early stages, where only wealthy people will be able to afford to go, the market may not suffer that much. Later on, when space is open to the middle class, economic downturns could depress ticket sales significantly. The question would be whether the industry could survive such dips.

AMERICANS' SPENDING HABITS

According to *American Demographics* (April 1999), the most important predictor of spending is where the consumer is in the life cycle. Most twenty-somethings spend less than the average on most products and services because their households are small and their incomes on the low side. Middle-aged people, who have larger families and incomes, spend the most. Older people spend less than middle-agers as their family size and income fall.

In the 1990s, major changes occurred in consumer spending patterns. Early middle-aged people thirty-five to forty-four—the largest share of American households and the largest share of most consumer markets—decreased spending dramatically in just ten years from 1987 to 1997. The reason? Economic insecurity caused by the recession of the early 1990s and falling incomes. During the same period, the number of households headed by forty-five- to fifty-four-year-olds increased 44 percent, and their spending rose accordingly. And spending by those sixty-five and over—an increasingly educated and affluent group—rose faster than that of any other segment. This trend will continue as Boomers move into that age group.

However, if the age of Medicare eligibility rises, then Boomers will have to spend a larger share of their money on medical care.

Average annual spending per household, 2000: $38,045

Food, both home and away, 2000: $5,158

Housing, 2000: $12,319

Transportation, 2000: $7,417

Entertainment, 2000: $1,863

Personal insurance and pensions, 2000: $3,365

Healthcare, 2000: $2,066

Apparel and services, 2000: $1,856

Other, 2000: $4,001

Numbers of SUVs sold in the U.S. in 2000: 3.35 million. Price roughly the same as that of a ticket to space.

Sources: American Demographics; Bureau of Labor Statistics.

HOW PEOPLE TRAVEL

Travel and tourism is one of the world's largest industries. That's very good news. Expensive luxury vacations do sell. That's also good news.

Here's some more possible good news. The Travel Industry Association of America reports that in 1997, U.S. residents made 1.3 billion person-trips. Seven hundred fifty-two million of those were vacation trips; 862 million were pleasure trips. (There is overlap between the two categories.) In order to fulfill Patrick Collins' predictions of one million space vacationers, the industry would have to turn only 0.1 percent of those trips into space trips. A 0.1 percent market share is a reasonable target. Of course, by making that statement I'm assuming that you can consider the current trips comparable to space tourism trips, which is not a given.

The "How the World Travels" sidebar shows that most people travel for pleasure. The fact that 72 percent of all vacation travel is family travel is a bit worrying at first glance because of the expense involved in taking so many people into space; however, combining that percentage with the number of vacation trips people take shows us that 211 million vacation trips were not family vacations. That's only U.S. residents, not the total market. Therefore, to reach Patrick Collins' estimate of one million tourists from this segment is also not a gargantuan goal.

The fact that Americans own 9.3 million recreational vehicles is a good and a bad sign. Recreational vehicles can cost quite a bit of money—amounts easily comparable to those of a space vacation. However, the types of people who buy and use recreational vehicles are probably not the best target market for space vacations as they tend to be frugal. (Buy an RV once, take vacations with it for years, not the one time that most people will probably go into space.)

As for convention delegates, forget it, unless their organizations pay their way, which is improbable. You probably won't see conventions in space for quite some time.

Findings from *The Wirthlin Report: Traveling in the '90s* are less rosy. People have a difficult time getting the money together for vacations, and an even harder time getting the time to take them.

HOW THE WORLD TRAVELS

In 2000, 75 percent of Americans' domestic trips were for pleasure. Leisure, recreation, and holidays accounted for 62 percent of trips worldwide, while business travel made up 18 percent of trips. Fourteen percent of American trips were for business.

In 1999, 66 percent of Americans traveled for pleasure. In 1998, leisure, recreation, and holidays accounted for 62 percent of trips worldwide, while business travel made up 18 percent of trips worldwide in 1998. In 1999, 17 percent of American trips were for business.

In 2000, family vacation travel accounted for about 30 percent of all U.S. vacation travel. School calendar changes, including snow days, starting school before Labor Day, and year-round schedules, had an effect on 15 percent of the vacations. Of those affected, 58 percent changed the time of their vacation to accommodate kids' changing school schedules. In 2000, 22 percent of parents who took a trip let their children miss school to go with them.

Most vacations in the U.S. are short. In 2000, 17 percent were day trips, 38 percent are one to two nights, and 45 percent are three nights or more. More than a third of those trips involve spending the night in a private home. The average traveler spends about three nights in a hotel or motel.

Delegates to conventions spend an average of four days and $862 to attend the events, according to the International Association of Convention & Visitor Bureaus.

There are 9.3 million recreational vehicles on U.S. roads, including motor homes, travel trailers, fifth wheels, truck campers, pop-up trailers, and van conversions.

According to *The Wirthlin Report*, forty-five percent of U.S. adults report that their major vacation planning challenge is saving money for the trip. The same poll found that 24 percent of vacation planners had trouble getting time off from work.

Sources: American Demographics, Travel Industry Association

Look at the Wealthy Americans and Travel sidebar. Wealthy Americans have both the time and the money to take space vacations. Good news.

WEALTHY AMERICANS AND TRAVEL

According to American Demographics in August, 1998, more than one in five people whose annual household income is $1 million or more take off at least two months per year (Lou Hammond and Associates survey). People with household incomes between $250,000 and $999,000 take quite a bit of time off as well: Fourteen percent take sixty-four days per year, 25 percent take thirty-six to sixty-three nights, and 48 percent take fifteen to thirty-five nights.

Those with incomes $1 million and over spend an average of $48,800 annually on vacations; households earning between $250,000 and $999,000 spend $23,200 on average.

The next sidebar shows that the average American spends a lot less for a vacation than it would take to go into space. But remember how averages are calculated: You add up all the spending and divide by the number of people. That means that there are a whole lot of people who spend more—sometimes way more—than a thousand dollars per year on travel. Travel and tourism wouldn't be the world's largest industry if that weren't the case. These numbers may be realistic from one angle, but they don't tell the whole story.

U.S. TRAVEL TRENDS

On average, Americans spend between $1000 and $2000 per trip. About 15 percent of those who travel for pleasure spend more than $2,500.

Older Americans, that is, those fifty-five and over, account for almost a third of resident travel within the U.S. This group has the highest net worth, lowest consumer debt, and highest discretionary income of all age groups. As Boomers age (as of this writing, the oldest Boomers are turning fifty-six), the travel industry may experience dramatic growth. Thirty-two percent of today's workers, regardless of age, say that the activity they most look forward to when they retire is travel (according to a Gallup survey).

Source: Travel Industry Association

Now here's the bad news: September 11, 2001. The terrorist attacks on the World Trade Center and the U.S. Pentagon severely undermined the willingness of people to travel by air, and airline bookings plummeted. Despite a government bailout, the industry sustained a heavy blow, and at press time, no one knew if or when air travel would return to pre-attack levels. If people are afraid to fly on airplanes because of possible terrorist attacks, surely they will fear space flight.

Has September 11th killed the prospects for near-term space tourism? It's tempting to say yes, but I think that's too facile. The answer is "Maybe." Under the circumstances, investment will be harder than ever to come by, particularly if demand no longer appears robust. Further, having already taken major hits from terrorist activity, insurers will be even less sanguine about taking risks than they are today. (I have always thought that risk avoidance by the insurance industry was a paradox, but maybe I'm missing something.) It goes without saying that if space tourism materializes, there will be questions about passenger screening that could turn even uglier than those involving the airlines.

On the other hand, it depends what happens in the world over the next decade. If people feel that efforts to stem terrorism have succeeded, the attacks will have minimal effect, especially in the U.S., where memories are short. In the U.K. and other places that live with terrorism on a daily basis, life goes on. People are used to seeing submachine-gun-toting soldiers patrolling at London's Heathrow Airport and feel safer for their presence. Life is tenacious. Humanity is a gritty species. I believe we will adapt.

It used to take a generation for people to recover from major stock market crashes. In the U.S., it now takes but a few years to rebound from major natural disasters. The attacks have set us back; any future ones will do so as well, but not for long, I think. I also believe that the future holds continual surprises for us, and that one or more of them might just lead to societal changes major enough to spark a space tourism industry at last.

Other Space Tourism Market Issues

What about the time to travel? If going into orbit is all the tourist is going to do, and training takes only a day, that may not be so bad. But if travel to a remote location is involved, that adds time and money to the equation, and it means that couples and families all have to be available at the same time in order to go together. *American Demographics* reports that most Baby Boomers love to travel, but they don't have the time to go for more than a quick getaway. Instead, people sixty and over are the ones with the time. And we know that those over sixty are less likely to want to take space holidays than younger people.

In order to garner the proposed number of tourists, there will have to be a massive advance global high-profile marketing effort comparable at least to that of a major feature film. That means hundreds of millions of dollars for marketing. Such a campaign will add significantly to the costs and lengthen the payback period.

There's also the issue of early feedback. If the first tourists rave about their experience, the floodgates are sure to open. But if they spend half their time being space sick, or if there are adverse incidents relating to safety, word of mouth could go a long way toward spoiling the market. All we have to go on now is reports from astronauts and cosmonauts, who travel under different circumstances with different training and who are not allowed to take motion sickness medicine because it interferes with their ability to perform their duties. On the other hand, even if we don't know what the first experiences will be like, operators can adjust the imperfections—probably. The speed with which they do it and get the word out about improvements will determine if the industry sinks or swims.

What about the investment required to develop a fleet of space tourism vehicles? If you take Collins' figures, we're talking about nearly $40 billion to develop and build them. That's a lot of money—three times NASA's budget. But if you look at just the initial $12 billion that Collins and the Japanese Rocket Society have estimated for development and certification alone, it's not that much money. Amounts in that range are raised and spent all the time. Compare costs with other expensive projects and say that since money is available for them, why not for space? For example, $9 billion is about to be spent to upgrade JFK Airport in New York—that's just one airport.

A new factor that's emerged with the U.S. presidency of George W. Bush is that of the military and space. The new U.S. Secretary of Defense, Donald Rumsfeld, has pledged that he will beef up military involvement with and spending on space. Like it or not, this trend will influence commercial space, probably positively. Military involvement with technology often serves as an incubator, after which the technology moves into mainstream markets. It happened with airplanes after World War I, it happened with global positioning satellites, it happened with computers after World War II, and it could happen again. Considering the vast amounts of money necessary for developing new kinds of space vehicles, it may be that only governments will be able to come up with them, especially since they don't have to worry about making a profit. So this is an area to watch.

Developing consumer markets is always a tricky thing. Consumers are much more fickle about spending their money than businesses are. They also have less money to spend in the aggregate. So attempting to develop a whole new industry around consumer demand may not be the easiest way of doing it. None of the industries we've looked at in this book—aviation, railroads, computers—started with or became strong by depending on consumers. It took a long time before the consumer aspects of those industries became at all significant. Yet here we are trying to start a tourism industry before we have a freight or industrial one.

On the other hand, freight has pretty much nowhere to go in space, other than *Space Station Alpha*. People don't need a destination as long as they can look at the Earth—at least in the early stages. So, maybe it will work.

— Appendix B —

Space Tourism Market Surveys

In 1993, the first major space tourism survey was conducted in Japan by Dr. Patrick Collins, while a Science & Technology Agency Fellow based at the National Aerospace Laboratory in Tokyo. Participants were given a brochure explaining that the service would be akin to that of airliners, and telling them that they would stay in a hotel orbiting the Earth. About 3,000 people were surveyed face-to-face; 90 percent of them responded, an extremely high rate. Here's what Collins found:

- Percentage of population that wanted to visit space at least once: 70–80
- Percent of those who claimed they'd pay up to three months' salary: 70
- Significant difference between the responses of men and women: none

One important factor in evaluating the data is that the Japanese personal savings rate is high. The Japanese also invest more than many other people, and their unemployment rates are lower than those in the U.S. and Europe.

In 1995, Collins conducted another survey, this time in the U.S. and Canada. About 1,000 people were questioned over the phone. Here's what he found:

- Sixty percent of the population wanted to visit space.

- In every age group, men were more interested than women, with the difference being about 10 percent.
- A majority said that they'd like to visit space for either several days, a week, or more.
- About 15 percent said they'd like to visit for a few hours.
- Two-thirds expressed a desire to go several times.
- 45.6 percent said they'd part with three months' salary.
- 18.2 percent said they'd pay six months' salary

At the time of the survey, average income for Americans was $2,000 per month, or $24,000 per year. That means that about 10 percent of the people would pay $24,000 for a trip, while about a fifth would pay $12,000 and close to half would pay $6,000. Obviously, because some of the recipients earn more than $2,000 a month, many people would pay more.

The survey did not spell out in detail what the space tourism experience would be. This omission is extremely important, and could influence the results. When questions arose, the researchers told the subjects that the trip would be to Earth orbit, not to other planets or solar systems, but they didn't describe these limits to all participants.

Another survey using the same questions was carried out in the two Berlin airports in 1994. This group of people obviously represents a narrower slice of the population than the samples in the Japanese and North American surveys. Air travelers tend to be of middle and high income. They also represent people who are willing to fly, even if they fear doing so. Nevertheless, the survey found that only 43 percent of those surveyed were interested in traveling to space.

All three surveys asked what people would like to do in space. Here's what people said:

- Most people in all three geographical areas said that they would want to look at the Earth.

- The next most popular activity was walking in space.
- About a third wanted to do astronomical observations.
- About a third of the Japanese respondents wanted to engage in zero-gravity sports, about 15 percent of the Germans, and just over a fifth of those in North America.
- A third of the Germans, a fifth of the North Americans, and just over a tenth of the Japanese wanted to do zero-gravity experiments.

Here's what the surveys revealed about length of stay:

- Between 5 percent and 7 percent wanted to take a day trip.
- More than 40 percent of Germans wanted to go for two to three days.
- Thirty-five percent of Japanese and 30 percent of North Americans found the couple-of-days option attractive.
- Forty-five percent of Japanese and 35 percent of the others wanted to go for a week. More than a quarter of North Americans wanted to go for several weeks. A fifth of Germans and about a tenth of Japanese wanted to stay for that long. This means that hotels will be necessary; orbiting vehicles alone will not be sufficient to satisfy demand.

Number of people willing to pay for a trip:

- One hundred percent of North Americans and 90 percent of others would pay a month's salary.
- Seventy percent of Japanese and about 45 percent of the others would pay three months' salary.
- One third of Japanese and one fifth of the others would pay six months' salary.

- 20 percent of Japanese and a tenth of the others would pay a year's salary.
- 10 percent of Japanese and a percent or two of the others would pay three years' salary.
- About 4 percent of Japanese and almost none of the others would pay five years' salary. Indeed, at first costs will be high, so that the numbers represented by those willing to pay higher prices help define early market demand.

Reasons that people did not want to go:

- 40 percent of North Americans who didn't want to go cited safety as the reason. Fifteen percent of Germans and about 27 percent of Japanese said the same.
- About 35 percent of the people who did not want to go said they'd prefer to dream about going. Thirty percent of Japanese and about 26 percent of Germans said the same.
- Between five and ten percent of those not interested said they didn't think the idea was realistic; the most skepticism was expressed by North Americans, with Germans second and Japanese last.
- About 18 percent of Japanese, 16 percent of Germans, and 3 percent of North Americans cited expense as the reason they wouldn't go.
- Other reasons offered for not going had to do with ecological concerns and age.

In 1998, another survey was conducted, this time in the United Kingdom by Olly Barrett at Bournemouth University, Dorset, England. Subjects were questioned face to face after the researcher explained the space tourism concept, but without detail:

- 34.7 percent said they'd like to take a trip into space.
- 23.6 percent said maybe.

- 4.17 percent said no.

Price findings:

- One hundred percent would pay one month's salary.
- Sixty percent would pay three months' salary.
- Twenty percent would pay six months' salary.
- About 10 percent would pay one year's salary.
- Forty percent would spend £700 (about $1,050).
- Twelve percent would pay £8,500 (about $13,000).

Length of stay:

- About 10 percent would go for a few hours.
- About 16 percent wanted to go for one day.
- Thirty-three percent wanted to go for two to three days.
- Twenty-eight percent wanted to go for a week.
- Eleven or 12 percent wanted to go for several weeks.

The U.K. survey asked respondents to rank the activities that they'd like to engage in:

- The most popular activity was viewing the Earth.
- Almost equally popular were looking deeper into space and walking in space.
- The next most popular activity was to be weightless.
- Scientific experimentation and reentry ranked fifth.
- Space sports ranked last.

Reasons for not going:

- 36.4 percent cited safety.
- 27.3 percent were uninterested.
- Expense was a concern in the under twenty and thirty-to-fifty age groups.
- Age was a barrier to those over forty.
- People over forty also thought space tourism was an unrealistic concept.
- Some other reasons were fear of flying, the desire to visit other places instead, better ways to spend one's money, and not healthy enough.

— Appendix C —

Regulatory Issues

Despite forty years of space flight, the U.S. lacks a regulatory framework that would support space tourism. The only commercial flights that launch from North America involve lobbing satellites and remote sensing equipment into orbit—not people, and not on reusable craft. This means that the Federal Aviation Administration (FAA), which oversees such flights, is concerned only with the safety of people on the ground and those involved in aviation flights that travel through the same airspace. So even if we had the vehicles today, we wouldn't necessarily be able to use them.

For safety reasons, launches on American soil are conducted from the two coasts—Cape Canaveral in Florida, Vandenburg Air Force Base in California, and Wallops Island, Virginia—though Kistler Aerospace is establishing a facility outside of Las Vegas, Nevada. A new spaceport in Alaska is dedicated largely to launching commercial vehicles. The FAA is worried about possible damage and liability if a craft were to crash on land. (The space shuttle does fly over land when it lands in Florida, but that coasting and landing phase of shuttle voyages is far less likely to come to grief than the takeoff phase.) These facilities cannot support high launch volumes. Therefore, changes in regulations will be necessary to allow launches from other locations, some involving flight over land.

Some of the other issues involved in regulating space tourist vehicles will be the same as those for aviation:

- Spaceport operation
- Traffic management
- Training and qualification of pilots and mechanics

- Safety surveillance
- Liability
- Environmental impacts and noise
- Hazardous cargo
- National security and foreign policy interests

But the space environment will require other issues to be addressed as well:

- Space debris
- Judicial jurisdiction for facilities in orbit
- Property rights and rights to noninterference with property in space
- Individual rights
- Taxation

We can draw upon existing codes as models for many of these items. The airline industry covers transportation aspects, building codes similar to those on land and for ocean liners can be used for orbital facilities, and maritime law and cruise industry regulations can be looked at for issues of personal behavior such as gambling and drinking.

Even though the FAA is working on some of these tasks, the only way that regulation really will work is to set worldwide standards. It's unlikely that suborbital vehicles will take off and land in the same country, especially if they carry packages and other cargoes around the world. (Currently, bilateral agreements exist between countries that certify commonly operated aircraft.)

Right now, two different parts of the FAA are involved in space launch and certification, the Office of Commercial Space Transportation and AVR (The FAA regulation and certification organization). AVR has developed a licensing procedure for reusable launch vehicles, but so far, no certification is actually required. The Office of Commercial Space Transportation believes that it can regulate passenger-carrying space vehicles on a case-by-case basis, but right now it doesn't have the staff or the funding to do so on a full-time, programmatic basis. AVR has the authority for ensuring safety within the atmosphere, so who has jurisdiction over space vehicles, which travel both through the atmosphere and beyond it? Part of the difficulty in solving the problems is that government officials at the top are not pushing for nor sustaining interest in doing so. Another part is that NASA is split over the whole idea of mass space transportation.

Suggestions from the industry for streamlining development of the regulatory environment include allowing that some voluntary personal risk be taken by space tourists and crew. Space trip operators would be legally able to accommodate passengers on space vehicles that have successfully completed an abbreviated (compared to today's commercial aircraft) certification test program when the passengers are acknowledged "adventure" tourists. Potential passengers could be provided with the reliability and safety information accumulated from the vehicle's test program so as to make their own decisions.

Obviously, regulators have to feel that the flights are safe enough to allow operators to run them. They also have to decide when that point has been reached, determine who decides that the conditions have been met, and define the level of physical qualification that passengers must meet.

Patricia Smith, FAA Associate Administrator of Commercial Space Transportation, explains that the approach to rule making will be for "a very high degree of demonstrable safety and stress the inherent right of free people to give informed consent to accept a certain amount of personal risk."

The FAA is looking at current regulations for aircraft in the attempt to determine how they can be applied to space vehicles. Some will be insufficient. For example, aircraft don't have to deal with the effects of the space environment and reentry into the atmosphere. Space vehicles may have to be designed with more redundancy: Should the systems be fail-safe or fail-operational?

Another issue that has to be addressed include instituting a space tourism regulatory framework that's consistent and compliant with international space treaties. One question is whether space tourists are considered "astronauts," a title that conveys certain rights.

There's no accepted international definition of where airspace ends and orbital space begins. These issues matter because of national security. Currently, U.S. airspace extends about 60,000 feet above sea level; the country has the right to deny access to its airspace. The kicker is that no one knows who has the authority over 60,000 feet, and outer space doesn't begin at 60,001 feet. By some definitions, it starts at fifty miles; by others, at sixty-two. The U.S. Air Force awards astronaut wings to pilots who've flown above fifty miles, while the Federation Aeronautique Internationale defines space as beginning at sixty-two miles. The Office of Commercial Space Transportation doesn't say officially, but it implies that space begins at sixty-two miles by exempting amateur rockets from the definition of space vehicles. (Amateur rockets can't exceed sixty-two miles, but rockets and satellites launched into space or suborbit must do so.) What happens between the end of air space—60,000 feet—and 327,360 feet (sixty-two miles)? If a failure occurs there, which treaty applies?

In the United States, private space launches are regulated by law and by red tape piled upon bureaucracy. As a result, there's an interminable lead time between applying for permission to launch and actually launching. This delay dampens the market for commercial launches.

Part of the issue is that there are thirteen agencies in the U.S. setting space policy. Some of that mess has grown out of requirements set by international treaty. Tort liability, that is, harm from products, is covered by the 1967 Outer Space Treaty and the 1972 U.N. Convention on Liability. Under the latter, it's not the civil spaceline operator who's liable for accidents—it's the government of the country of registry. The former requires that all space activity be conducted only by states, which must authorize and supervise any nongovernmental organization operating in space. So states assume a huge responsibility for everything that happens out there and must constantly hover over anyone who's working there.

Because of these treaty requirements, the Office of Commercial Space Transportation was established within the U.S. Department of Transportation. The office created regulations covering safety inspections, licenses, clearances, permits, and approvals for launches, and invoked the authority to inspect all launch vehicle and payload builder and operator facilities.

Every airplane that crosses an international border must carry a certificate of airworthiness. New designs take money and at least a year—often several years—to obtain certification from the FAA. So far, no regulation allows a space vehicle to be certified. The FAA is writing stringent medical requirements for spacecraft; some in the space community are trying to get the idea accepted of accredited passengers who will accept risks.

One of the issues is that of launch pads. Commercial companies don't necessarily want to launch from government-controlled facilities, so some of them are looking at new possibilities. Kistler, for example, is examining a possible launch site in West Texas as well as locations in Nevada and Australia.

What you're allowed to do when launching varies with the spaceport and the country. With some spaceports you can go polar, east, to *Space Station Alpha*. Whether you can depends on range safety and location. Whatever country you launch from has a liability to other countries for your activities, which means that individual countries have the right to set the laws of who can launch and who can land.

—— Appendix D ——

Propulsion

Primer on Propulsion

Efficiency. There are two key factors under our control in making rockets efficient:

1) Maximize the proportion of the rocket that's propellant. The more mass you throw backwards the faster you can move forward. Rockets typically put less than 1 percent of their mass in orbit. (This is why they're so expensive.)

2) Create a system that throws mass backward more quickly. Specific impulse (Isp) measures how much propulsive force you're getting from the propellant. The higher the specific impulse, the faster the mass being ejected is moving and the farther you can go on a given load of fuel.

Typical Isp numbers:

35–65 seconds (exit velocity about 1,000 mph) for rockets with stores of pressurized gas, usually helium or nitrogen, expelled through nozzles. Same Isp for propulsion packs on astronauts' space walking suits.

250–400 seconds with exit velocities over 8,000 mph for chemical systems that burn propellant combinations, including shuttle rockets.

Over 2,000 seconds and exit velocities greater than 40,000 mph for electrically driven rockets that accelerate ions in an electric field. Propellant usage efficiency is high but thrust is low (less than that of a house fly). This type of propulsion applied over many months in space can take a spacecraft beyond the solar system.

Propellants. All rockets use either a nonreactive propellant like nitrogen gas or a combination of chemicals that react with each other. Electric propulsion usually uses a single nonreactive propellant combined with a lot of electric power. Liquid rockets generally carry their propellants as liquids in separate fuel and oxidizer tanks. Solid rockets premix the fuel and oxidizer into a solid crystalline or rubbery material that is molded and bonded into the rocket motor casing.

Choosing a technology. Launching into orbit from near Earth requires a lot of thrust and huge flow rates from liquid propellants. It also requires big, fast pumps and a fuel injector and chamber that can handle it all. All that heavy hardware makes it even more difficult to get off the ground.

Solid rockets require no pumps, valves, or tanks. They're ideal for weapons. Solids have lower Isp than liquids because they're larger and heavier. They're dangerous because they can be inadvertently ignited and cannot be shut off. Solids also tend to be toxic. They give a high thrust, which can be good except that it subjects the payload to large acceleration loads.

The most common configuration is a mixture of liquid, which helps attain a high Isp, and solid, which provides extra thrust for the liftoff.

Once you're in orbit, you need a high Isp to change to another orbit or change direction. Both solids and liquids are used for this.

Sailing on the solar wind requires no propellant.

Paraphrased from Rick Fleeter's excellent and funny (yes, funny) book, *Micro Space Craft*. Reston, VA: The Edge City Press, 1995.

ALTERNATIVE METHODS OF PROPULSION

Magnetic levitation (maglev). A maglev system involves the use of magnetic fields generated electrically to levitate and accelerate a craft along a track at very high speeds. Once the vehicle reaches 600 mph, it switches over to rocket engines to launch it into orbit. The cost of electricity per launch: $75. Advantage: No propellant means 20 percent lower weight than that of a typical rocket.

Laser. This method is in the research stage. A craft ascends on a laser beam.

Microwave power. This method would require a microwave power station in orbit.

Appendix E

Bibliography

Ashford, D. M. (1997, February). Space Tourism—How Soon Will It Happen? Paper presented at IEEE Aerospace Conference, Snowmass, CO.

Bond, P. (1999). *Zero G: Life and Survival in Space.* London: Cassell.

Collins, P. (1995). Benefits of Commercial Passenger Space Travel for Society. Paper presented at 5th International Space Conference Of Pacific Basin Societies.

Collins, P. (1999, April). Space Activities, Space Tourism and Economic Growth. Paper presented at 2nd International Symposium on Space Tourism, Bremen, Germany.

Collins, P. "The Space Tourism Industry in 2030." Proceedings, Space 2000, ASCE, 594-603. Presented at Space 2000, Albuquerque, March, 2000.

Collins, P. Q. (1990).The Coming Space Industry Revolution and Its Potential Global Impact. *Journal of Space Technology and Science, 6,* 21-33.

Collins, P., & Ashford, D. M. (1988). Potential Economic Implications of the Development of Space Tourism. *Acta Astronautica, 17,* 421-431.

Collins, P., & Isozaki, K. (1997, July). The JRS Space Tourism Study Program Phase 2. Paper presented at Nagasaki: 7th International Space Conference of Pacific Basin Societies (ISCOPS), Nagasaki, Japan.

Collins, P, Iwasaki, Y., Kanayama, H., & Ohnuki, M. (1994). Commercial Implications of Market Research on Space Tourism. *Journal of Space Technology and Science, 10,* 3-11.

Dickinson, B., & Vladimir, A. (1997). *Selling the Sea: An Inside Look at the Cruise Industry.* New York: Wiley.

Fawkes, S., & Collins, P. (1999, April). Space Hotels: The Cruise Ship Analogy. Paper presented at the 2nd International Symposium on Space Tourism, Bremen, Germany.

Fleeter, R. (1995). *Micro Space Craft.* Reston, VA: The Edge City Press.

Gordon, S. H. (1997). *Passage to Union: How the Railroads Transformed American Life,_1829-1929.* Chicago: Ivan R. Dee.

Fountain, Henry. Shoot It, Howl at It, Buy It. *New York Times.* July 1, 2001.

Gump, D. P. (1990). *Space Enterprise: Beyond NASA.* New York: Praeger.

Heppenheimer, T.A. (1995). *Turbulent Skies: The History of Commercial Aviation.* New York: Wiley.

Koman, V. (1998). *Kings of the High Frontier.* Centreville, VA: Bereshith Publishing.

Lewis, J. S. (1997). *Mining the Sky: Untold Riches from the Asteroids, Comets, and Planets.* Reading, MA: Helix Books.

Matsumoto, S., Amino, Y., Mitsuhashi, T., Takagi, K., & Kanayama, H. (1989). Feasibility of Space Tourism "Cost Study for Space Tour." Paper no. IAF-40, presented at Proceedings 40th International Astronautical Federation Conference, Malaga, Spain.

Nagatomo, M., & Collins, P. (1997, July). A Common Cost Target of Space Transportation for Space Tourism and Space Energy Development. Paper presented at Nagasaki: 7th International Space Conference of Pacific Basin Societies (ISCOPS), Nagasaki, Japan.

National Research Council (2000). *Radiation and the International Space Station: Recommendations to Reduce Risk.* Washington, DC: National Academy Press.

Norris, D. A. (1999, July). So Where's the Moon Base? Paper presented at Apollo 11 Launch Anniversary Breakfast Address, Commercial Lunar Base Development Symposium, League City, TX.

O'Neil, D. (Compiler). Mankins, J., Bekey, I., Rogers, T., Stallmer, E., & Piland, W., (Eds.) (1999). *General Public Space Travel and Tourism—Volume 2 Workshop Proceedings.* Huntsville, AL: NASA Marshall Space Flight Center.

O'Neill, G. K. (1989). *The High Frontier: Human Colonies in Space.* Princeton, NJ: Space Studies Institute Press.

Robinson, K. S. (1993). *Red Mars.* New York: Bantam.

Rodgers, E. (1996). *Flying High: The Story of Boeing and the Rise of the Jetliner Industry.* New York: The Atlantic Monthly Press.

Rogers, T.F. (1997, June). The Decreasing General Public Support for Our Federal Civil Space Program and How Two "Outreach" Activities Could Reverse This Trend. Paper presented at the 12th Man in Space Symposium, Washington, DC.

Stine, G. H. (1985). *Handbook for Space Colonists.* New York: Holt, Rinehart and Winston.

Stine, G. H. (1996). *Halfway to Anywhere: Achieving America's Destiny in Space.* New York: M. Evans & Co.

Stover, J. F. (1997). *American Railroads* (2nd ed.). Chicago: University of Chicago Press.

U.S. Public Interest Research Group Education Fund (1999, October and 2001, April). *Flirting with Disaster: Global Warming and the Rising Costs of Extreme Weather.* www.pirg.org.

Wasser, A. (1997, January). Staking a Claim on the Moon and Mars: Property Rights in Outer Space. *SpaceViews.* Available: http://seds.lpl.arizona.edu.

White, F. (1998). *The Overview Effect: Space Exploration and Human Evolution.* Reston, VA: American Institute of Aeronautics and Astronautics, Inc.

Wichman, H. (1992). *Human Factors in the Design of Spacecraft.* MIT Press and McGraw-Hill.

Wirthlin Worldwide. (1998). American Travelers Planning Better, Vacationing More. *The Wirthlin Report, 8,* 1-4. Available: http://wirthlin.webaxxs.com/publicns/report/wr9808.htm.

Zubrin, R., & Wagner, R. (1996). *The Case for Mars: The Plan to Settle the Red Planet and Why We Must.* New York: Simon and Schuster.

Glossary

Aerospatiale. A French aerospace company.

Apollo Program. The humans-to-the-Moon program instituted by U.S. President John F. Kennedy in the 1960s. The *Apollo 11* mission marked the first time a human set foot on the Moon.

Ariane launch system. A rocket made by a European manufacturer (Arianespace).

Arianespace. A European launch vehicle manufacturer.

Atlas. An American ballistic missile that was converted into an expendable launch vehicle.

Backward contamination. Bringing alien life to Earth that disturbs our ecosystem.

Boeing. One of the two big U.S. aerospace giants. The other is Lockheed Martin.

Bootstrap. A process of building up financial resources and self-financing little by little. Companies that bootstrap use the proceeds of small projects to do larger and larger projects.

CATS. Cheap access to space. The Holy Grail of those wishing to develop space and establish a permanent human presence. Currently, it's too expensive to launch people and other things into space economically, so there aren't very many launches.

Circadian rhythm. A cyclical variation in the intensity of bodily processes or behaviors over a 24-hour period.

Cosmic rays. Constant emissions from other galaxies. Composed primarily of ionized particles, they travel at near-light speed. Cosmic rays are a health risk for human space travelers. They ram themselves deep into the body, causing a nuclear reaction in the tissuesone that causes massive damage to our DNA and puts us at great risk for cancer.

DC-X. An experimental single-stage-to-orbit (SSTO) spacecraft built and flown by McDonnell Douglas in 1993.

Delta Clipper. See DC-X.

Europa. A Moon of Jupiter featuring a frozen ocean that may harbor life.

EVA. Extra-vehicular activity. When an astronaut goes outside sealed spacecraft and space stations to effect repairs or install something, he or she is engaging in an EVA. EVAs are highly dangerous.

Expendable launch vehicle (ELV). A rocket that puts something into space and is then discarded.

External tanks. Fuel containers on the outside of the space shuttle.

Flyback booster. A rocket fuel container that returns to Earth once it is empty. Advantages of flyback boosters are that discarding them reduces the weight of a vehicle and that they may be reused.

Forward contamination. Invading and disturbing the ecosystem of another planet.

Future-X. A NASA/U.S. Air Force program that develops experimental spacecraft.

G force. The acceleration of gravity.

Gagarin, Yuri. The first human to orbit the Earth (1961). He was a cosmonaut working for the USSR's space program.

Geostationary orbit. The point 24,000 miles above the earth at which an object in orbit hovers over a single point on the earth.

Gravity well. The area around the Earth in which gravity exerts its pull.

HUD. The U.S. Department of Housing and Urban Development.

Isp. (Specific impulse). Specific impulse measures how much propulsive force a propellant yields. The higher the Isp, the faster the mass being ejected is moving and the farther it is possible to go on a given load of fuel.

In-situ resource use (ISRU). Making propellant, air, water, and other essentials out of indigenous materials, primarily the atmosphere of a celestial body.

Jet Propulsion Laboratory (JPL). The NASA/Cal Tech facility in Pasadena, California from which planetary exploration is carried out.

Johnson Space Center (JSC). The NASA facility at which Mission Control for human spaceflight missions is located. JSC is in Houston, Texas.

Kankoh-maru. A passenger-carrying space vehicle designed by the Japanese Rocket Society used as a reference for various passenger craft designs.

L5 Society. The original American space advocacy organization. In the mid-eighties, it merged with the National Space Institute to become the National Space Society.

Liquid rockets. Launch vehicles that use liquid fuel.

Lockheed Martin. A large American aerospace company formed by the merger of Lockheed and Martin Marietta. Lockheed Martin is one of the big two American aerospace companies, the other being Boeing.

Magnetic field. An area of space in which magnetic force can be detected at all points.

McDonnell Douglas. An American aerospace company that built and flew a successful experimental single-stage-to-orbit (SSTO) vehicle in 1993 called the Delta Clipper, or DC-X. The vehicle was never made into a production model.

Microgravity. Near-weightlessness caused by free fall.

Moon Treaty. A treaty dating from the early 1980s that bans individuals from claiming private property on the Moon and makes it illegal for any earth country to claim sovereignty over a celestial body. No major spacefaring nation signed the Treaty.

National Space Society. A space advocacy organization formed by the merger of the L5 Society, a grass-roots organization that came out of the ideas of the late Princeton University professor Gerard O'Neill, and the National Space Institute, founded by rocket pioneer Wernher von Braun.

Outer Space Treaty. A 1967 United Nations document ratified by ninety-one countries including the U.S. and the USSR. The treaty bars claims of national sovereignty in space and forbids countries from claiming land or objects there. (The full name is Treaty on Principles Governing the Activities of States in the Exploration and Use of Outer Space, including the Moon and other Celestial Bodies.) No nuclear weapons or other weapons of mass destruction are to be placed in space, and countries are not to interfere with other countries' exploration and/or use of space. A document signed by more than ninety nations including the United

States that says that spacefaring nations will avoid "harmful contamination and also adverse changes in the environment of the Earth resulting from the introduction of extraterrestrial matter."

Pathfinder. A Mars probe launched by the U.S. in 1996, arriving at the Red Planet in 1997.

Payload. The deliverable carried by a launch vehicle or spacecraft. A payload can comprise equipment, people, and cargo.

Photometric. Of the measurement of the properties of light.

Photon. A unit of light.

Planetary Society. An organization founded by Carl Sagan to push for government funding of manned space projects.

Progress vehicle. An automated version of the Russian Soyuz spacecraft used to supply the Salyut space station, Mir, and *Space Station Alpha*.

Rand Corporation. A think tank in Santa Monica, California.

Re-entry. The process of returning to Earth from beyond its atmosphere.

Regolith. The upper layer of the Moon's surface.

REM. Roentgen Equivalent Man. A unit of X-ray or gamma ray dose.

Reusable launch vehicle (RLV). A rocket that puts something into space and is then reused.

Saturn V. The rocket used to launch Apollo moon missions.

Single-stage-to-orbit (SSTO). A space vehicle composed of one stage.

Solar flares. Sporadic bursts of radiation from the sun.

Solid rocket booster. A casing that carries solid rocket propellant. Solid rocket propellant is a rubbery mixture of fuel and oxidizer, highly flammable and toxic.

Soyuz. A Russian spacecraft.

Space Frontier Foundation. A space advocacy organization that believes near space should be developed by the private sector, while far space (everything beyond the Moon) should be explored by governments.

Space Station Alpha. The International Space Station constructed in the late nineties and early oughts by sixteen nations.

Spectroscopic. Of the study of the energy emitted by a radiant source.

Sputnik. The first artificial satellite lobbed into space by humans. Sputnik, launched by the Russians in 1957, inspired a space race between the USSR and the U.S. that culminated in the first humans (the Americans) setting foot on the Moon in 1969.

Stage. A section of a rocket that may be kept or discarded while the vehicle is in flight, depending on the design. In two- and three-stage rockets, one of the stages carries the payload, while the other(s) carry fuel. When the fuel is spent, the stages are shed to decrease the vehicle's weight.

Suborbital. A point above the earth below orbital height. In order for a vehicle to reach orbit, it has to be at least 100 miles up.

Subselene. Beneath the surface of the Moon.

Titan. An American ballistic missile that was converted into an expendable launch vehicle.

U.S. Space Foundation. An organization whose members hail primarily from large aerospace companies and government. The group's purpose is to bring together the sectors of the space community and serve as a credible source of information about the practical use of space.

Von Braun, Wernher. A German-born American rocket engineer who died in 1977.

X-15. An experimental American rocket plane of the 1950s and 1960s. Three were built and flown by NASA and the U.S. Air Force.

X-33. A single-stage-to-orbit (SSTO) advanced technology demonstrator commissioned by NASA and built by Lockheed Martin. The project was cancelled in 2001 due to cost overruns and technical difficulties.

***Zenit* rocket.** An expendable rocket made in Ukraine.

About the Author

Paula Berinstein has been told she asks too many questions, but she doesn't care. She really wants to know what's out there. The way she sees it, when you stop being curious, you might as well just fade away.

Ms. Berinstein has asked questions about space ever since she can remember. (For a time she also worked in the space industry—eight years as a programmer/analyst at Rockwell International's Rocketdyne division, now Boeing, where they design and make the Space Shuttle main engine.) What would it be like to go into space? What will we find there? What will happen if we never get there? She still asks those questions, but she's added some more immediate and practical ones to her list: How much will it cost to establish a permanent human presence in space? Where will the money come from? Is there a mass market for space tourism, and if so, how do we meet the demand? What happens to the human body in space, and what do we need to do to help the body function there? Who's doing what to get us there, and how promising are the results?

Faced with these questions and more, Ms. Berinstein decided to write her sixth book—one that explores and answers them as far as possible. *Making Space Happen: Private Space Ventures and the Visionaries Behind Them* is the happy result.

Ms. Berinstein has exercised her curiosity in the writing of five other books, including *Alternative Energy: Facts, Statistics, and Issues, Finding Statistics Online,* and *The Statistical Handbook on Technology.*

Ms. Berinstein lives in Los Angeles with her husband Alan, their cat, Speckles, and occasional teenagers running through. She spends about two months a year in the couple's second home, Cambridge, England.

Index

C

D

E

F

G

H

I

L

M

N

P

Q

R

S

T

U

V

X

Y

Z

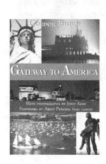